Scientific Crosscurrents between Italy and England

EUROPA PERIODICA
STUDIES ON PERIODICALS AND NEWSPAPERS

Edited by
Patrizia Delpiano, Fabio Forner, Giovanni Iamartino,
Sabine Schwarze and Corrado Viola

VOLUME 2

Lucia Berti

Scientific Crosscurrents between Italy and England

Italian Contributions to the *Philosophical Transactions of the Royal Society*, Seventeenth to Nineteenth Centuries

Bibliographic Information published by the Deutsche Nationalbibliothek
The Deutsche Nationalbibliothek lists this publication in the Deutsche Nationalbibliografie; detailed bibliographic data is available online at http://dnb.d-nb.de.

Library of Congress Cataloging-in-Publication Data
A CIP catalog record for this book has been applied for at the Library of Congress.

Cover image: © The Royal Society.
Extract of an edited translation of part of a letter from Paolo Mattia Doria (Naples, 1731), to Paolo Rolli. In the letter, Doria seeks the Royal Society's opinion on his book *Duplicationis Cubi Demonstratio*.

ISSN 2749-215X
ISBN 978-3-631-88809-4 (Print)
E-ISBN 978-3-631-88812-4 (E-PDF)
E-ISBN 978-3-631-88813-1 (EPUB)
DOI 10.3726/b20113

© Peter Lang GmbH
International Academic Publishers
Berlin 2023
All rights reserved.

Peter Lang – Berlin · Bruxelles · Lausanne · New York · Oxford

All parts of this publication are protected by copyright. Any utilisation outside the strict limits of the copyright law, without the permission of the publisher, is forbidden and liable to prosecution. This applies in particular to reproductions, translations, microfilming, and storage and processing in electronic retrieval systems.

This publication has been peer reviewed.

www.peterlang.com

To my sons, Marcus and Arthur

Abstract: Soon after its foundation in the 1660s, the Royal Society became the centre of an international network of scientific correspondence and many of the learned letters it received were subsequently published in its journal, the *Philosophical Transactions of the Royal Society* (*PTRS*). The relations of Italian learned men with the Society allowed several discoveries – such as Marcello Malpighi's anatomical studies and Alessandro Volta's invention of the battery – to receive the attention of the international scholarly community and become the subject of further research that allowed medical, scientific, and technological advancement. The *PTRS* papers and the letters preserved in the Society's archives provide precious insights into the making of science and the workings of the Republic of Letters in the early and late modern periods. This book investigates the Royal Society's relations with Italy through a socio-historical and critical linguistic analysis of the papers concerning Italian research published in the *PTRS* and of the epistolary exchanges between the Society's Fellows and Italian scholars. The aim from the linguistic perspective is to describe the features and development of Italian-research-based papers in the *PTRS*, as well as the discursive aspects that characterise the exchanges between the two countries. Ultimately, from the historical and cultural point of view, the study will provide a picture of the development of Anglo-Italian relations in a scientific context from the mid-17th to the end of the 19th century.

Keywords: Anglo-Italian relations, scientific journals, Republic of Letters, history of science, Critical Discourse Analysis, early and late modern periods.

Contents

List of abbreviations 13

Chapter 1 Backgrounding the study 17
   1.1 Introduction 17
   1.2 The socio-historical contexts 22
      1.2.1 Notes on the history of the Royal Society and
          Philosophical Transactions, 17th–19th centuries 22
      1.2.2 The Italian states and their academies 32
   1.3 The Royal Society and Italy: A review of relevant literature ... 40

Chapter 2 Methodological framework 47
   2.1 The groundwork 47
      2.1.1 Critical Discourse Analysis 47
          2.1.1.1 The Discourse-Historical Approach (DHA) 48
          2.1.1.2 *Discourse and social change* – Fairclough's
                approach 50
          2.1.1.3 Politeness 56
      2.1.2 A socio-historical approach to scientific writing 58
   2.2 Approach to text analysis 62
   2.3 Collection of the papers and analysis 65

Chapter 3 Social, cultural and historical insights 73
   3.1 17th century 73
      3.1.1 Fellows and contributors 74
      3.1.2 Creating a discourse community: Correspondence
          and information exchange 79
      3.1.3 Fellows and Englishmen in Italy 83

3.1.4 Topics 86
3.1.5 Landscapes of northern and central Italy 95

3.2 18th century 99
3.2.1 Fellows and contributors 101
3.2.2 Fellows and Englishmen in close contact 108
3.2.3 A complex network of exchanges 112
3.2.4 Topics 114
3.2.5 Beauties of northern, central and southern Italy 122

3.3 19th century 126
3.3.1 Foreign Members and contributors 128
3.3.2 Fellows and Englishmen in Italy 134
3.3.3 Topics 137

3.4 Languages of the papers and translation practice 138
3.4.1 In the 17th century 138
  3.4.1.1 English'd out of the *Giornale de' Letterati* 141
3.4.2 In the 18th century 141
3.4.3 In the 19th century 145

Chapter 4 Discourse features 149

4.1 17th century 150
4.1.1 Textual dimension 151
  4.1.1.1 Macrostructural features 151
  4.1.1.2 Language use 152
4.1.2 Discursive practice 158
  4.1.2.1 Discourse representation 158
  4.1.2.2 Meeting minutes and letter exchanges 161
  4.1.2.3 Evaluation in the discourse of the *PTRS* 163
4.1.3 Reporting disputes and disagreements 168
4.1.4 Witnessing 173
4.1.5 Toponymy 174
4.1.6 Interdiscursivity and intertextuality 176

                4.1.6.1 Dialogicity in the discourse on *amianthus* 177
        4.2 18th century 181
            4.2.1 Textual dimension 182
                4.2.1.1 Macrostructural features 182
                4.2.1.2 Language use 184
            4.2.2 Discursive practice 188
                4.2.2.1 Discourse representation 188
                4.2.2.2 Evaluation in the discourse of the *PTRS* 190
                4.2.2.3 Original, translation and publication:
                         A brief comparison 193
            4.2.3 Reporting disputes and disagreements 195
            4.2.4 Witnessing 202
            4.2.5 Toponymy 206
            4.2.6 Interdiscursivity and intertextuality 208
        4.3 19th century 210
            4.3.1 Textual dimension 210
                4.3.1.1 Macrostructural features 210
                4.3.1.2 Language use 211
            4.3.2 Discursive practice 214
                4.3.2.1 Discourse representation 214
                4.3.2.2 Evaluation in the discourse of the *PTRS* 216
            4.3.3 Witnessing and toponymy 216
            4.3.4 Interdiscursivity and intertextuality 218

Chapter 5 General conclusions 219
    5.1 Development of Anglo-Italian socio-cultural relations 219
    5.2 Languages of international scientific communication
         and linguistic consequences 221
    5.3 Development of Italian and English *PTRS* papers 223
    5.4 Development of Italian discourse representation 224
    5.5 Concluding remarks 224

Appendix 227
  Tables of Fellows and contributors 227
  Loanwords in the *PTRS* 253

Bibliography 261
  Primary sources 261
    17th century 261
    18th century 269
    19th century 287
  Secondary sources 292

Index of names 307

List of abbreviations

art.:	article(s)
Eng.:	English
For. Sec.:	Foreign Secretary
Fr.:	French
FRS:	Fellow of the Royal Society
GdL:	*Giornale de' Letterati*
Lat.:	Latin
Phil. Trans./ *PTRS*:	*Philosophical Transactions of the Royal Society of London*
PRS:	President of the Royal Society
Sec.:	Secretary

Archival material

AP:	Archived Papers (unpublished at the time of receipt)
CMO:	Council Meetings
EC:	Certificates of Election and Candidature
EL:	Early Letters
JBO:	Journal Books
L&P:	Letters and Papers
LBO:	Letter Book Original
MS:	Manuscripts general

Acknowledgements

I would like to wholeheartedly thank my mentor, Giovanni Iamartino, for his continued guidance and encouragement, selflessly supporting me even when I had to considerably reduce my workload to give birth and look after my two children during the period that led to the publication of this book.

I would also like to express my gratitude to the members of staff at the Royal Society of London, who were always very helpful while I worked on my research at the library; to Manuela D'Amore, Felicity Henderson, Fabio Forner, Laura Pinnavaia and Elisabetta Lonati for their varied and extremely valid input; and to the scientific committee of the Research Network for the History of European Periodicals for welcoming this book into the *Europa Periodica* series.

Chapter 1 Backgrounding the study

1.1 Introduction 17
1.2 The socio-historical contexts 22
   1.2.1 Notes on the history of the Royal Society and *Philosophical Transactions*, 17th–19th centuries 22
   1.2.2 The Italian states and their academies 32
1.3 The Royal Society and Italy: A review of the relevant literature 40

1.1 Introduction

> *The Royal Society has so just a regard & Veneration for ye memory of ye Galilei, the Borelli, Malpighi, and Bellini, y^t she can never be incurious of what is doing in a Country, y^t produced those Great & Excellent Genii.*[1]
>
> *'Tis hardly to be believed, what a high esteem all, where I have passed [in Italy], have for the Royal Society and the universal knowledge and learning of the Britons.*[2]

The UK's national science academy – the Royal Society of London – and its journal – *The Philosophical Transactions of the Royal Society* (hereafter *PTRS*) – were created by a group of learned gentlemen in the 1660s. The Society's main source of inspiration was FRANCIS BACON's idea of natural philosophy based on an empirical approach to the study of nature. A few years after the Society's founding, the *PTRS* started being published (1665) and soon became the leading scientific journal of the time. Thanks especially to its first secretary HENRY OLDENBURG,[3] the Royal Society became the centre of an international network of scientific correspondence with many learned letters being subsequently published in the journal. Indeed, foreign natural philosophers became aware of the Royal Society's prestige and Baconian agenda and wrote to the Society

---

1 JURIN to DEREHAM, 1722, in Rusnock (1996: 91).
2 ROBERT BALLE, 1721, in Fisher (2001: 356).
3 OLDENBURG was appointed secretary in 1662, very likely thanks to his "scientific contacts and communicative skills" (Atkinson 1999: 19). The Society generally had two secretaries, who recorded what was discussed during meetings and represented the Society in foreign correspondence with "natural philosophers", physicians, and other learned gentlemen.

offering scientific information and hoping to receive approval and possibly publication in the *PTRS*. Publication in the Society's journal meant not only gaining international visibility and contributing to a collective enterprise of science, but also establishing priority of one's findings. It is moreover well known that the development of the *PTRS* itself traces the history of the modern scientific journal; this will also be seen in the changing nature of the contributions sent from Italy to the Royal Society and published in the *PTRS*, which gradually develop from personal letters to becoming more and more like present-day scientific papers.

Hence, the *Philosophical Transactions*, the world's longest-running scientific journal, represents an invaluable repository of historical and linguistic material for scholars to investigate. Further, the Fellows of the Royal Society preserved originals and copies of most of their correspondence and bureaucratic documents together with instruments, portraits, natural specimens and other curiosities, which can be found in the Royal Society's archives in London. A notable example of the historical worth of the Society's treasures is the recently discovered holograph letter from GALILEO to BENEDETTO CASTELLI (21 Dec. 1613) where he first set out his ideas on the relation between science and religion, and defended Copernican astronomy from charges of being contrary to the Holy Scriptures. This letter is of primary importance for the history of GALILEO's relations with the Church and had severe consequences; namely, the suspension of COPERNICUS' *De Revolutionibus* (Nuremberg, 1543) and the warning to GALILEO to abandon Copernican astronomy, which was seen as a threat to the traditional interpretation of the Bible. Up until this discovery, historians had relied on manuscript copies, which differed between each other. The letter has shed new light on our knowledge of GALILEO and displays a more daring and compromising wording compared to its copies (Camerota et al. 2018: 1). GALILEO and his letter are but one example of a long line of instances in which the Royal Society saw the value of Italian men of science and treasured material related to them.[4] From the outset, the Fellows had shown great interest in the Italian Peninsula and the research of its scholars. A great deal of Italian-research-based

---

4 Although the present study will frequently adopt the terms *science* and *scientists* to refer to the scholars of the three centuries under study, it ought to be reminded that these words started being used regularly only in the 19th century. The present meaning of science did not exist back in the 17th and 18th centuries, where those who studied nature empirically defined themselves as "natural philosophers". However, starting from the scientific revolution, it is common to refer to natural philosophers as scientists, in that this is the time where modern science started developing. Similarly, it ought to

papers have been published in the *Transactions* and many letters to and from Italy are preserved in the Royal Society's archives. These resources can provide new insights into the history of Anglo-Italian relations, which have hitherto been studied primarily from a cultural and literary perspective. But as it will be seen in the course of this study, British interest in Italy was not limited to Italian literature and natural and cultural curiosities. The aim of this study is thus to investigate the Royal Society's relations with Italy through an analysis of the Italian contributions to the *Philosophical Transactions* and the letter exchanges between the Fellows and the Italians.

The Royal Society and the *Philosophical Transactions* have been extensively researched by historians and linguists. Linguistic analyses carried out so far on *PTRS* articles have focused, for instance, on the development of scientific writing in general;[5] on more specific fields and genres such as medical writing[6] and reports of scientific experiments;[7] and on select linguistic features such as stance.[8] Nevertheless, many linguistic analyses carried out so far on the *PTRS* have focused on the English language and its stylistic development with less regard for the fact that many papers published in the journal came from foreign countries and were the result of translation.[9]

---

    be noted – especially in view of the series on the history of European periodicals of which this book is a part – that, for economy purposes, throughout this study I often refer to the texts published in the *PTRS* as *papers* or *articles*, despite the fact that the journal article was not yet an established genre in the early modern period and that the published texts are varied in nature (full letters, letter extracts, news notices, book reviews, lists, extracts from other journals etc.). For more on the debate on the history, forms and development of Europen periodicals see Forner et al. 2022, Peiffer et al. 2013, and Generali 2012.

5 For linguistic studies on historical scientific writing see for instance Atkinson (1996 and 1999), Banks (2008a and b, 2009a and b, 2010a and b, 2012, 2017), Bazerman (1988), Gotti (2006), Gross et al. (2000 and 2002), and Gunnarson (2011).

6 For linguistic studies on historical medical discourse see, among others, Atkinson (1992), Berti (2019), Canziani et al. (2014), Gotti (2011), Pahta and Taavitsainen (2011), and Lonati (2013).

7 See, for instance, Bazerman (1988), Biber and Finegan (1989), and Gotti (2014).

8 Gray et al. (2011) and Banks (2008b).

9 Studies on translation practices in the field of science and at the Royal Society have also been carried out, see for instance: Beer (1990), Boschiero (2010), Henderson (2013 and 2017), Knowles Middleton (1969), Olivari and Torna (2012), Plescia (2011), Turner (2008), and Vicentini (2019).

From a historical and cultural point of view and relevant to the purpose of the present research, only a few studies have focused on the Royal Society's relations with Italy. For instance, Knowles Middleton (1979 and 1980) and McConnel (1986 and 1993) focus on a number of Italian Fellows of the Royal Society; Cavazza (1980 and 2002) and Cook (2004) focus on the Society's relations with specific Italian intellectual communities (from Bologna and Rome); Hall (1982) examines the role played by Italy for the Royal Society up to the 18th century; Gomez Lopez (1997) explores the correspondence between Italians and the Royal Society in the first decades of the Society's existence; and D'Amore (2015 and 2017) shows how the journal's papers on the Italian south were in harmony with the literary and cultural trends of the 17th and 18th centuries and contributed to increase English interest in Italy.[10] These studies, however, are generally focused on individual Italian natural philosophers or limited to specific geographical areas and periods (especially the early Royal Society). Only a glimpse can be caught of the relations between the scientists of the two countries and of the scientific contributions that the Italians made to the *Philosophical Transactions*.

As anticipated earlier, research has also been carried out on Italian and English socio-cultural relations.[11] The general picture that emerges – for the period considered here, 17th–19th century – is that after a period of decline of English interest in Italy in the 17th century – when Italy came to be seen as the country of Catholicism as opposed to the puritanism of the Commonwealth – Italy regained popularity in the 18th century and through the 19th as one of the favourite destinations of the *Grand Tour* and for a rediscovery of classic Italian literature and the Italian opera. From the Italian side, in the course of the 18th century England was becoming increasingly favoured in Italy as *the* enlightened country, appreciated for its institutions, economy, liberal thinking and literature. However, the above-mentioned studies generally focus on literary sources, while scientific exchanges offer a new perspective. For instance, religious and political views hardly interfered with the relations between Englishmen and Italians, in that the focus of the Royal Society was on experimental philosophy, which was seen as independent of metaphysical and political thinking. In actual fact, Italian religious restrictions enabled the Society to be the first to publish Italian research

---

10 See also Bertucci (2013), Cavazza (2010), Cook (2002 and 2004), Eccles (1975), Ferrari and McConnel (2005), Findlen (2009), Fisher (2001), Hunter (2014), Knowles Middleton (1979 and 1980), Quinn (2005), Waller (2012), and Wis (1996 and 1970).
11 See among others Costa (1968), Crinò (1971 and 1972), Giannini (1936), Pesaresi and Ascari (2015), Praz (1944), and Rebora (1936 and 1938). From a linguistic perspective see also Iamartino (2001 and 2002), Pinnavaia (2001), and Praz (1939).

that met with criticism and rejection in Italy, such as the studies by the Italian physician MARCELLO MALPIGHI.

Other than the *PTRS* papers, Italian interest in English science and culture – and, vice versa, English interest in Italian research – is perceived by the considerable amount of Italians who were elected Fellows of the Royal Society – 135 between the 17th and 19th centuries – and who were given the opportunity to attend the Society's meetings. Among them were men of science such as astronomers, physicists, mathematicians, botanists, and many physicians but also humanists such as historians, philosophers and poets. A considerable number of Italian Fellows were also statesmen and diplomats. Further, this study reveals that contributions made by non-elected Italians who had relations with the Society were often more significant than those made by elected Italian Fellows.

The present study goes beyond the research topics considered in the above-mentioned literature by carrying out a historical and critical linguistic analysis on *PTRS* articles written by Italians or based on Italian research and by analysing English and Italian relations through the papers and the epistolary exchanges of the scientists from the two countries. The aim from the linguistic perspective is to describe the features and development of Italian and Italian-research-inspired scientific writing in the *Transactions;* and ultimately, from the historical and cultural point of view, to provide a picture of Anglo-Italian relations in scientific context. The critical linguistic analysis of the primary sources here becomes functional to a more precise analysis of cultural relations. It moreover adds to the existing research on the development of scientific writing by providing a study that is focused on a culturally-restricted group of papers and which considers the original sources of the publications. Comments and descriptions on editorial and translation practices will also be provided. The period considered starts with the birth of the *Philosophical Transactions* in 1665 and finishes in 1900; the temporal limit is related to the great socio-historical changes that occurred in the 20th century and which subsequently influenced the Royal Society's activity and its publications.

It is hoped that the present study will be of interest both to scholars interested in the history of science and/or the development of Anglo-Italian scientific relations, and to linguists interested in historical scientific writing. In this view, the results have been organised into two main chapters, the first focusing on the social, cultural and historical insights that arose from the analysis (Chapter 3), and the second focusing on the results of the more purely linguistic analysis (Chapter 4).[12]

---

12 The division into a cultural-historical part and a linguistic part of the results chapters allows the reader, who may be interested only in specific parts of this research, to skip

The general organisation of the study is as follows: the rest of the present chapter draws on the existing literature to provide a brief historical and socio-cultural background to the research. The information provided below is considered an integral part of the following analysis in that it provides a context for it. The central chapters will in fact present and try to account for the results by contextualising them. The following sections will thus report on the history of the Royal Society, the *Philosophical Transactions*, and relevant notes on Italian history and scientific academies (Section 1.2); while Section 1.3 provides information on aspects of Anglo-Italian scientific relations that have already been dealt with in the literature. Chapter 2 presents linguistic research that has inspired the methodology developed for the analysis (Section 2.1 and subsections); the approach adopted for the present piece of research (Section 2.2); and a description of how the analysis was carried out step by step (Section 2.3). The central Chapters, 3 and 4, present and discuss the results of the historical and critical linguistic analysis of the papers and letters collected. The two chapters are further divided by century and feature case studies on different aspects that were perceived to be relevant to each period. Chapter 5 provides some general conclusions to the study, adding some considerations on Italian scientific borrowings in the *PTRS* as a linguistic result of contact between the two cultures. Finally, the appendix provides tables of the Italian Fellows and contributors divided per century reporting brief biographical notes and the number of contributions to the journal. The second and last table lists the borrowings found during the analysis with information on their entry into the English language.

1.2 The socio-historical contexts

1.2.1 Notes on the history of the Royal Society and *Philosophical Transactions*, 17th–19th centuries[13]

The roots of the Royal Society are generally traced back to an informal meeting that was held at Gresham College in London in November 1660, when a group

---

any sections they might not want to read. However, the historical and linguistic analyses are connected and functional to one another and cannot therefore be completely separated. Further historical detail is provided with the results of the linguistic analysis, especially in the sections focusing on discursive practice. Intratextual references are thus provided to redirect to sections where a given topic is further discussed.

13 Where not otherwise specified, this account is based on Atkinson (1999), Fyfe et al. (2015), Hall (1975, 1991 and 1992), and Rusnock (1996).

of gentlemen discussed the formation of an organisation for the promotion of experimental philosophy on the pattern of Continental academies.[14] The main source of inspiration for the group was BACON's empirical approach to the study of nature. Regular meetings were held from the start, a constitution was drawn up, and the members paid weekly subscriptions. In July 1662, the Society was given chartered status by King Charles II and was officially named "the Royal Society" and, later, in a second charter of 1663, as "the Royal Society of London for improving naturall knowledge". While remaining a private Society focused on experimental research, the chartered status meant that the Fellows were granted privileges such as direct patronage from the king, permission to print without government censorship, and freedom to correspond with other countries. From the beginning the Society had an elaborate organisational structure with a president, a treasurer and a register-keeper at its top. The Society also appointed two secretaries who were to record what went on in the meetings, manage correspondence with outside parties, and read a selection of their correspondence at the meetings. The Fellows were generally gentlemen – or *virtuosi*, as they were to call themselves –[15] who pursued "natural knowledge" as a pastime and not as professionals. Religious and political opinions were to be left behind, with the sole subject of debate being natural philosophy.[16] The concept of

---

14 This is where the first instances of Italian influence on the Royal Society can be seen. Indeed, when LAWRENCE ROOKE, professor of astronomy at Gresham, spoke of "a designe of founding a Colledge for the promoting of Physico-Mathematicall Experimentall Learning", which was to be done by debating things according to "the manner of other countries", the Florentine Accademia del Cimento was one of the models in the Fellow's mind. Like the Royal Society, the Cimento academy worked under private patronage and control and, unlike other European Societies, was experimental in concept (Hall 1991: 9). Later, instead, Italian academies such as the Istituto delle Scienze e delle Arti in Bologna (1714) and the re-founding of the Academia degli Investiganti in Naples, were inspired by the Royal Society, and FRANCESCO NAZZARI's *Giornale de' Letterati* by the *Philosophical Transactions*.
15 Indeed, the Fellows of the Royal Society were men of diverse interests and not necessarily all learned. Hall describes the virtuosi – which she separates from those who studied the more physical branches of learning – as "men with an interest in the world of nature who enjoyed discovering what others were doing in the investigation of nature but who were not mentally or temperamentally equipped to investigate it themselves, except possibly in its simplest form, namely by collecting rarities or observing what went on around them" (1991: 10).
16 Indeed, the Society's neutrality in terms of politics and religion was one of the features that the Italians admired in it. Later, politics did however sometimes interfere with the

natural knowledge was loose, and many of the interests of the Fellows pertained to disciplines which would not in the present age be considered scientific, such as archaeology, numismatics and antiquarianism.

BACON's programme of natural philosophy[17] conceived the study of nature as a cooperative endeavour with the aim of creating a "Natural and Experimental History such as may serve to build philosophy upon".[18] This approach included minute recordings of experiments; the circumstances in which they were performed; the presence of eyewitnesses; and the consideration of the works of other researchers. In line with this programme, the plans of the Royal Society were thus to cultivate "a sound and useful philosophy" through the "joint labours of the industrious and wise men of the whole world in mutual co-operation" and by a "diligent and unremitting examination into Nature through observation and experiment, carefully and frequently performed".[19] Hence, the research carried out by the Society's Fellows needed to be public, and natural philosophers from different countries should have communicated their findings to each other. To this objective, a "Paper of advertisements" was devised, in which the substance of the Society's inquiries, their progress, and information on what the Fellows received from their contacts was to be published. This paper was later to become the *Philosophical Transactions of the Royal Society* first published 6 March 1665 with the subtitle "Giving an Accompt of the Present Undertakings, Studies, and Labours of the Ingenious in many considerable Parts of the World".

Just a few months earlier, on 5 January 1665, the French *Journal des Sçavans* started being published in Paris. It was printed on a weekly basis and contained reviews on books on theology, history, medicine, and natural philosophy. HENRY OLDENBURG, the creator of the *Transactions*, had also been in contact with the authors of the French journal who asked him to contribute information and book reviews from England. OLDENBURG's project was very similar to this, as he himself was to acknowledge, "but much more philosophical in nature" (Fyfe et al. 2015: 7).[20]

Society's business, and in the 18th century an Italian candidate to election was rejected for his appreciation of French republicanism (see Section 3.2.1).

17 Discussed in his *Novum Organum* (1620) and *New Atlantis* (begun 1624 and never completed).
18 BACON, *Novum Organum*, 1620, cited in Atkinson (1999: 18).
19 OLDENBURG in Hall and Hall (1966: 620–621).
20 For comparisons between the *PTRS* and the French *Journal* see Banks (2009b, 2010 and 2017), McCutcheon (1924), and Turner (2008).

The *PTRS*'s character was thus thereby shaped by OLDENBURG, who is today considered the inventor of the scientific journal. OLDENBURG was made Fellow of the Society in 1660 and appointed secretary in 1662. Already before this time, he began the creation of his extensive network of correspondence on scientific matters. OLDENBURG tactfully employed his linguistic – he wrote English, Dutch, French, Italian, German and Latin with fluency – and rhetorical skills to encourage individual scientific activity and stimulate discussion. He would then translate the received letters into English and edit them for publication in the *Transactions*. The first issues of the journal contained adapted extracts from OLDENBURG's correspondence, accounts of books, and reports of experiments.[21]

The Royal Society and the *Philosophical Transactions* were always inextricably associated in the public eye; however, until 1752 the journal was run as a private endeavour. Indeed, OLDENBURG published the journal at his own expense through the Society's printers and, after his death, the *Transactions* continued to pass through the hands of a series of editors who were generally also secretaries.[22] Under OLDENBURG the journal was published mostly on a monthly basis until his death in 1677,[23] after which publication became less regular and frequent.[24]

Although the *PTRS* remained a generalist publication until the end of the 19th century, featuring articles from a wide range of disciplines, the Society's presidents and secretaries played an influential role in the contents of the journal. Hence, the editorship of Sir HANS SLOANE, from 1695 to 1713, marked

---

21 None of the first pieces published in the *PTRS* would today be considered articles, but rather short news items all heavily marked by OLDENBURG's adaptations.
22 OLDENBURG's successors were NEHEMIAH GREW; then ROBERT HOOKE – from 1679 to 1683; during this period the journal was published as *Philosophical Collections* –; ROBERT PLOT, until 1687, at which point the journal was not published for a year and later continued to be published with less frequency; RICHARD WALLER, until 1695; and Sir HANS SLOANE until 1713, who restored the journal to a healthy condition (Atkinson 1999: 21); The journal has been in continuous publication since the 1690s.
23 Under OLDENBURG's editorship 136 issues of the *Transactions* were published. Minor interruptions in this period were given by the plague outburst of 1665, the Great Fire of 1666, and the second Anglo-Dutch war of 1667, during which OLDENBURG was briefly imprisoned in the Tower of London suspected to pass information to the Dutch enemy (Fyfe et al. 2015: 8).
24 According to Atkinson (1999: 22), the period before SLOANE's office as secretary was marked by a decline in foreign contributions; articles by non-Englishmen amounted to 40 % in the 1660s but fell to 20 % in the 1690s. As far as Italian contributions are concerned, however, the presence of Italian papers in the *PTRS* continues regularly until the end of the 18th century.

both a period of stability for the journal but also a period of discontent over his editorship. SLOANE was in fact a physician, natural historian and collector of plant specimens, whose interests were reflected in the *Transactions*. During this period there was a decline in experimental practice in favour of more theoretical and less experimental subjects such as natural history, medical curiosities, case histories and geology (Hall 1991: 120). During SLOANE's office the *Transactions* were publicly attacked as a "collection of outlandish miscellanea, written in a confused style" (Atkinson 1999: 23).[25] Also, this period was marked by the presidentship of Sir ISAAC NEWTON and many Fellows of the Society were ardent Newtonians, who had different views from SLOANE on what the Society should concern itself with. Eventually, SLOANE was replaced by EDMOND HALLEY as secretary and editor (1713–1721), marking the beginning of a period in which the Society was controlled by Newtonians. During HALLEY's editorship foreign correspondence was more focused on astronomy, while medical and biological subjects were neglected. HALLEY was also less devoted to the journal compared to his predecessor and successor and during his tenure the *PTRS* was not published for two consecutive years (Rusnock 1996: 16). However, although the Newtonian period was marked by an increase in physical and experimental interests, attention for non-experimental subjects also continued and NEWTON himself showed an interest for medical reports, case histories and other curiosities (Hall 1991: 121).[26]

JAMES JURIN succeeded HALLEY in 1721 and reinvigorated both the atrophied *Transactions* and foreign correspondence. He wanted to make the Society's library more cosmopolitan by furnishing it with foreign books and copies of scientific journals. He therefore extended the Society's contacts abroad by relying on Englishmen and diplomats residing in foreign countries. JURIN was a physician and his interests in medicine (especially the inoculation of smallpox) and in meteorology were reflected in the *Transactions*. He was succeeded by a series of medical secretaries: WILLIAM RUTTY (1728–1730), CROMWELL MORTIMER (1730–1752), and MATTHEW MATY, all of whom continued contacts with Italy. In the early 18th century an increased number of foreigners were elected to

---

25 The attacks were published by WILLIAM KING in an anonymous pamphlet entitled *The Transactioneer with some of his Philosophical Fancies: In Two Dialogues* (1700) (Atkinson 1999: 22).

26 Hall reminds us that the Society's aim was to improve *all* natural philosophy, and not only its physical branches (1991: 120).

the Society and their presence at meetings (including several Italians) became regular (Hall 1991: 137).

Up until this point the main subjects appearing in the *Transactions* had been medicine and astronomy. Under the editorship of MORTIMER and the presidentship of MARTIN FOLKES (1741-1752)[27] – a mathematician with strong archaeological and literary interests – natural history and antiquities gained new emphasis in the journal. The Society was once again sharply attacked and the *Transactions* was depicted as "a catalogue of futility, error, and triviality" (Fyfe et al. 2015: 10).[28] Consequently, it was decided that the *Transactions* should be officially taken over by the Society. Financial and editorial responsibility were no longer in the hands of the secretaries, but in those of the Society's governing Council, who appointed a Committee of Papers (members of the Council themselves) to vote on each paper proposed for publication and read before the Society. This is generally seen as the first step into the development of the modern journal-referee concept (Atkinson 1999: 26). Greater attention was placed on the singularity and utility of the subjects, and many papers were now being refused publication. The election of the Earl of Macclesfield GEORGE PARKER as president (1752-1764), a mathematician and astronomer, contributed to strengthening the new editorial policy. During his term the first government-funded expedition to the Islands of the South Atlantic for the observation of the 1761 transit of Venus was launched (Atkinson 1999: 28). More scientifically oriented was also his successor, the Earl of Morton JAMES DOUGLAS (1764-1768), but antiquarian interests were soon to be revived under the presidentship of JAMES WEST (1768-1772). WEST was then followed by the physician JOHN PRINGLE (1772-1778). Whether the Presidents were men of science or not, science continued to be represented by a minority in the Council and scientific Fellows were only a third of the whole Fellowship; hence, dissatisfaction continued (Lyons 1944: 164).

After this series of short-term Presidents, in 1778 came the election of Sir JOSEPH BANKS who had been a Fellow since 1766, had served on the Society's Council, had taken part in four voyages of discovery, and was both a wealthy landowner and tropical botanist. He served as president for nearly 42 years, until his death in 1820. During his term the Society lived a period of stability

---

27 After NEWTON (PRS 1703-1727), interim presidentship was given to Sir HANS SLOANE (1727-1741), who had remained active in the Society after losing his office as secretary.
28 The attacks came from JOHN HILL, an apothecary and naturalist, who had been refused Fellowship to the Society. He thus published over the course of two years three works in which he ridiculed the Society, its president and the *Transactions*.

and a leisurely intellectual atmosphere. BANKS was a socially and politically active man; he was very familiar with the Society's business and became very controlling of the activities of the Council; he appears to have personally selected members of the Council and surveyed the acceptance and rejection of candidates (Lyons 1944; Hall 1984). BANKS also devoted himself to the improvement of the Society's administration. The *Transactions* continued steadily during this period and so did the work of the Committee of Papers, although here too there were ways in which the President could bypass the Society's publishing procedures; for instance by preventing papers from being read at the meetings. There is also evidence of BANKS' editorial intervention in the papers, such as proposals for cuts and emendations. Informal evaluation of papers also occurred during social gatherings – Sunday evening receptions, breakfast parties, dinners before the Thursday evening meetings, and after-meeting tea-table conversations – organised and frequented by BANKS (Fyfe et al. 2015: 15). BANKS represented the more conservative side of the Royal Society's Fellowship in the continuing underlying struggle between scientists and genteel members. His political influence ensured a close relationship with the government[29] and friendship to politically powerful individuals, who, despite the opposition raised by those devoted to the hard-line sciences, were necessary for the Society's survival. BANKS also began a close association with the British Museum which continued well into the 19th century (Hall 1984: 2).[30]

BANKS was very controlling outside the Society as well. The 19th century was in fact the time when the sciences were becoming more specialised, and various scientific societies were founded during BANKS' presidentship. While in some cases he supported the foundation of new institutions, in other cases he attempted to suppress them fearing that they would attract members away from the Society and influence publications in the *Philosophical Transactions*,[31] The Society however was little affected by the existence of other specialist Societies. The *Transactions* continued its publications regularly; frequent topics for this period were medicine, astronomy, natural history and electricity.

After BANKS's death, the Society found itself in an unstable situation, and the struggle between professional and amateur scientists was to re-emerge. After a

---

29 He moreover became an indispensable government advisor on colonial affairs and an authority on imperial trade (Atkinson 1999: 30).
30 Already in the 18th century a few of the Society's secretaries had been officers of both institutions, such as MATTHEW MATY and JOSEPH PLANTA (Hall 1991: 136).
31 Some of these Societies, such as the Astronomical and the Geological Society, printed their own journals.

brief interim presidentship assigned to WILLIAM HYDE WOLLASTON (1820) came the election of HUMPHRY DAVY (1820–1827), who represented a compromise between the wealthy and the professional, being a self-made man, who earned his living as surgeon and by lectureship at the Royal Institution, but who also relied on the patronage of several influential gentlemen (Atkinson 1999: 35). His tenure was characterised by a cooperative relationship with the specialist societies and the inclusion of an increased number of scientific members in the Society's ruling body. During this period new membership was also restricted, showing the Society's development towards a more scientific orientation: literary and antiquarian interests disappeared in the evaluation of a candidate's worth in favour of a "devotion to literature and natural knowledge, or science" (Hall 1984: 23). Gradually stress was also placed on the fact that Fellows and candidates should also make contributions to the *Transactions*, which had often not been the case. The quality of the papers proposed for publication was to meet higher standards, and new roles were created such as a junior secretary, who was to sort bureaucratic details such as dating, sending rejection letters, and preparing abstracts for the Committee of Papers; a Sub-committee of papers, which decided on the papers to be read at meetings; and specialised sub-committees for the revision of papers within specific fields. Interestingly, although there was a dominant portion of medical men among the Society's Fellows in the 1820s, during this period the physical sciences thrived – in fact, many medical Fellows such as WOLLASTON appeared to be more interested in the physical rather than the biological sciences (Hall 1984: 44).

Opposition, however, continued as DAVY was seen as sympathising with the conservative side of the Society. He also suffered from ill health and on these grounds resigned in 1827.[32] The 1820s in general were marked by dissatisfaction and private criticism within the Society. Gradually, towards the 1830s, discontent became public; starting from a series of incidents, more generalised attacks on

---

32 Successive 19th-century presidents were DAVIES GILBERT (1827–1830); AUGUSTUS FREDERICK, Duke of Sussex (1830–1838); SPENCER COMPTON, second Marquis of Northampton (1838–1848); WILLIAM PARSONS (1848–1854); JOHN WROTTESLEY (1854–1858); Sir BENJAMIN COLLINS BRODIE (1858–1861); General Sir EDWARD SABINE (1861–1871), the last president to hold office for more than five years; Sir GEORGE BIDDELL AIRY (1871–1873); Sir JOSEPH DALTON HOOKER (1873–1878); WILLIAM SPOTTISWOODE (1878–1883); THOMAS HENRY HUXLEY (1883–1885); Sir GEORGE GABRIEL STOKES (1885–1890); WILLIAM THOMSON (1890–1895); JOSEPH LISTER (1895–1900).

the Society and the supposed decline of science in England were published.[33] Points of criticism concerned the presence of too many medical men among the Society's members and the lack of specialisation of the Society and of many of its members. The Society needed reform and greater selectiveness in election procedures. Little criticism was this time directed at the Society's journal, although it was emphasised that there were many Fellows who contributed no papers, and that the Committee was composed by members who lacked expertise on the topics of the papers presented, a consequence of which was that many valuable papers were rejected without an explanation.

The period of criticism was necessarily followed by a period of reform (1830s–1850s), which included a restriction in elections with a maximum of fifteen Fellows to be elected per year – possibly a way to reduce the possibility of admitting non-scientists within the Society (Atkinson 1999: 39). After 1840, prior publication in the *PTRS* became another important criterion for membership. The selection procedure for papers also became stricter and made greater recourse to outside referees (1831); and later seven Sectional Committees covering the main areas of scientific activity were appointed to referee papers (1838), which were then to be submitted to the Committee of Papers. However, this system only lasted a decade and in 1847 the Sectional Committees were dissolved, while the practice of sending most papers to outside parties for the review remained (Fyfe et al. 2015: 15). Another major innovation of the period was the creation of a new journal in 1832, *The Proceedings of the Royal Society*, which was initially devoted to publishing paper abstracts, short news notices, meeting minutes, obituaries and medal awards, but later became a journal in its own right featuring full (shorter) articles. While initially a paper could be published in both journals – in abstract form in the *Proceedings* and in full in the *Transactions* – later paper-length became the main criterion for inclusion in either one or the other journal, although the *Proceedings* continued to feature paper abstracts and to act as a support to the *Transactions* until the 20th century (Atkinson 1999: 40–43; Fyfe et al. 2015: 18).

Another major change in the Society's publications took place in 1887, when the *PTRS* was split into two journals, *Philosophical Transactions A* and *Philosophical Transactions B*, specialising in the physical and biological sciences

---

33 Most influential criticism in this period came from CHARLES BABBAGE's *Reflections on the Decline of Science in England* (1830) and AUGUSTUS BOZZI GRANVILLE's *Science without a Head, or the Royal Society Dissected* (1830 and a second edition was published in 1836).

respectively. Further, in 1896 the Sectional Committees were reinstalled and in 1898 the names of paper referees were no longer written in the Society's register books in order to ensure their anonymity.

As to the Presidents, they continued to feature both professional scientists and gentlemen up until 1885. Since then, the Royal Society's Presidents have all been professional scientists. The number of scientific Fellows also increased and gradually overtook the non-scientific side. Foreign members have regularly continued to be elected to the Society with the numbers gradually decreasing. The period in which the greater quantity of Foreign Members were elected was between 1730 and 1780 with an average of 28 foreigners elected per year – against an average of 69 home Fellows. From that moment on, numbers of Foreign Members decreased to less than 10 Fellows a year, while home Fellows ranged between 90 and 150 (19th century).[34] Today, the Royal Society elects a maximum of 52 ordinary Fellows and 10 Foreign Members a year. From its early years the Royal Society had been in cooperative relations with both individual scholars and broader institutions. As far as Italy is concerned, it will be seen that the Society held mostly one-to-one relations with Italian men, while also showing an interest towards larger Italian academies. In the 19th century especially, new links for cooperation with Italian academies were created.[35]

---

34 The averages here provided were calculated from a table reported in Lyons (1944: 343) providing average numbers of ordinary Fellows and Foreign Members for each five year period between 1665 and 1940.
35 Hall (1984: 195) notes the same in speaking of the Royal Society's foreign relations in general; namely that broader cooperation with foreign scientific academies and societies came only in the last quarter of the century.

Finally, although the 20th and 21st centuries are not here dealt with as they exceed the scope of the present research, relations with Italy after the 19th century continued and the following Italians were elected Foreign Members: AUGUSTO RIGHI (1907), physicist; Sir VITO VOLTERRA (1910), mathematician and physicist; LUIGI LUCIANI (1918), physiologist; TULLIO LEVI-CIVITA (1930), mathematician; ENRICO FERMI (1950), physicist; EDOARDO AMALDI (1968), physicist; RENATO DULBECCO (1974), biologist and Nobel laureate; GIUSEPPE OCCHIALINI (1974), physicist; CARLO RUBBIA (1984), particle physicist and Nobel laureate; LUIGI LUCA CAVALLI SFORZA (1992), geneticist; RITA LEVI MONTALCINI (1995), neurobiologist and Nobel laureate; UGO FANO (1995), physicist; MICHELE PARRINELLO (2004), physicist; LUCA CARDELLI (2005), computer scientist; DARIO ALESSI (2008), biochemist; MAX PETTINI (2010), astronomer; EZIO RIZZARDO (2010), polymer chemist; MARIA GRAZIA SPILLANTINI (2013), molecular neurologist; TULLIO POZZAN (2018), professor of General Pathology; RINO RAPPUOLI (2016), microbiologist; and FABIOLA GIANOTTI (2018), physicist.

1.2.2 The Italian states and their academies

While the English end of the Anglo-Italian scientific exchanges is focused on one society, the Italian end is given by many individual scholars and various Italian academies based in different parts of the Peninsula. Italy's geo-political history changed considerably over the course of the three centuries under study, and the Kingdom of Italy did not in fact exist until 1861. Hence, while it was common to refer to the inhabitants of the Peninsula as *Italians*, even when its territory was divided into a number of political entities, the main historical events that characterised Italy up to its unification and the political division of its territory must be borne in mind. Indeed, the Italian political boundaries, the Church, state sovereigns, and wars, influenced possibility of travel, correspondence and publication.

Briefly, in the 17th century the Italian Peninsula comprised the Duchies of Savoy, of Milan, of Modena, and of Parma and Piacenza; the Republic of Venice, which included the cities of Padua, Bergamo and a number of ports on the opposite coast of the Adriatic Sea; the Republic of Genoa (inclusive of Corsica until 1768), and of Lucca; the Grand Duchy of Tuscany; the Papal States, which comprised the regions of Umbria, Marche and the cities of Bologna and Ferrara; and, in the south, the Kingdoms of Sardinia, of Naples,[36] and of Sicily. Moreover, a considerable part of the Italian territories were under Spanish rule until the early 18th century.[37] Only the Papal States and the Republic of Venice maintained their own independence. At this time, Italian scientific activity was related to its universities and academies. The most active intellectual circles were in Tuscany, Rome, Naples, Bologna and Padua. However, Italian science, even outside the Papal States, was heavily constrained by the controlling rule of the Catholic Church and the Inquisition, especially after the condemnation of GALILEO for his support of heliocentrism in 1633. The activity of the Italian academies generally revolved around Princes and individual patrons of learning; as a consequence, their existence was conditioned by the political and financial interests of their patrons.[38]

---

36 The Kingdom of Naples included the regions of Abruzzo, Molise, Campania, Puglia, Basilicata, and Calabria.
37 The Duchy of Milan and the southern Kingdoms of Sardinia, Naples, and Sicily.
38 The following description of Italian academies in the 17th century largely draws on Clericuzio (2013). All of the academies here mentioned were known to the Royal Society and while some of them had close connections with it, others were mentioned in their epistolary exchanges and *PTRS* articles. For a general study on Italian academies see Testa (2015).

Academies with scientific objectives included the Accademia dei Lincei, founded in Rome in 1603 by FEDERICO CESI. The members of the Lincei were devoted to the study of nature and mathematics. CESI had moreover planned to create branches of the academy throughout Italy, Europe and other continents. One of these branches was created among the intellectuals of Naples, although this circle was never very active. In 1611, GALILEO was made a member, and he further created links with the academy among his friends in Tuscany. Eventually, however, after CESI's death in 1630 and the condemnation of GALILEO, the first Lincei came to an end. Although the Academy no longer existed by the founding of the Royal Society, their history and the work of various members were known to the Fellows in London. Moreover, later, in the 19th century, a new version of the Academy was created as the Accademia Pontificia dei Nuovi Lincei (1847) and cooperative relations between the two institutions were created.

In Rome, scientific activity was also carried out at the Jesuit Collegio Romano, which had been in existence since the 16th century, and in the Accademia Fisico-Matematica, generally associated with Queen CHRISTINA of Sweden, who acted as its patron. The Roman Accademia was founded in 1677 by GIOVANNI GIUSTINO CIAMPINI who, together with MICHELANGELO RICCI and FRANCESCO NAZZARI, also created the Roman *Giornale de' Letterati*, on the model of the *Philosophical Transactions*. This academy was devoted to the study of mathematics, medicine and mechanics. Among its members were also GIOVANNI ALFONSO BORELLI, the telescope maker GIUSEPPE CAMPANI, and the astronomer FRANCESCO BIANCHINI.

Another important scientific academy which was to have close connections and several members in common[39] with the Royal Society was the Accademia del Cimento. The Cimento was founded in Florence in 1657 by Cardinal LEOPOLD DE' MEDICI and his brother the Grand Duke of Tuscany FERDINAND II. A branch of the Cimento was also active in Pisa with an informal group of physicians and naturalists connected with GIOVANNI ALFONSO BORELLI (these included CARLO FRACASSATI, MARCELLO MALPIGHI, LORENZO BELLINI, and NICOLAS STENO). The MEDICIs had already made of Florence and Tuscany a primary centre of Italian science by collecting ancient scientific texts for their library and by employing mathematicians and architects in the construction of ports and fortresses, as

---

39 Members of both societies were for instance GIOVANNI ALFONSO BORELLI, VINCENZO VIVIANI, MARCELLO MALPIGHI, and LORENZO MAGALOTTI. A few members of the Cimento, such as FRANCESCO REDI and CARLO FRACASSATI, were not elected but corresponded with the Royal Society.

well as the enhancement of the botanical garden of the University of Pisa. The Cimento's motto was "provando e riprovando" (trying and trying again), which emphasised the experimental character of the academy. Its members concerned themselves with various branches of learning, including mathematics, geometry, physics, chemistry and medicine. However, the researches were generally carried out at the request of the Prince and the Grand Duke, and the members followed them in their various places of residence in Tuscany (Palazzo Pitti in Florence, Livorno, Pisa, and on the Elba Island). Some of the research of the Cimento – on air pressure, thermometry, phase transition, acoustics and more – was published in the *Saggi di Naturali Esperienze fatte nell'Accademia del Cimento*. The book was written by the Academy's secretary, LORENZO MAGALOTTI, and it took several years to be completed as it was frequently restyled both for linguistic reasons and to avoid censorship from the Inquisition. Hence, the *Saggi* were published only at the end of the Cimento's existence in 1667, by which point experiments reported in the book had already been carried out in other European countries and no longer represented a novelty.[40] The book moreover had only reported some of the researches of the Cimento; studies carried out on planet Saturn, for instance, were not included, probably because they would have revealed an inclination towards Copernican ideas (Clericuzio 2013). Worthy of notice is that, like the Royal Society, the Cimento chose to present their findings to the public in their own vernacular (Tuscan), rather than in Latin.

In Naples TOMMASO CORNELIO and LEONARDO DI CAPUA founded the Accademia degli Investiganti (ca.1663). CORNELIO was professor of mathematics and medicine at the University of Naples and was inspired by European experimentalism and the medical and scientific ideas of PIERRE GASSENDI, WILLIAM HARVEY, THOMAS WILLIS and ROBERT BOYLE. The patron of the Investiganti was ANDREA CONCLUBET, Marquis D'Arena, who wanted to promote scientific innovation in the Kingdom of Naples along the lines of the Royal Society. A number of English Fellows kept contacts with and visited the Academy, such as JOHN FINCH, THOMAS BAINES, JOHN RAY, FRANCIS WILLOUGHBY, and PHILIPP SKIPPON. The Investigantes concerned themselves primarily with medicine and chemistry, but also with physics and atmospheric pressure. Their existence was characterised by conflict with traditional galenic physicians and eventually the Academy was closed down in 1668, although its members continued to carry out their research elsewhere.

---

40 On the reception of the *Saggi* see Boschiero (2010), Gomez Lopez (1997), and Knowles Middleton (1969).

In Bologna ANTONIO FELICE MARSILI, both a powerful clergyman and an experimental philosopher, recreated in 1687 the Accademia dell'Arcidiacono (previously founded in 1656), which was divided into two distinct branches: the one religious, focused on a critical study of the Church; the other focused on experimental philosophy. The Academy also had relations with BENEDETTO BACCHINI and his *Giornale de' Letterati* of Parma (1686), where some of the discourses of the Academy were published. Among the members of this Academy was DOMENICO GUGLIELMINI, who became a Fellow of the Royal Society in 1698. Also active in Bologna, from the late 16th century to the end of the 17th was the Accademia dei Gelati founded by MELCHIORRE ZOPPIO with the brothers BERLINGERO, CAMILLO and CESARE GESSI, which was more of literary bent. Also to be mentioned is the Accademia degli Inquieti (ca. 1694), which was an informal group of scholars – among which was EUSTACHIO MANFREDI – with both literary and scientific interests. The group initially met in the residences of its members and later, in 1704, they moved to the palazzo of Count LUIGI FERDINANDO MARSILI. Other Bolognese academies to follow experimental pursuits between the 1680s and the 1690s were the Accademia delle Traccia and the Accademia del Davia (Cavazza 1980: 106).

What transpires from the Italian scientific context of the 17th century is that most of the individuals active in experimental philosophy were either directly or indirectly connected with one another; in their movements across the Peninsula they met, discussed science and promoted natural philosophical learning. At the same time, they were often forced to dissimulate or self-censure their writings for fear of the Inquisition. Many of them, starting from GALILEO, tried to live their religious and scientific interests in harmony, rather than in opposition.

In the wider European context, other academies that played an influential role in the history of science were the French Académie des sciences, founded in Paris in 1666; the German Academy of Sciences Leopoldina, also known as the Academy of the Curious of Nature, founded in 1652; and the Berlin Academy of Sciences (1700). Among the most widely circulating scientific journals were – other than the *Transactions*, and the Italian *Giornali de' Letterati* – the French *Journal des Sçavans* (1665) and the German *Acta Eruditorum* (1682–1782).

In the 18th century Italy was still fragmented into a number of states, duchies and republics; the major changes concerned the Kingdom of Sardinia, which was now ruled by the Savoyard dynasty, with Turin as its capital, and comprised Piedmont, the Aosta Valley, Nice, and Sardinia (after 1720). The Grand Duchy of Tuscany, while maintaining its independence, passed from the now extinct House of Medici to the control of the House of Hapsburg-Lorraine in 1737 – and would remain under their influence, with a brief interruption during the

Napoleonic period, until the unification of Italy. Lombardy, Trentino-Alto Adige and Venezia Giulia were also under Austrian control and part of the Austrian Empire (1700–1796). The Duchy of Parma and Piacenza came under the control of the Italian Bourbons in 1731. And the southern territories were under the Bourbon dynasty for most of the century.[41]

Before moving into the description of the relevant 18th-century Italian academies, an important aspect that influenced European and Italian science in the late 17th and 18th centuries was the reception of Newtonianism. One of the main scientific areas of study in the northern parts of Italy was the study of waterways (hydrology), which exploited mathematical techniques to learn about and control physical phenomena related to rivers and irrigation systems (Mazzotti 2013). News of Italian research in hydrology, such as VALLISNERI's studies on fountains and rivers, also reached the Society through its contacts. The economic, social and political importance of controlling waterways frequently led to disputes between the Papal States, the Republic of Venice and the Empire. Hence, in this context NEWTON's mathematical principles found favourable reception, especially in Bologna, Padua, Venice and Ferrara. Scholars such as GIOVANNI POLENI, JACOB HERMAN, JACOPO RICCATI, and STEFANO DEGLI ANGELI carefully studied the *Principia mathematica* (1687) and critically considered them for their own research interests. Not all Italians however favourably received Newtonianism; an example was GIOVANNI RIZZETTI who was to criticise NEWTON's theory of colours.[42] Newtonian ideas started being more openly discussed among the learned only in the 1730s; MARIA GAETANA AGNESI, for instance, openly defended NEWTON's theories of colours, magnetism and tides. Newtonianism was also favourably received at the Accademia Fisico-Matematica and the Accademia degli Antiquari Alessandrini in Rome, and in Naples with CELESTINO GALIANI and NICOLA CIRILLO among others.

---

41 Following the war of Spanish succession and the treaty of Utrecht, the Kingdom of Sicily saw brief interruptions to Bourbon control between 1713 and 1720, when the House of Savoy took control, and between 1720 and 1734 when Sicily passed under the House of Hapsburg. The Kingdom of Naples was also under Austrian rule between 1713 and 1720. When the two Kingdoms were reconquered by the House of Bourbon in 1734 (Battle of Bitonto) they were united into the Kingdom of Naples and Sicily, although the unification was officialised only in 1816, when the two formerly distinct Kingdoms became the Kingdom of the two Sicilies.

42 NETWON's theory of colours was presented in his *Opticks* (1704). On the dispute see Section 4.2.3.

The socio-historical contexts 37

In the 17th century Italy had created more scientific academies than any other part of Europe; yet none of them remained active by 1700 (Findlen 2009: 4). Nonetheless, Italian academies still played a primary role in Italian science, and the 18th century saw not only the birth of new academies but also of new branches of learning and, geographically speaking, of new scientific centres, especially towards the end of the century.

In Bologna, the Count LUIGI FERDINANDO MARSILI, brother of ANTONIO FELICE, founded the Istituto delle Scienze e delle Lettere in 1714,[43] which was closely connected with the University. MARSILI designed the Institute inspired by the Royal Society and wanted it to be not only a centre of research but also of experimental teaching, complete with library, museum of natural history, laboratories of physics and chemistry, and an astronomical observatory (Cavazza 2002). Towards the end of the century the Institute became one of the main centres for the study of electricity.

Florence, thanks to the previous work of the MEDICIS, continued to treasure scientific books and collections of natural history. Worth mentioning is the Real Museo di Fisica e Storia Naturale (1775–1878). The museum, based in the Torrigiani palace, was created at the will of the Grand Duke of Tuscany PETER LEOPOLD and housed the natural historical collections of the Medicis, which were formerly spread among various Florentine palaces (Palazzo Pitti, the Imperial gallery, and the Uffizi gallery). The cataloguing of the collections was carried out by GIOVANNI TARGIONI TOZZETTI and the museum was directed by FELICE FONTANA; both men and their work were known to the Royal Society. In 1780, FRANCESCO ZACCHIROLI, FONTANA, and GIOVANNI FABBRONI[44] proposed the transformation of the museum into an academy, but the proposal was rejected by the Grand Duke on the grounds that the project would have required incredible resources without guarantee of success (Borelli 1997: 573). The museum however provided a place for these and other intellectuals to gather and discuss science.

Naples housed the Reale Accademia delle Scienze e delle Belle Lettere (1778).[45] The Reale Accademia administered a number of institutions: a museum, a botanical garden, an astronomical observatory, the school of medicine of

---

43 It incorporated two academies, the Academia degli Inquieti and the Academy of Arts ("Clementina").
44 FABBRONI sent a paper to the Royal Society, "Chemical actions shown by simple contact to two metals" (1794, L&P/10/69), which was read but not published.
45 The founding of the Reale Accademia has its roots in a number of previous academies: the Accademia Palatina (1698); the Accademia delle Science, founded by CELESTINO GALIANI (FRS 1735) in 1732; and the Reale Accademia Ercolanense (1778).

the Ospedale degli Incurabili, and the Academy of painting, sculpture and architecture. While some have judged Neapolitan science to be in decline at the end of the 18th century, others have revaluated it; the Kingdom not only was not lagging behind in the fields of medicine and mathematics, but played a relevant role in the spreading of technical-scientific knowledge (Borelli 1997: 573–574).

In Verona, in 1782, ANTONIO MARIA LORGNA founded the Società italiana delle scienze, also known as the Society of the Forty because it came to include among its members forty of the most eminent Italian men of science. LORGNA had conceived the academy much earlier, in 1766, and he wanted it to be an Italian scientific Society above the political divisions of Italy. Among its supporters and early members were ALESSANDRO VOLTA, LAZZARO SPALLANZANI and RUGGERO GIUSEPPE BOSCOVICH.[46] The Society is today the Italian national academy of science, under the name of Accademia nazionale delle scienze, based in Rome.

Turin became an important centre for the study of medicine and electricity for medical purposes. Among the most innovative medical scholars of Piedmont was CARLO ALLIONI (FRS 1758). Moreover, in 1757 a private society, the Società Scientifica Privata Torinese, was created by the Count GIUSEPPE ANGELO SALUZZO di Monesiglio (FRS 1760) in collaboration with GIUSEPPE LUIGI LANGRANGIA (FRS 1761). The society was devoted to the study of mathematics, mechanics and physics. In 1783 it was officially recognised as the Academia delle Scienze di Torino.

Finally, Hapsburg Lombardy had as its main cultural centres Milan, Pavia (with the University), and Mantova. Milan was home to various academies. Worth mentioning is the Accademia Clelia de' Vigilanti, conceived around 1722 by CLELIA BORROMEO in collaboration with ANTONIO VALLISNERI (FRS 1703). The Royal Society, through the mediation of THOMAS DEREHAM, had shown interest in the Clelian Academy from its inception. BORROMEO's plan was to create an academy that not only served Milan, but the whole of Italy and wanted to revive and improve the greatness of the Florentine Cimento. The interests of the Clelian academy included the liberal arts but were primarily focused on natural history, mathematics, mechanics, physics, botany, medicine, anatomy and chemistry. While a detailed plan for the academy, its objectives and members had been created, and a lively international network of correspondence was set up in its

---

46 BOSCOVICH (FRS 1761) was a Jesuit priest, and a polymath from the Republic of Ragusa, in Dalmatia, which was under the influence of the Republic of Venice. BOSCOVICH studied in Italy and France and eventually settled in Rome. His main studies were in the fields of astronomy and physics.

favour, the actual realisation of the project proved very difficult and the academy never quite took off.⁴⁷ Moreover, the Austrian empress MARIA THERESA founded in Milan the Società Patriottica (1776–1798) for the promotion of agriculture, fine arts and manufactures; and in Mantua she founded the Accademia di Scienze e Lettere (1768), which in the early 20th century was in contact with the Royal Society. The Mantuan academy – still existing today as the Accademia Nazionale Virgiliana – concerned itself with most branches of learning, including the natural, physical and mathematical sciences. While under Spanish control, Milan had perished, however, under the Habsburg dynasty – when a series of reforms were promoted by MARIA THERESA and her son JOSEPH II and involved all cultural activities from the humanistic and scientific to the technical and administrative (Berzolari 2002: 48) – Milan saw the improvement of its botanical gardens, the astronomical observatory, and libraries. The libraries, such as the Braidense and the Ambrosiana, were reorganised in order to respond to modern requirements, and it is in this period that the role of the professional librarian started developing (Borelli 1997: 577).

At the opening of the 19th century, most of the Italian Peninsula was under NAPOLEON's control (1796–1815) and its geo-political division had been greatly altered. After the Congress of Vienna and the recreation of the Ancien Régime, Italy's territories were once again modified.⁴⁸ However, the French revolution and the oppression of the great powers after the Congress contributed to the enhancement of nationalistic views and the desire for independence and unity. Hence, Italy was soon to take part in the revolutionary waves of the 1820s and 1830s. After these unlucky revolutions, the first and the second wars of Independence, and the final movements to free Italy from foreign hegemony, eventually led to the Italian unification in 1861 and the creation of the Kingdom of Italy.⁴⁹

---

47 See Findlen (2009) for a detailed account of the Clelian academy.
48 Post-congress Italy comprised: the Lombardo-Venetian Kingdom, which was part of the Austrian Empire and comprised the former Republic of Venice, the whole of Lombardy, Friuli and Valtellina; the former Kingdom of Sardinia, which now also included the territories of the previous Republic of Genoa; the Grand Duchy of Tuscany continued to be ruled by the House of Hapsburg-Loraine; and under Austrian influence were also the Duchies of Modena, Parma, and Piacenza, while after 1847 they returned to the Bourbons of Parma; the Kingdoms of Naples and Sicily were united in the Kingdom of the two Sicilies (1816) with the Bourbon FERDINAND IV as King; and the Pope was restored in the Papal States, although these too lost some of their territories.
49 Veneto, Friuli, Mantua (controlled by the Austrians) and Lazio (Papal States) were not part of the Kingdom as early as 1861, they were annexed later in 1866 and 1870.

Despite the turmoil that characterised Italy in the 19th century, Italian scholars, who were often also active patriots, continued to pursue their scientific interests. Among the most significant aspects of 19th century science in Italy were the first conferences of Italian scientists held annually between 1839 and 1847 (the first was held in Pisa), and then again in 1862 (Siena), 1873 (Rome) and 1875 (Palermo) (Ingaliso 2011). In Lombardy, biology and experimental physics gained primary importance through the work of LAZZARO SPALLANZANI and ALESSANDRO VOLTA; and the late 18th-century studies of the Bolognese LUIGI GALVANI in animal electricity continued to influence the research of physiologists and physicists throughout Europe. In the Papal States, science was still influenced and limited by religious control. In Naples, the first centre for research in volcanology, the Osservatorio Vesuviano, was established in 1845 under the direction of the physicist MACEDONIO MELLONI. Another important contribution to the advancement of science came from the chemist STANISLAO CANNIZZARO, who was able to convince the international scientific community of the distinction between the atom and the molecule, which would set the basis for modern atomic theory. The Milanese CARLO CATTANEO founded the scientific journal *Il Politecnico* in 1839 for the benefit and intellectual growth of all. These are only a few examples of Italy's 19th-century scientific contributors and their activities. What characterises the new men of science is a greater specialisation, compared to the previous centuries, and their activeness in Italian political history.

Pre- and post-unification Italian scientists also took part in the intellectual circles of the academies, some of which were to establish cooperative links with the Royal Society, especially at the close of the century. Worthy of mention is the Istituto Lombardo Accademia di Scienze e Lettere, founded by NAPOLEON in 1797 and initially based in Bologna. The aim of the Istituto Lombardo was the investigation and improvement of the arts, the sciences, and philosophical thought. Among the first 31 members worth mentioning are ALESSANDRO VOLTA, ANTONIO SCARPA, and BARNABA ORIANI. After 1810 the headquarters of the Institute were moved to Milan, with branches in Venice, Bologna, Padua and Verona. The Roman Lincei were also revived in the Accademia Pontificia dei Nuovi Lincei (1847) and new links were created between the Accademia dei Fisiocritici of Siena and the Royal Society.

1.3 The Royal Society and Italy: A review of relevant literature

Among the studies on the relations between the Royal Society and Italy some are more specialised and focus on individual Fellows of the Royal Society and their

Italian connections and, vice versa, on Italian scholars and academies and their connections with the Society; while others provide a more general picture. This section will report on those studies of more general character.

Hall, in her paper "The Royal Society and Italy 1667–1795" (1982) provides a picture of the Society's connections with Italy in the 17th and 18th centuries. Here, Hall highlights a few key points that ought to be borne in mind when dealing with the Society's relations with foreign members and correspondents, namely:

1. that the importance of individual scientists and their writings lasted for much longer than their own life time. Nowadays, researchers may consider a twenty-year-old piece of writing dated, and this can be reasonable in our rapidly moving society. Yet, it must not be forgotten that the further one goes back in time the slower and more difficult to achieve scientific advances were. Thus an important discovery could remain fresh and unsurpassed for a very long time. Hall exemplifies this by demonstrating how a few 17th-century Italian scientists, who were very influential in the shaping of the Society itself, were still being read and cited well into the 18th century. Among them were GALILEO GALILEI, promoter of the new experimental science and, as a Fellow, MARCELLO MALPIGHI (FRS 1669), physician and "perhaps the leading scientist of the thriving circle centred around the University of Bologna" (1982: 64). Indeed, it will be seen that MALPIGHI was still being referenced and praised even in the 19th century;

2. that there were severe communication issues, given by the difficulty in transmitting letters and books between the two countries. This was especially true in the case of books, which required trustworthy travellers (either merchants or diplomats) who were willing to take the books from one country to another. Even more troublesome was finding learned men with whom the Society might exchange scientific knowledge;

3. that the Royal Society had an unbalanced knowledge of what went on in the Italian scientific world. This is partly due to the communication issues reported above, but especially to the scientists' individual interests and their efforts in keeping correspondence and collecting and reporting information. For instance, the predominance of a topic in a given period often reflects the interests of the president in charge. Thus, during FOLKES' and PRINGLE's presidencies, (1741–1752 and 1772–1778 respectively) a relatively conspicuous number of Italian papers on classical subjects and archaeology appeared in the *Transactions* reflecting the presidents' taste for "antiquaries" (1982: 73);

4. finally, that very often Italians or Englishmen were not made Fellows because of their scientific achievements but for their potential role as correspondents and intermediaries.

But how were contacts with Italian academics established and how was the Society able to learn about the Italian scientific advances? In the case of notable men it was generally the Society's secretaries who got directly in contact with them. In other cases, it was the Italians who contacted the Society, either because they were interested in receiving information on the Society's doings – as in the case of FRANCESCO NAZZARI, who regularly published extracts from the *Transactions* in his *Giornale de' Letterati* (founded 1668) – or because they sought approval for their own work. Generally Englishmen residing in Italy were very helpful to the Society in reporting the doings of the academic circles in Venice, Florence, Rome and Naples. Hall mentions for instance Dr HENRY NEWTON (FRS 1709) who took up residence in Florence between 1709 and 1714; the baronet Sir THOMAS DEREHAM (FRS 1720), who lived in Rome in the first part of the 18th century; and the English plenipotentiary Sir WILLIAM HAMILTON, who was in Naples from 1764 to 1800.

Finally, Hall mentions the main topics of Italian research that interested the Society in the first 130 years of its existence; namely medicine, botany, astronomy, meteorology, earthquakes and volcanic eruptions, mathematics, archaeology, classical studies, and, later, experimental physics, electricity and animal electricity. Her paper also includes a list of all Italian Fellows between 1667 and 1795 for a total of 110 members, four of which were resident in England (GREGORIO LETI, DOMENICO FERRARI, PAOLO ROLLI, General PASQUALE DE PAOLI and TIBERIO CAVALLO).

Cavazza, in two papers published in 1980 and 2002, reports on the relations between Bologna and the Royal Society in the 17th and 18th centuries. She explains that 17th-century Bologna, while in social and economic decline and heavily controlled by the Inquisition (being part of the Papal States), was still able to host a number scholars who attempted to disseminate the aims, methods and organisation of the Florentine Accademia del Cimento, among which were MALPIGHI, GEMINIANO MONTANARI, and GIOVANNI DOMENICO CASSINI. The Bolognese scholars became aware of the Royal Society and its experimental programme and soon echoed its principles in their own academies and writings. Cavazza also provides examples of how the Italians were eager to receive the Society's judgement and how they asked for the Society's help in solving controversies. The Society inspired the Bolognese with its political and religious neutrality; they realised that science needed to become independent.

Cavazza then focuses on the Istituto delle Scienze e delle Arti (1714), which was founded on the model of the Royal Society and of the French Academie des Sciences. The following paper (2002) provides more detail as to the relations between the Institute and the Royal Society in the 18th century. Cavazza reports on the reception of Newtonianism and the censorship exerted by the Inquisition on scientific books suspected of Copernicanism. Hence, the Bolognese, rather than openly expressing Newtonian ideas, showed their appreciation through experimental contributions, i.e. by repeating and corroborating NEWTON's experiments and findings. The members of the Institute sent their own scientific writings to the Royal Society; EUSTACHIO MANFREDI for instance sent his astronomical observations, while FRANCESCO MARIA ZANOTTI sent the first publication of the Institute, the *Commentarii*, which he had edited himself as the Institute's secretary. The Royal Society reciprocated with the *Transactions* and the Catalogue of its Fellows. A number of Fellows (mainly English ambassadors at Naples) were also given membership to the Institute. The paper concludes with a focus on the Bolognese studies in electricity and the Fellows' reception of them.

A more negative picture of the early Royal Society's relations with Italians is provided by Gomez Lopez in her paper "The Royal Society and post-Galilean science in Italy" (1997). Differently from Cavazza, who focuses on the examples of the Society's appreciation towards Italian science, Gomez Lopez points out "the feelings of suspicion towards the scientific contributions of the Italians" (1997: 37). Moreover, contrary to the previous studies, Gomez Lopez claims that new science and Protestantism "were inseparable elements" and that the Society's suspicions towards Italian science were likely due to religious differences. She provides some evidence for this by highlighting that only two Italians were elected to the Society in the 1660s.[50]

In his paper "Rome and the Royal Society, 1660–1740" (2004), Cook reminds us that several of the Englishmen that would later form the Royal Society travelled to the Papal city and maintained relations with the Romans even before the founding of the Society. They were attracted to Rome for its antiquities – which learned travellers already knew through their classical education – but they also met with the intellectuals creating links with the Accademia dei

---

50 The number of Italian contributions to the journal in the first years is however considerable. Moreover, while Gomez Lopez provides an instance in which OLDENBURG stated his doubts as to FRANCESCO NAZZARI's intentions to cooperate, it will be seen that initially it was especially OLDENBURG who sought contact with Italians.

Lincei and later with the Accademia Fisico-Matematica. In Rome moreover the English could count on the presence of other fellow countrymen residing there, especially around the Venerable English College. Cook explains that GALILEO's trial had discouraged speculative natural inquiry in the capital; however, in the Collegio Romano the Jesuit professors still pursued astronomy and after 1664 Queen CHRISTINA of Sweden created a lively circle of intellectuals and would later extend her patronage to the Accademia Fisico-Matematica. Towards the end of the 17th century, after the death of Queen CHRISTINA and the outbreak of the Nine Years War (1688–1697) contacts with Rome had become more difficult and appear to have been interrupted for some time. However, later, especially after the Peace of Utrecht (1713) correspondence and visits were resumed. Cook then provides detail into the visits of individual Fellows such as JOHN RAY, EDMOND HALLEY, GILBERT BURNET, and GOTTFRIED W. LEIBNIZ, explaining that natural philosophy was a minor concern for most of these travellers who were far more interested in the remains of the imperial past and the great churches and palaces of the Papal city. They did however meet notable natural philosophers and their reports to the Society led to further contact between London and Rome.

Rusnock (1999) focuses on the Royal Society's correspondence network in the first half of the 18th century. While her paper is not limited to Italian relations with the Society, a considerable part of it focuses on the Italian correspondence, which is indicative of the weight of Italian contributions to the Royal Society at this time. Rusnock explains that the Royal Society was seen as an arbiter to which natural philosophers turned for judgement and approval. However, despite the authoritative reputation that the Royal Society had gained, it tried to avoid judgements and limited itself to selecting writings for publication, leaving the judgement part with the reading public. In various instances the Italians insisted on receiving the Society's opinion,[51] and at a certain point the secretary JAMES JURIN became somewhat exasperated and in a letter to DEREHAM wrote:

> The Italian Virtuosi always expect, & you seem to require, that I should send you ye Opinion of ye Society upon ye several Papers you transmit to me. But this is what ye R.S. never gives. They pronounce no Judgement upon what comes before them, but only return their thanks to ye Authors. (JURIN to DEREHAM, 1729, in Rusnock 1999: 162)

Her study also emphasises the important role of correspondence and how relations with foreign natural philosophers highly depended on the individual efforts of the secretaries in encouraging exchange and cooperation.

---

51 See also Cavazza (1980 and 2002) and Section 3.2.3.

D'Amore in *The Royal Society and the discovery of the two Sicilies. Southern routes in the Grand Tour* (2017), provides a new perspective on *Grand Tour* literature and on Anglo-Italian relations by investigating the Fellows' interests and travel itineraries in the south of the Peninsula and by assessing the impact that the Royal Society and its publications had on the *Grand Tour* experience. D'Amore traces the Fellows' paths in their Italian travels "starting in Restoration times and culminating with the Fellows' 'discovery' of mysterious Sicily at the end of the eighteenth century" (2017: 2). The interests that brought the Fellows to the Italian south were connected to their Baconian principles and desire of studying and controlling nature. Italy was moreover rich in antiquities, which made it even more attractive, especially in the period in which modern archaeology was born and the Neoclassical mode was developing. Archaeological findings – most importantly the discovery of the buried Roman city of Herculaneum in 1738 – and the devastating eruptions of Vesuvius and Etna opened a new chapter in the *Grand Tour* of Italy. Previously, the final Italian destination had been Rome, now Grand-tourists were realising that the Kingdom of Naples and the "wild and mysterious" Sicily were full of wonders to discover. In describing and discussing the Fellows' writings about the Kingdoms and their explorations, D'Amore also provides details into the relations between the travelling English Fellows and their Italian hosts.

Knowles Middleton (1979) instead focuses on "Some Italian visitors to the early Royal Society". By reporting about the visits of LORENZO MAGALOTTI and PAOLO FALCONIERI, FRANCESCO RICCARDI and ALESSANDRO SEGNI, COSIMO DE MEDICI, ERCOLE ZANI, and LORENZO PANCIATICHI he reminds us that the admiration of the Italians was such that a visit to the Royal Society became a *must* for Italian travellers. And indeed travelling Italians continued to pay their visits to the Royal Society and to attend its meetings well into the 18th century.

Finally, since this study is also interested in language use and translation practices, worthy of mention is Knowles Middleton's analysis (1969) of the translation of the title of the *Saggi di Naturali Esperienze*, the book published by the Accademia del Cimento in 1667. It was previously stated that the *Saggi* were not so well received in England, after LORENZO MAGALOTTI and PAOLO FALCONIERI had personally delivered them to the Society in 1668. Yet in 1683 RICHARD WALLER was appointed to make an English translation of the book.[52] The translation was published in 1684 as *Essays of Natural Experiments, made in the Accademie del Cimento &c.* The Italian noun *saggio* (pl. *saggi*) however

---

52 On the translation of the book see Boschiero (2010).

presents an ambiguity, in that it can both mean *essay*, as WALLER understood it, but it can also mean *assay*, as in testing a substance, or *sample*, as in a part or an example of a whole. The preface to the later Italian editions of the book would seem to confirm that the meaning of *Saggi* was *samples*, in that the writer of the preface explains that the title is an announcement that the experiments described in the book represent "a selection of a few among many". This brief analysis is worthy of notice in that it further highlights the problems that could arise from international exchanges and miscommunication; had the Fellows been aware that the book only contained a minor portion of the Cimento's experiments, their expectations and final judgement might have been different.

To conclude, the studies described here provide relevant insights into the Fellows' connections with the various Italian States, especially in the first century of the Society's existence. It is hoped that this book will complete the picture not only by providing new detail, but also by drawing a full map of the Fellows' interests in and connections with the whole of the Peninsula and the development of their relations through the three centuries. Indeed, it will be seen that in the 19th century the Royal Society's interest in Italy did not wane; the Fellows continued to visit Italy both for cultural and scientific reasons and maintained friendly relations with Italian scientists. The forms of cooperation developed, becoming more scientifically oriented. Also, while in the 17th and 18th centuries Italians who visited the Society had the privilege to attend meetings and discuss natural philosophy with the Fellows, later they were granted more prestigious roles by participating in projects and expeditions and holding lectures at the Royal Society.

Chapter 2 Methodological framework

2.1 The groundwork	47
2.1.1 Critical Discourse Analysis	47
2.1.1.1 The Discourse-Historical Approach (DHA)	48
2.1.1.2 *Discourse and social change* – Fairclough's approach	50
2.1.1.3 Politeness	56
2.1.2 A socio-historical approach to scientific writing	58
2.2 Approach to text analysis	62
2.3 Collection of the papers and analysis	65

This chapter starts by introducing major approaches towards the study of discourse – in CDA, pragmatics and socio-historical linguistics – that have inspired the methods and procedures adopted in the present study (Section 2.1 and subsections). Subsequently, Section 2.2 will present the approach developed for the present analysis and the main analytical tools employed, which are strictly related to the objectives of the study; while Section 2.3 will provide more detailed information as to how *PTRS* papers related to Italy and Italian research were retrieved and will outline the various steps of the analysis.

2.1 The groundwork

The approach to the analysis of the papers was inspired and shaped by a series of theoretical works and analyses in critical discourse studies, pragmatics and socio-historical linguistics. However, given the very different purpose of the present study from that of the works that will now be briefly introduced, none of them can be considered the sole or main inspirational source but all of them have contributed valuable tools for a discourse analysis aimed at gaining insights into Anglo-Italian relations.

2.1.1 Critical Discourse Analysis

Making discourse analysis critical means not only exploring language use in connection with the social and political contexts in which it occurs, but also trying to reveal hidden or simply out-of-sight values, positions and perspectives. By carrying out critical discourse analysis (CDA) the analyst goes beyond the level of description and attempts to explain why the text is as it is and what it is aiming to do. The main interest then is not on the linguistic unit *per se* but rather analysing, understanding and explaining social phenomena.

Although this is not the case in the present study, CDA tends to be problem-oriented, i.e. it generally focuses on social and political issues. However, Wodak and Meyer (2016) also point out that any social phenomenon lends itself to critical investigation and not only "exceptionally serious social or political experiences or events" (2–3).

Two critical approaches that are relevant to the present analysis – the discourse-historical approach by Wodak and Reisigl (2016), and Fairclough's approach as explained in *Language and Social Change* (1992) – will now be outlined.

2.1.1.1 *The Discourse-Historical Approach (DHA)*

The DHA as a branch of CDA is an interdisciplinary approach that combines linguistic analysis with historical and sociological approaches. What distinguishes it from other approaches is its special focus on the historical embedding of the object under study. At the heart of DHA, as well as of most critical approaches, lie the concepts of *critique*, *ideology* and *power*, although not all of them will be equally relevant to the present piece of research.

The linguistic interest in ideology – i.e. a worldview and a system composed of related mental representations, convictions, opinions, attitudes, values and evaluations, which are shared by members of a specific social group. […] (Wodak & Reisigl 2016: 25) – lies in the way ideologies are created, maintained, enhanced, negotiated and changed through language use. Discourses[53] do not only mediate ideologies but also help gain, maintain or lose *power*, which is defined as "an asymmetric relationship among social actors who have different social positions" (26).

The discourse historical approach focuses on discourse as strongly related to its context and thus distinguishes four levels of context: (1) co-text, i.e. contextual linguistic information provided within the same text; (2) the intertextual and interdiscursive relationships between texts, i.e. explicit and implicit links between texts and genres (see also Fairclough's approach below); (3) the social

---

53 *Discourse* is treated in CDA and DHA as language in use. It is moreover seen as socially constituted and socially constitutive and is thus considered a form of social practice. This view of discourse is the one adopted in the present study.

variables and institutional frames of a specific "context of situation"; and (4) the broader socio-political and historical context (31).

In analysing the texts the following three steps from DHA were followed: (1) identifying the specific content or topic(s) of the sampled piece(s) of discourse; (2) investigating the presence and use of any discursive strategies; and (3) examining the employed linguistic means and context-dependent realisations. Wodak and Reisigl (2016: 32–33) provide five examples of discursive strategies related to salient features that were investigated when analysing the sampled texts, i.e.:

- Nomination, related to the way in which social actors, objects, phenomena, events, processes and actions are linguistically constructed;
- Predication, related to the qualities (positive or negative) and characteristics attributed to social actors, objects, phenomena, events, processes and actions;
- Argumentation, the way in which validity claims are justified through argument schemes (for instance through topoi).[54]
- Perspectivisation, the perspective adopted in the text, the author's point of view (signalled for instance by deictics, use of direct free indirect and indirect speech etc.);
- Intensification or mitigation related to the way in which utterances are put forward, e.g. whether they are mitigated or intensified (thorough diminutives and augmentatives, modal expressions, vagueness, question tags, indirect speech acts etc.)

Analysis was carried out at the level of the text's microstructure (e.g. individual utterances, linguistic devices and discursive strategies), macrostructure (e.g. structural and argumentative organisation) and context analysis. Context analysis does not only consider the socio-historical embedding of the text but also other texts that are intertextually related to the text under study. Hence, for instance, in analysing a *PTRS* paper based on an Italian scientist's letter, the analysis also included an investigation of the original letter.

Finally, reproduced below is the eight-step programme to carry out DHA, which was also adopted for the present analysis:

---

54 Topos is a rhetorical notion referring to content related warrants which connect premises to conclusions. They are a type of stereotypical arguments based on socially shared opinions generally implying common sense reasoning schemes for the sake of persuasion (Kader 2016: 34).

1. Activation and consultation of preceding theoretical knowledge;
2. Systematic collection of data and context information;
3. Selection and preparation of data for specific analyses, macro-analyses and micro-analyses (e.g. selection and downsizing of data according to relevant criteria);
4. Specification of the research questions and formulation of assumptions (on the basis of a literature review and first skimming of the data);
5. Qualitative pilot analysis, including context, macro and micro analysis;
6. Detailed case studies;
7. Formulation of a critique, i.e. interpretation and explanation of results;
8. Practical application of analytical results.[55]

2.1.1.2. Discourse and social change – *Fairclough's approach*

Fairclough (1992) provides an interdisciplinary approach to investigate social and cultural change. He sees discourse as a form of social practice. And, in this view, discourse – i.e. spoken and written language use – not only represents the world but also acts upon it. Hence, discourse is not only shaped and constrained by the socio-historical context in which it is produced but it is also socially constitutive, that is, it contributes to maintaining or changing the social reality.

In Fairclough's approach to discourse analysis any *discursive event* (i.e. any instance of language use) is treated three-dimensionally as a (1) piece of text (spoken or written); (2) an instance of discursive practice; and (3) an instance of social practice. The textual dimension relates specifically to the linguistic components of the text. The view of the discursive event as discursive practice instead focuses on the processes of text production, distribution and consumption. And, finally, by viewing text as social practice the focus is on the social circumstances of the language product and how this is both shaped by and shapes the social structure at all levels. During analysis, the textual dimension corresponds to a more descriptive stage, while the dimensions of discursive and social practice are interpretative and explanatory. The advantage of this multidimensional approach lies in allowing the analyst to assess the relationships between discourse and social change.

---

55 The final point, application of results, means publishing the piece of research not only to inform the scholarly and general public, but also to make readers more aware of and resistant to what they are reading. This is done in order to avoid, for instance, being manipulated or persuaded through false arguments. As far as the present research is concerned, the final step is fulfilled mainly from an informational point of view.

Fairclough's approach is also interdisciplinary in that it combines various linguistic theories and approaches to the study of language with different social, political and psychological theories. Among the most influential linguistic theories are Systemic Functional Linguistics (SFL); and pragmatic theories such as presupposition, politeness (Brown and Levinson, see below), and speech acts. Major influences from social and political theorists have been Foucault and his ideas on the construction of power, control, social subjects and knowledge through discourse; Bakhtin, Kristeva and their work on intertextuality; and Hegel, Marx, Gramsci and their theories on capitalism and hegemony. Of course, 19th- and 20th-century political theories that discuss issues of their time cannot be much of use for 17th- and 18th-century papers, but as it will be later seen, the present analysis will consider all the sampled discursive events within *their own* social structures. Finally, an influential social psychological approach to discourse was that of "speech accommodation theory", which deals with register changes in relation to the social situation.

Fairclough further defines his method as being critical, i.e. wanting to show "the connections and causes which are hidden" and "providing resources for those who may be disadvantaged through change" (1992: 9).[56]

Fairclough's checklist adopted for the present analysis consists of four ascending text-related headings – vocabulary, grammar, cohesion and text structure – and three further headings related to discursive practice – force of utterances, coherence and intertextuality. Tab. 2.1 below, summarises the stages and sub-stages of the analysis:

---

56 The latter point – that is the action or praxis (drawing on Marx) as the final stage of CDA – will be more strongly emphasised in his later works (see Fairclough 2010 and 2015).

Tab. 2.1: Fairclough's framework for discourse analysis (Adapted from Locke (2004: 46))

Text			
Vocabulary	Grammar	Cohesion	Text structure
Deals mainly with individual words.	Deals with words combined into clauses and sentences	Deals with how clauses and sentences are linked together	Deals with large scale organisational properties
• word meaning • wording vs alternative wordings • metaphor	• modality • transitivity, theme, nominalisation and voice	• connectives and argumentation	• interactional control[57]
Discursive practice			
• Force of utterances			
• Coherence			
• Intertextuality			
• Ethos			
Social practice			
• Social matrix of discourse • Orders of discourse • Ideological and political effects of discourse			

Since words and longer stretches of language use can have multiple meanings, and meanings can be assigned to more wordings, two stages are distinguished in the analysis of the wording-meaning relation. In analysing word meaning the emphasis is upon key words, their meanings and meaning potential. In the case of wording, the focus is on how a particular meaning has been assigned a particular wording. Choosing one wording rather than another means making choices about how to signify or construct social identities, social relationships, knowledge and belief (1992: 76). Hence, considering the particular wording choices that have been made, especially in contrast to alternative wordings, can provide further useful insight into what may have been the authors' thoughts and intentions and into their cultural and ideological meaning. The choices behind the use of particular metaphors, and the connotations of meaning they evoke, are also considered. Modality reveals the degree of commitment between speaker and utterance – for instance compare "you are right", categorical simple present, strong commitment, and "you may be right", weak commitment, the speaker is not sure about the truthfulness of their utterance (as signalled by possibility modal

---

[57] Since interactional control is more relevant to conversation, it shall not be dealt with here. For further information see Fairclough (1992) or Locke (2004).

The groundwork 53

may). But, most importantly, modality has to do with interpersonal meaning, since it can reveal insight into how social relations and identities are presented (1992: 28). Transitivity refers to the types of processes – relational, action, event, and mental – and participants contained in clauses, while theme represents the initial part of the clause. What is "thematised" is generally the taken-for-granted (Fairclough 1992: 178).[58] In the case of a written text especially, the frequent occurrence of a given theme can show what is the perspective adopted in the text. Nominalisation is another feature of transitivity and consists of:

> the conversion of processes into nominals, which has the effect of backgrounding the process itself – its tense and modality are not indicated – and usually not specifying the participants, so that whom is doing what to whom is left implicit. Medical and other scientific and technical language favours nominalisation, but it can be abstract, threatening and mystifying for 'lay' people. (179)

Voice represents the distinction between passive and active. According to Fairclough, active is the unmarked choice, since it can be used with no particular intentions, while the passive voice is generally chosen for various reasons, such as wanting to omit the agent (agentless passive, e.g. "the paper was written"). In looking at cohesion, the focus is on means of linkage between clauses. This can be achieved, for instance, through repetition of vocabulary from the same semantic field (lexical cohesion), through substituting devices (e.g. pronouns), and conjuncts. Different cohesive strategies create different argumentative structures and provide evidence of the different modes of rationality adopted for the production of a given discursive event (77). The overall structure of the text, i.e. its organisational properties, are also important in the analysis of argumentative strategies.

Moving on to the analysis of a discursive event as discursive practice, the pragmatic concept of force refers to the actional component of each utterance, that is what the utterance *does*. The force of an utterance is not always straightforward; for instance: "are you watching the football match?" may be a simple question or it may be a complaint, if uttered in a specific context and with a specific tone of voice.[59] Context then becomes fundamental for decoding this

---

58 Although the starting point of the clause frequently corresponds to given information, Systemic Functional Linguistics (SFL) distinguishes between the two by using the term Theme to refer to the starting point of the clause and, Given to refer to the given information, in that the Given may not necessarily be placed at the beginning of the clause. See also the section on SFL below.
59 Consider for instance someone who dislikes football, going home wishing to watch something on television and finding a relative engaged in a football match.

ambiguity of meaning. Coherence is found when the parts of a text (sentences) are meaningfully related, and the overall reading "makes sense" to the reader, rather than eliciting doubt and confusion as to what the author/speaker's points are. Ethos concerns all the features that contribute to constructing the "social self". Finally, Fairclough devotes much space to intertextuality, which will also be an important point in the present study. The term "intertextuality" accounts for

> the property texts have of being full of snatches of other texts, which may be explicitly demarcated or merged in, and which the text may assimilate, contradict, ironically echo, and so forth. In terms of production, an intertextual perspective stresses the historicity of text: how they always constitute additions to existing 'chains of speech communication' consisting of prior texts to which they respond. In terms of distribution, an intertextual perspective is helpful in exploring relatively stable networks which texts move along, undergoing predictable transformations as they shift from one text type to another (for instance, political speeches are often transformed into news reports). And in terms of consumption, an intertextual perspective is helpful in stressing that it is not just 'the text', not indeed just the texts that constitute it, that shape interpretation, but also those other texts which interpreters variably bring to the interpretation process. (1992: 84–85)

Fairclough distinguishes two types of intertextuality: "manifest intertextuality", where the intertextual relations are explicitly marked – through reference, for instance – and "interdiscursivity" or "constitutive intertextuality", which is how texts appear to subscribe to other discourses in their underlying structural features, properties and boundaries.

The way in which texts are transformed is also part of intertextuality. Moreover, Fairclough expands on features of intertextuality by discussing: discourse representation, i.e. the explicit incorporating of other texts, the focus being on *how* other discourses are represented; presupposition, i.e. what is tacitly assumed to be true in a sentence; negation (negative sentences, often used polemically); metadiscourse, for instance speakers can distance themselves from their text by means of linguistic devices (such as hedging), therefore creating different levels within the same text (1992: 122); and irony that can display the speaker's attitude towards someone or something.

Returning briefly to discourse representation, Fairclough includes under this heading the analysis of attribution strategies; i.e. how represented discursive events are indicated as deriving from a source. This concept has been dealt with in detail by media discourse studies (see for instance Bednarek 2006; Martin & White 2005; White 2004). Indeed, attribution is a constitutive feature of both contemporary and historical media. During the analysis of the sampled *PTRS* papers it was observed that attribution strategies played a subtle but influential role in the presentation of Italian discourses. It will be seen that attribution strategies not only appeared to reveal author stance but also implicitly attempted to influence reader opinion.

Finally, in analysing a given discursive event by viewing it as social practice, the idea is to interpret and explain the underlying reasons that account for the text being as it is and the effects it has on the social context in which it is produced.[60]

Before moving on to the most relevant analytical tools exploited within the above-described critical approaches, some further notes are required. It is important to highlight that certain aspects of CDA differ considerably from the objectives of this study in analysing scientific discourse. Focus on the interpersonal function of language in Fairclough's model, for instance, is part of the analysis but not the ultimate goal. In the present research instead, understanding, describing and trying to explain the interpersonal meanings and functions of language, and thus also how social relationships are constructed, changed and maintained, is the principal aim.

Fairclough's *Discourse and Social change*, which is an earlier work of his, was chosen as a model, because his most recent books – at least those considered for the present study (2010, 2015) – as well as other works on CDA (Wodak & Meyer 2016) become too strongly involved with political and ideological issues. And although discourse as a mode of political and ideological practice is said to be one of the principal concerns of *Discourse and Social Change* (1992:67) as well, Fairclough here gives ample space to the analysis of interpersonal features (in Chapter 5) giving momentary breaks to his concern with political and ideological issues. Of course, politics and ideology can also be implicated in scientific discourse, but in the present piece of research they play a minor role.

Moreover, most critical discourse analyses, even those in DHA, focus on the present or recent past due to the practical, socially improving, objectives of CDS. Whereas this study focuses on an earlier and much broader time span (second half of the 17th century up to the end of the 19th), which also entails that the practical application of results, at least those concerning the first centuries under study, is less achievable. However, it is hoped that this study will prove that CDA can also be carried out on ancient history and not just contemporary history.

---

60 Since Fairclough's concepts of "social matrix of discourse", "orders of discourse" and "Ideological and political effects of discourse" have not been considered for the present study, they shall not be dealt with here (see Fairclough 1992 for more).

2.1.1.3 Politeness

Politeness can be considered as a socially and culturally defined behavioural system. In pragmatics, it is seen as "a set of strategies on the part of discourse participants for mitigating speech acts which are potentially threating to their own 'face' [i.e. one's public self-image] or that of an interlocutor" (Fairclough 1992: 162).[61] Politeness theory distinguishes between two types of "face": "positive face", when the individual wants to be appreciated by other individuals; and "negative face", when the individual gives more importance to their own freedom and independence rather than to being liked and admired by others. Fairclough moreover adds that politeness can be seen as a means of acknowledging and maintaining social and power relations (1992: 163).

Since the way we behave with others can readily affect the opinions they have of us and vice versa, politeness should be considered when dealing with the Royal Society and its foreign exchanges. Indeed, in its first centuries of existence papers were framed by numerous polite conventions. The difficulty, however, lies in the fact that politeness conventions also change diatopically – and thus, broadly speaking, Italian conventions as opposed to English conventions – and diachronically – hence, the way we interpret an 18th century utterance today may not be the same as the way in which it would have been interpreted in the 18th century. This means that in interpreting texts and utterances there will always be the issue of the subjectivity of the interpretation.

A key concept of the theory is that of "face-threatening acts" (FTAs). If we were to ask someone to help us fix a broken vase, we could risk threating their negative face (i.e. their want to be free and not impeded by others) by putting them in the position of having to respond to our request. At the same time, the way in which we make our request can also threaten our own positive face (i.e. our want to be liked). This may therefore lead one to make the request by adopting different politeness strategies in order to save both their own face and that of their interlocutor. Hence, FTAs are speech acts that can potentially damage the speaker's and the interlocutor's positive or negative face and they are often inevitable if one wants to be considered part of society and interact with its members. The desire to maintain positive face will elicit the use of politeness strategies in order to minimise face damage both to the speaker and the interlocutor.

Politeness theory distinguishes four strategies for putting forward an FTA, which will be here explained by exploiting an example from a *PTRS* paper,

---

61 Fairclough is here drawing on the famous work on politeness theory by Penelope Brown and Stephen C. Levinson: *Politeness, Some Universals in Language Use* (1987).

whose author informs the public that an experiment on blood made by the Italian physician CARLO FRACASSATI in Italy had been previously performed in England by the author of the same paper. This author would thus seem to claim the originality of their experiment, however, doing so may bring about the inference that the Italian physician *copied* his experiment making the claim of originality highly face threatening both to the author, who may damage his own positive face by appearing as a person who makes accusations, and to the face of the Italian physician since plagiarism was and is commonly perceived as negative behaviour. The following examples are based on the author's original wording, which is reported at point 3; the examples represent four different ways, corresponding to four different strategies, in which the author could have presented his claim:

1. Bald or on record: *Fracassati heard about our experiment and took a hint to make and publish what now is English'd in the* Transactions. In this example there is no attempt to mitigate what the author of the paper believes FRACASSATI has done. The author does not therefore care about reducing the threat to the reader;
2. Positive politeness: *I am flattered that the esteemed Fracassati was inspired by my experiments to make and publish what now is English'd in the* Transactions. In this case the author mitigates the FTA by showing admiration towards the Italian physician. The general idea is to make the reader or hearer feel closer to the speaker who is producing the FTA;
3. Negative politeness: "[FRACASSATI] *may have had some imperfect Rumour of our Experiment without knowing whence it came, and so may, without any disingenuity [i.e. dishonesty], have thence taken a hint to make and publish what now is English'd in the* Transactions"[62] In the original wording the author mitigates the claim of authorship through extensive hedging given by the modal verb *may* and the use of negation found in *imperfect* and *without*.

---

62 *Phil. Trans.* 1666–1667: 551–552. This paper may also serve to briefly exemplify the concepts of *manifest intertextuality* and *interdiscursivity*. The text is manifestly related (through direct reference) to a paper on one of FRACASSATI's blood experiments published in the *PTRS* in 1666 and it can also be interdiscursively related to other texts on blood researches that were published before and after it and which together are part of the discourse of blood experiments. Common features, which are not explicitly addressed as such, that show how this paper, and other sampled papers, pertain to the discourse on the nature of blood and blood experiments are for instance the subject, the procedures followed in carrying out the experiments, the interest in understanding how blood behaves and reacts to stimuli and so on.

Moreover, by taking a closer look at the chosen wording as opposed to alternative wordings, it can be seen that the author uses words that have a connotation of smallness. Compare for instance *an imperfect rumor of our experiment* with *an account of our experiment*. The author's wording suggests that FRACASSATI only heard a little of the experiment, while the alternative wording suggests that the author heard all the details of the experiment. Also *a hint* has a connotation of smallness, which the author could have avoided, if he had wanted to be more direct, for instance by saying *...about our experiment, which led him to make ....* The point of negative politeness is to produce the FTA in a more indirect form in order to show respect for and be less imposing upon the hearer/reader;

4. Indirect or off-record: *is it not amazing that the experiment I devised three years ago is so similar to the one that now is English'd in the* Transactions? In this case the utterance does not literally express the FTA but only alludes to it and it is up to the reader to infer the FTA from what is being said/written.

Both strategies 3 and 4 exploit indirect speech acts. Although indirect speech acts can be less clear, they are generally preferred to direct speech acts, which can often sound imposing, harsh or, in this case, accusatory.

Finally, it is interesting to notice that this paper is entitled "A Confirmation of the Experiments made in Numb. 27. to have been made by Signor Fracassati in Italy, by injecting Acid Liquors into Blood". The optimistic approach given by the noun *confirmation* can again be seen as an example of positive politeness.

The example of the paper above also goes to show that politeness strategies play an important role when managing social relationships and especially in delicate cases of disagreement, misunderstanding and the like.

2.1.2 A socio-historical approach to scientific writing

The applied study by Dwight Atkinson (1996 and 1999) has a rather different focus from the theoretical works presented so far. Atkinson is concerned with the historical development of scientific writing as a genre. Starting from the assumption that linguistic registers derive their distinctive characteristics from inherent co-occurrence patterns of linguistic features (Biber 1988), Atkinson's focus is on the changing presence of said features in order to describe how scientific writing in the *PTRS* developed between 1675 and 1999 – that is most of the journal's existence. Moreover, the discourse analysis is situated in its socio-historical context in order to view relations between the changing nature of scientific discourse and of the socio-historical context and scientific discourse community.

Atkinson's study represents a comprehensive account of the development of scientific writing in diachronic perspective. His findings (and those by previous research in the same field) are very relevant to, and provide a sociolinguistic background for, the present study. A brief account of his study and main findings is therefore needed.

Atkinson carried out qualitative rhetorical text analysis (based on Bazerman 1988) and quantitative sociolinguistic register analysis (through Biber's Multidimensional approach, 1988) on a corpus of 202 *PTRS* papers collected at seven 50-year intervals.

Rhetorical analysis is based on examining texts for genre features that signal attributes of the rhetorical situations that led to the production of the texts. Hence, analytical categories emerge from the engagement with the texts themselves (Atkinson 1996: 337). Multidimensional analysis (hereafter MD analysis), on the other hand, is based on a set of empirically predetermined discourse functions or dimensions of variation,[63] which are given by sets of co-occurring linguistic features. These dimensions of variation are conceived as continuous scales with opposite poles (signalling opposite discourse functions) along which a given discursive event may be placed according to the co-occurring linguistic features that form it. Dimensions that are most relevant to the present study are:

1. *Involved vs. Informational production.* Involved denotes production of language, with an affective and/or interactive focus, which is generally found in spontaneous language use. Informational production instead denotes planned communication of highly integrated prepositional content (Atkinson 1996: 351). Linguistic features which are associated with involved production are: private verbs (e.g. think, feel), that-deletions, contractions, present tense verbs, second-person pronouns, demonstrative pronouns, general emphatics (e.g. a lot, for sure, such a), the pronoun it, BE as main verb, causative subordination, general hedges, amplifiers (e.g. absolutely, completely, altogether), WH-questions, discourse particles (e.g. now, anyway), place and time adverbials and possibility modals. Informational production is characterised by a relative absence of the former features and an increased presence of nouns (including nominalisations), prepositions and attributive adjectives;

---

63 The discourse functions (or dimensions of variation) were elaborated from a quantitative study specifically aimed at determining which linguistic features tend to predict different register properties and the goal was to provide comprehensive descriptions of patterns of register variation. For more, see Biber (1988).

2. *Narrative vs. Non-narrative concerns*. Past tense verbs, third-person pronouns, perfect-aspect verbs and public verbs (e.g. say, tell) are forms that characterise narrative writing. Since there is no opposite discourse function to narrativity, its opposite pole has simply been defined as non-narrative.
3. *Abstract vs. Non-abstract information*. When information is presented mostly through the use of passives (both agentless and *by*-passives), conjuncts, past participial WHIZ-deletions[64] and adverbial subordinators (e.g. because, although) it is considered abstract.

Briefly, results of MD analysis showed: (1) a general shift from involved discourse to informational discourse; (2) a decrease in the level of narrativity; and (3) a movement from a moderately abstract style to a highly abstract one. Similarly, rhetorical analysis revealed a general shift from an author-centred to an object-centred approach, that is, a change in the place occupied by the author within the discursive event. In the first years of the journal's existence authors played a central role in the study being presented, but throughout the centuries their presence gradually decreased, allowing the object of study to be foregrounded. Nowadays an "effaced" or "distanced" author is considered a typical feature of scientific writing. In the *PTRS*, author-centredness was mostly signalled by first-person pronouns, active verbs and language showing the author's affective states and psychological processes. Typical features related to the initial author-centred period were also: *witnessing*, that is, the naming of other people (generally notable gentlemen) who witnessed what the author was writing about; *indexes of modesty and humility* in presenting one's work,[65] and *elaborate politeness*, which was found especially in the opening sections with displays of encomia towards the Society, its secretaries and fellow researchers (1996: 339).

Moreover, rhetorical analysis focused on genres and discourse structures. Atkinson reports that letters and experimental reports were the two most stable genre forms across most periods. Papers in letter form ceased to be published from 1875 onwards. The discourse structure of letters was initially digressive and unorderly and was generally framed by encomiastic openings. Gradually letters became longer, more elaborate and detailed. Towards the end of the 18th century letters were often accompanied by cover letters, and some articles were made up

---

64 That is, reducing relative clauses by omitting relative pronouns, e.g. the man (whom) I spoke to yesterday.
65 Signalled by stance markers or hedging devices such as possibility modals, adjectives and adjectives showing probability (likely, probably, possibly), and distancing verbs (appears, seems etc.) (Atkinson 1996: 368).

of multiple letters. In the 19th century articles in letter form were still present; however, forms of encomia were dismissed and after the introductory salutation authors would go straight to the point.

While letters gradually decreased and were eventually dropped, experimental reports were found to increase and became the most frequent type of article by the second half of the 19th century. At the beginning of the period under study, experimental reports varied considerably in form and structure; however, gradually terminology became more specific, and authors started showing greater concern with describing experimental conditions and making clear and precise descriptions of methods (19th century). Discourse structure became more organised with articles making use of sectioning and tables. Finally, in the 20th century experimental reports started being organised according to what is today known as the IMRD structure (introduction, methods, results, discussion).

Atkinson also searched the articles for information regarding the discourse community. In the 17th and 18th centuries, articles tended to display a cooperative dialogic relationship with one another. This is signalled for instance by answer lists and articles written in response to other articles. Atkinson points out that this cooperative behaviour seems to "fit in with the Baconian programme of empirical scientists, i.e. the cooperative activity in the service of constructing an enormous base of natural knowledge" (1996: 369). In the 18th century, a new more polemical stance was also displayed by authors, who would often also anticipate criticism. However, this oppositional period does not seem to last and a cooperative tone was soon recovered. Gradually authors became concerned with presenting the information more abstractedly, being clear and precise, and placing their work within larger community research contexts.

Atkinson contextually situates the results of his analyses[66] by highlighting how features characterising the first centuries of *PTRS* papers may be seen as an expression of the British genteel culture of the time. The discourse community of texts users and producers was formed by the highest ranks of the social hierarchy. These gentlemen adhered to conventionalised moral and social qualities such as self-reliance, modesty, civility and honesty; and in writing they "traded on this conventional social image of the gentleman for rhetorical purposes" (1996: 362).

---

66 The results of rhetorical analysis and MD analysis were correlated as follows: decline of author-centred rhetoric and linguistic shift towards and informational production; development of an object-centred rhetoric and increase in the level of text abstractedness; decline in the detailed description of experiments and the decrease in the level of narrativity.

Hence, elaborate politeness, modesty, letter writing, and cooperation between authors may be seen as an expression of a way of life in which gentlemen-researchers really stood at the centre of events. And the rhetorical effectiveness of discursive events strongly depended on the trust accorded to their authors; in other words, the person's credibility substantially determined the credibility of the account (1999: 148–152).[67] Atkinson also points out that the 19th century tendency to meticulously describe methods and materials co-occurs with the growth of specialist and scientific disciplines in Britain (1996: 364). New specialised societies, for instance, were born in opposition to the eclectic science of the Royal Society (see Section 3.3).

2.2 Approach to text analysis

The above studies from the fields of CDA, functional linguistics, pragmatics, sociology, and sociolinguistics have all been useful in providing a methodological background for the present study. Comparisons and considerations made in the previous sections should thus be kept in mind throughout the whole reading. One feature that is common to the linguistic studies presented above, and which underlies this analysis, is that discourse is strongly related to the social reality in which it is embedded; hence it not only reflects its social circumstances but also contributes to making, maintaining and changing reality. Discourse is therefore *functional*, and a detailed analysis of language use both at the micro- and macro-level – from words, to sentences, to whole texts – and at the interpersonal level can provide more precise descriptions and interpretations of human history. Having said this, the method followed in carrying out the research will now be described in the present and following sections.

The project started with a rather standard research procedure, which is also present in the DHA eight-step plan reported above (Section 2.1.1.1). Firstly, existing literature in the subject areas of the history of the Royal Society and *PTRS*, Anglo-Italian relations, scientific writing, and the history of science both in England and Italy was collected and investigated. All relevant historical information was recorded (see Chapter 1) for later contextualisation of results. From the very beginning, the primary sources, i.e. *PTRS* papers, were collected,

---

67 This point however will be further discussed in the following chapters. It will be seen that while one's status was generally pointed out in the narratives, credibility did not depend on this aspect alone, but rather on the person's skill, the presence of witnesses who had an understanding of the subject, and, later, on the methodological procedures and possibility of successfully reproducing the experiments and observations.

and so were available private letters, meeting minutes and reports to be integrated with the analysis of the papers and historical context.

Once all of the papers were collected, a first skim through the whole corpus was carried out in order to organise the sampled articles according to period, text types, languages (i.e. the language in which they were written: English, Latin, Italian and French), and topics. The number of papers per each category and period was counted – 102 papers for the 17th century, 185 for the 18th and 52 for the 19th, for a total of 339 papers. Moreover, this first view of the texts allowed us to identify topics and scientists that were particularly prominent within the corpus, and periods of apparently more assiduous cooperation between scientists from the two countries. Consequently, it was possible to narrow down focus onto specific periods and paper groups to which special attention could be dedicated and on which more specific case studies would be carried out.

Later, further manuscript material and contextual information related to specific areas of focus was collected at the library of the Royal Society.

The analysis was carried out by treating the texts three-dimensionally. Hence, the sampled discursive events were firstly analysed as *text* considering: text structure, word meaning, wording, nominalisations, modality,[68] hedging devices, and further stance – voice, transitivity and theme/rheme, personal pronouns, private verbs, negation, positive and negative evaluation, irony etc. – and politeness features.

Moving into the analysis of *discursive practice* and *social practice*, focus was put on intertextuality, interdiscursivity and the representation and transformation of texts. Indeed, most of the papers that relate to Italian studies and findings represent the reception of said studies and the further transformation and spreading of them. Hence, considering the sampled papers in relation to the texts that preceded and followed them is of primary importance for an assessment of the reception of Italian research in England and the management of Italy-Royal Society relations. The reception of one's studies, i.e. the negative, positive or neutral opinion that one "scientist" (or more) could form of another's work, was critical for further relations between the two scientists and future interest (or lack of it) in each other's work. In actual fact, the esteem built between two scientists, "A" and "B", could also be reflected on third parties, "C" "D", "E", "F" etc., if proposed by "A" or "B". Finally, results – as in the salience or the frequent

---

[68] By modality it is meant any linguistic item that has a modal function. Hence not just modal verbs, but also nouns, adjectives, adverbs, particles and whole sequences that have a modal effect – i.e. showing commitment, distance or marking politeness.

occurrence (or absence) of particular features in a given period or subject area – were interpreted in relation to the available contextual information. By exploring manifest intertextual relations between papers, motivations behind the texts and effects on subsequent texts were also assessed.

The present analysis is mostly qualitative based, although some quantitative, manually collected, data will also be provided. When the project was first conceived, the idea was to create a computer-readable corpus of all the texts and to exploit the tools of corpus linguistics (hereafter CL) for the counting and analysing of recurrent linguistic features, searching for keywords (unusually frequent words) for different text types, and more. However, although analysing the texts through the quantitative tools of CL would have greatly complemented the study, it was eventually decided to prioritise a qualitative-based analysis over a quantitative one. The reasons behind this shall now be explained.

Corpus linguistics exploits computational techniques to collect quantitative information about discourse. Among the advantages of this methodology is the possibility to carry out research on a much larger corpus of texts and the preciseness and objectiveness of the results. Corpora of scientific and *PTRS* papers have already been created for quantitative discourse analyses such as Atkinson's (1999),[69] and the use of statistical methods to evaluate the weight of results has proved very fruitful in the tracking of changes in discursive practices. A computer readable corpus of the texts collected would indeed be helpful, for instance, for a quantitative tracking of a given keyword and the development of its use through time. For instance, an unusually frequently occurring word in the corpus under study was the verb *pretend*, which in the early *PTRS* was used to attribute what was said to the original Italian author. The verb frequently, but not always, took on a slightly negative connotation. By doing a keyword search, the exact number of occurrences of this verb can be counted throughout the whole period and across all the different types of papers, and any variations in the frequency of its use can be tracked. However, to evaluate the meanings and the effects of this verb in the particular discursive event, the analyst still needs to integrate their qualitative, i.e. subjective, interpretation by reinserting the verb in its original context. Fairclough (2015: 20–23) is rather sceptical about the importance attributed to corpus linguistics, since it "omits essential properties of actual language use, notably the fact that actual language users interpret and contextualise language as an inherent part of using it". According to Fairclough, corpus linguistics is indeed useful but it is merely a tool that can *serve* analysis and is not the analysis itself:

---

[69] See also Taavitsainen and Pahta (2011).

> [Corpus linguistics] can be useful in checking out impressionistic conclusions about which words and co-occurrences of words are most significant [...], in alerting analysts to words and co-occurrences which they had not noticed, and in providing statistical information about certain features of e.g. New Labour discourse. Above all it can stimulate new ideas which might lead to new directions of investigation and analysis. But it is best regarded as part of the preparation from which the real work of analysis and critique can begin. The danger is in attributing to corpus linguistics a more elevated place in one's critical discourse analysis than it actually has, perhaps because one is in awe of the power of the technologies which are drawn upon (21).

The reader is being warned that CL is a helpful tool in critical discourse studies but only to a certain extent. Fairclough reminds the reader that the focal point of the analysis is the qualitative part. For instance, returning to the example of the verb *pretend*, while the corpus can provide the exact number of times the verb was used, its forms, the types of texts in which it appeared and its collocates; it cannot explain *what pretend means* or *why* it is used in the particular discursive event in which it occurs.

The meaning, the reasons and the effect of words are the main purpose of the present study. Hence, although a computer-readable corpus would have been a valuable aid in carrying out the present analysis, the actual creation of it would have meant manually transforming the scanned pages of 339 *PTRS* articles into computer-readable files and tagging all of the data contained within them in order to obtain satisfying results. But such a process would have taken away too much time from the most important part in this particular instance of discourse analysis, which focuses especially on the interpersonal function of the texts, i.e. how personal and social relationships are enacted through discourse (Halliday & Matthiessen 2014: 30). Moreover, the analysis of the texts through a corpus only allows focus on individual words, sentences or paragraphs, but a critical approach to the study of language goes beyond words and paragraphs: the entire discursive event, its organisational properties, its intertextual relations with previous and forthcoming events, and its social and historical context are all fundamental aspects of the analysis. Hence, due to time limitations and the incredible amount of time already taken for the actual collection of the papers, it was eventually decided to leave aside, at least for this first analysis, the addition of a computer-readable corpus and a corpus linguistic approach.

2.3. Collection of the papers and analysis

Following the initial consultation of existing research on the history of science, Anglo-Italian relations and linguistic approaches to DA, CDA and case studies, the first two steps to start the present study were (1) the finding and noting down all of the names of Italians elected to the Royal Society's membership; and (2) the collection of the *PTRS* papers and intertextually related primary sources. The identification of Italian Fellows was eased by an existing list published by the Royal Society of all of its Fellows; however, the list does not distinguish the Italian Foreign Members (FMs) from other Foreign Members and "home" Fellows,[70] and had to be read from beginning to end in order to find Italian Fellows. A total of 135 Italian Foreign Members of the Royal Society were identified – 20 for the 17th century, 103 for the 18th, and 12 for the 19th. Moreover, the search for biographical information and the first skimming through papers in order to establish whether they were suitable for insertion in the corpus, revealed the presence of many other Italians who, though never elected, held relations with the Society and published in the *PTRS*. These Italians have been classified as "contributors". The total number of Italian contributors identified was 62 – of which 29 in the 17th century, 22 in the 18th, and 11 in the 19th century. The Italian Fellows and contributors are not exhaustive of all the Italians who held relations with the Society, as many more names were found in the Society's archives, but only those who held correspondence and contributed papers were included in the study. The names of Italian FMs and contributors were later searched in the database of the Society's archive to find letter exchanges and mentions in other archival material.

The papers were collected by inserting keywords in the *Philosophical Transactions* online database. All *PTRS* papers have been scanned and digitalised and could therefore be downloaded in PDF form. The keywords used to search for papers were mainly the Fellows' and contributors' names, Italian place names, names of institutions and names of important Italian researches. The search for papers by keywords entails that not all of Italian-research-based papers may have been found, especially book reviews, which were often entitled simply as "an account of books..." and did not always report the titles of the reviewed books. Moreover, when the search was carried out, the database mainly only found the searched keywords in the titles of papers and not in their body. Later the database was improved by the Royal Society, and keywords can now also be found in the bodies of the papers, which allowed the collection of further material.

---

70 The distinction between Foreign Members and Fellows did not exist during the early years of the Royal Society, it was introduced later in the 19th century – although some Italians had been referred to as Foreign Members already in the late 18th century. Therefore, the first Italian members are defined as Fellows, while the 19th-century members are defined as Foreign Members (FMs).

Collection of the papers and analysis   67

Papers that were included in the corpus met at least one of the following conditions:

- papers originally written by Italians and translated into English;
- papers originally written by Italians and published in the original language (Latin, French, Italian, or English);
- papers reporting about Italian news, research and books;
- papers reporting the experiences of English Fellows travelling or resident in Italy;
- papers referencing Italian research and providing some detail about them.

The differences between the papers were kept in consideration throughout the analysis as a paper written by an Italian entails the representation of Italian ideas, opinions, procedures and results, and also, even if translated – as translations mostly reflected the originals –, they portray the Italian writing style and linguistic strategies. Instead, papers written by English Fellows may arguably be said to represent the English writing style and opinions. Hence, this distinction is necessary for a study in which the objective is that of analysing Anglo-Italian relations. It moreover represents a novelty in discourse analysis of historical scientific writing, in that most of the literature consulted does not focus on the original sources of the *PTRS* papers. The focus of these studies is generally on the development of English scientific writing and does not consider the translated nature of a very great number *PTRS* papers.

These first steps into the research also enabled the identification of a number of English Fellows who contributed to the transmission of Italian research and news to the Royal Society and of the spreading of English news and books in Italy. The English contributors will be reported about in the results chapters.

Once most of the papers had been collected, the issue arose as to how they could be quickly and easily consulted, together with how to write down the results of the qualitative analysis and render them countable for quantitative analyses. Hence, a spreadsheet of all the sampled *PTRS* papers was created. For the creation of the file, the papers were read for the second time. The spreadsheet includes fourteen columns, to which another thirteen were added when the critical discourse analysis was carried out. Hence, each paper was reported in the spreadsheet with the following features:

1) publication year; 2) volume number; 3) Issue number (where present); 4) Journal name; 5) reference pages; 6) reference author; 7) original author (when identifiable); 8) translator; 9) communicator of the paper; 10) topic of the paper; 11) references; 12) title of the paper; 13) language of the paper; 14) translated, not translated, or reported nature of the paper.

While some of these headings are clear, other require some explanation. The journal name was noted in that not all of the papers collected for the corpus come from the *Philosophical Transactions* – although the great majority do. Some of the 19th-century papers are taken from the *Proceedings* journal, as some research from this century was published in this journal only. The *Proceedings* moreover, which was supposed to publish abstracts of papers, sometimes actually briefly reported and reviewed the papers, and therefore represents a relevant source for the analysis of the Royal Society's impressions on Italian research. The "reference author" (column 6) represents the name provided by the journals' database as the "author" of the paper; however, in the early *PTRS* especially, the names provided are in most cases not the names of the actual authors, but mere names reported in the title. In many cases the name of the author was anonymous (mainly in the 17th century). This is also the reason why in the course of this study, *PTRS* papers will be referenced with their journal name, year, and page numbers and not according to author name. To provide the reference names given by the Royal Society's database would be misleading, especially in a study in which *who* the authors were matters. Column 7 thus provides the correct names of authors, whenever these were identified. Column 11 was exploited to note down any names of other scholars who were referenced in the papers. Instead, in the topics column the broad subject areas of the papers were noted. The classification of papers according to subject areas was not an easy process since the concept of science and sciences, as they are known today, did not exist until the 19th century when they gradually started being defined and distinguished. Moreover, in the early *Transactions*, some papers covered more than one subject area.

The advantage of a computer-readable spreadsheet is that this enables the researcher to apply filters to each category in order to find the total numbers of a given feature. For instance, in order to know how many papers were published in English all that was necessary to do was to apply a filter on the language column and select only the English papers. This way the total number immediately appeared. Further, by filtering the papers, the other features reported in the neighbouring columns were also closely consultable without the interference of papers that were not relevant in the specific case. For instance, if then the researcher wanted to view whether the English papers were reported (basing on an Italian piece of research), translated, or fully based on the English author's writing, the translation column could be looked at, or even better, another filter could be applied on the translation column, in order to find the exact numbers of how many papers of those written in English were translated. Spreadsheets can moreover be consulted by keyword searches just like a computer-readable corpus. Hence, for instance, if the researcher wanted to know how many paper titles made use of the word *experiment*, a word search could be put in and the spreadsheet would find all instances of the word – including variants, such as *experiments*, if they included the search term.

Once the preliminary spreadsheet of the 339 *PTRS* papers was completed, the first results, divided per century, were noted. These concerned the languages used to write the papers, the major topics that interested the Society as far as Italy was concerned, and translation practices.

At this point the discourse and critical discourse analysis could be carried out and the papers were consulted for the third time and more minutely analysed. While some research questions were posed beforehand, other aspects arose during the analysis of the first century that were then kept in consideration for the following centuries. Hence, the analysis of the papers was partly built during the analysis itself. The features that were analysed were noted down in further columns in the spreadsheet:

> 15) General linguistic and sociocultural aspects of relevance; 16) Italian place names; 17) Narrative or non-narrative; 18) Place of the author; 19) Letter form?; 20) Text type(s); 21) Features of intertextuality and dialogicity; 22) Paper length; 23) Discourse representation; 24) Positive evaluation; 25) Negative evaluation; 26) Borrowings; 27) Structural features.

The more purely linguistic analysis, at the level of text, focused on the macrostructure of the papers – titles, structural organisation, broad text types (e.g. observations, experimental reports, travel accounts, book accounts etc.), presence of tables and drawings, and page length – and, at the microstructural level, on the recurring discourse features according to which the papers were marked as tending either towards author-centredness and involvement or towards informativity and abstractedness, i.e. object-centredness. In this study, Biber's (1988) dimensions of variation and Bazerman's (1988) categories of authorial presence are correlated: involvement corresponds to author-centredness; informativity alone marking an absence of author-centred features and an increase in informational ones, and abstractedness marking high object-centredness. The analysis of recurring linguistic features also enabled the classification of papers as being either narrative, descriptive, narrative and descriptive, or neither of the former. During the analysis of the textual dimension of the papers, wording choices were also considered and any unusually frequent wording choices and linguistic strategies (such as humble presentations of papers, witnessing strategies, attribution strategies, and encomia) were noted down in column 15. While in column 26 any Italianisms and borrowings were noted for a subsequent analysis of the presence of Italian scientific borrowings in the *Transactions* and in the English language, which will be briefly discussed in the conclusions as another important indicator of the quality of Anglo-Italian scientific relations and their development.

For the more critical analysis of discursive practice, the features considered and noted down were explicit and implicit evaluation strategies, politeness strategies, argument strategies and how Italian discourses were represented in English writings and English discourses in Italian writings (column 23 discourse representation). Papers were marked as "neutral" when there was no kind of judgement or bias towards what was being reported or, in the case of translated papers, when there was no form of framing and commentary, that is, the paper was only a direct translation of the Italian original. However, translated papers could themselves represent other discourses, both Italian and foreign. Papers were marked as "neutral-positive" when they were mainly neutral but contained some minor forms of positive evaluative language; while they were marked as "neutral-negative" when they contained minor negative or disagreeing comments, which were however irrelevant to the representation of the discursive event as a whole. Papers were marked as "negative" when they contained criticism and/or disagreement that could negatively influence the readers' opinion of the represented discursive event. In column 21 any features that marked intertextual and interdiscursive relations with other papers, together with features that showed a dialogic relation between authors and papers were noted. In column 16 any Italian place names mentioned in the papers were written down. The count of the frequency of toponyms made it possible to confirm what had appeared, from the socio-historical analysis, as the main Italian areas of contact with the Society.

The *Philosophical Transactions* was also the place where disputes were sometimes reported, at least up until the 18th century, as it will be seen. When delicate matters arose from the papers, these were considered into more detail and all the available complementary primary sources specifically collected for the purpose were analysed for a more precise and impartial analysis.

The results of the analysis will be presented and discussed in Chapters 3 and 4. Chapter 3 focuses on the results of the historical analysis dividing them per century, while Chapter 4 focuses on the results of the linguistic analysis. The results are arranged according to Fairclough's (1992) three-dimensional approach to analysis, i.e. *textual dimension*, *discursive practice* and *social practice*. However, the analysis of *social practice* is here reported before the other two dimensions (Chapter 3). Hence, results are divided into two main parts: the first that reports social, cultural and historical insights that arose from the content analysis of papers and complementary sources (*social practice*); and a second which reports the more purely linguistic and critical linguistic results of the analysis of the *textual* and the *discursive* dimensions of the papers and related primary sources (Chapter 4). The division of the results per century was a choice made for practical reasons, i.e. to view the gradual changes that took place both in the Society's foreign relations and in the development of the *PTRS*. It should not be forgotten however that historical and linguistic developments are not marked by dates; they took (and take) place slowly. Moreover, Fellows often lived over the course of two centuries and the contributions of Fellows considered in one century often continued into the next. In the course of this study, the terms *paper*, *article*, and *text* will be used interchangeably to refer to the sampled discursive events.[71] It will be seen, moreover, that the tools of CDA will sometimes be more intensively exploited than on other occasions, where a comparison between the intertextually related sources (papers and letters), added to the analysis of discourse representation, appeared to be sufficient.

---

71 See footnote 4, chapter 1.

Chapter 3 Social, cultural and historical insights

3.1 17th century 73
   3.1.1 Fellows and contributors 74
   3.1.2 Creating a discourse community: Correspondence and information exchange 79
   3.1.3 Fellows and Englishmen in Italy 83
   3.1.4 Topics 86
   3.1.5 Landscapes of northern and central Italy 95
3.2 18th century 99
   3.2.1 Fellows and contributors 101
   3.2.2 Fellows and Englishmen in close contact 108
   3.2.3 A complex network of exchanges 112
   3.2.4 Topics 114
   3.2.5 Beauties of northern, central and southern Italy 122
3.3 19th century 126
   3.3.1 Foreign Members and contributors 128
   3.3.2 Fellows and Englishmen in Italy 134
   3.3.3 Topics 137
3.4 Language of the papers and translation practice 138
   3.4.1 In the 17th century 138
      3.4.1.1 English'd out of the *Giornale de' Letterati* 141
   3.4.2 In the 18th century 141
   3.4.3 In the 19th century 145

3.1 17th century

This section reports and discusses the results of the historical analysis carried out on 17th-century *PTRS* papers, dating from 1665 to 1699. For this period, a total of 102 papers were collected, 20 Italian Fellows were counted and over 29 Italian contributors – that is non-Fellows who sent the Society news and writings on Italy and Italian studies – and 11 English contributors were identified.

3.1.1 Fellows and contributors

> *And your own intelligence will spur you on, without the urging of others, to inform yourself about these matters; in the same way you will be led, without doubt, to encourage all the keen minds of Italy to employ their talents in advancing the sciences and the arts by observations and experiments faithfully and diligently performed. We hope that that great prince of the Roman Church, Cardinal de' Medici, will never leave off philosophizing, or making his academicians philosophize, nor that those celebrated men Rucellai, Ricci, Capponi, Cassini, Viviani, Rinaldini, Dati, Redi, Borelli, Fabri, del Bono, de Angeli, Settalla, Magalotti, Falconieri, Manfredi, Travagino, etc. will ever cease to contribute their knowledge and diligence to increasing the glory of this century, so exalted already by the growth of knowledge and useful discoveries.* (OLDENBURG to AZOUT, 1668, Hall and Hall 1967: 482–483)
>
> *I took the freedom to recommend to you the promoting of the dessein and concerns of the Royal Society, partly by procuring us the best Philosophical correspondencys, you could all over Italy, with the Excellent Virtuosi in Florence, Rome, Naples, Bononia, Venice, Milan; partly by communicating to Us, what is transacted where you are in matters of Experimentall Philosophy [...]* (OLDENBURG to FINCH, 1667, Hall and Hall 1966: 618)

As anticipated above, the number of Italian Fellows for the 17th century, i.e. 1665–1699, amounts to a total of 20.[72] The Fellows are listed in Tab. 6.1 in the appendix in order of their election together with brief biographical notes. Among them the most numerous were researchers in the biological sciences, especially in medicine. They were eight in total – MARCELLO MALPIGHI, FRANCESCO TRAVAGINO, GIACOMO PIGHI, JACOBUS GRANDI, DOMENICO BOTTONI, FRANCESCO SPOLETI, GIORGIO BAGLIVI and SILVESTRO BONFIGLIOLI – some of them being better known to the Society than others. Another relatively large group was formed by Italians who may not have been necessarily interested in

---

72 Hall's (1982) list totals 17 Fellows, her list does not include the military diplomat Sir BERNARD GASCOIGNE (Bernardo Guasconi), the astronomer GIOVANNI DOMENICO CASSINI (1982:73) – because he had moved to Paris –, and the abbot and historiographer GIOVANNI BATTISTA PACICHELLI.

"natural philosophy", that is BERNARDO GUASCONI, a soldier and diplomat known in England as Sir Bernard Gascoigne and who assisted early Italian visitors to the Society; the Count CARLO UBALDINI of Montefeltri, who had converted to Protestantism;[73] GIOVANNI AMBROSIO SAROTTI, the son of PAOLO SAROTTI, a Venetian ambassador in England; the courtier and envoy TOMMASO DEL BENE; and the barrister and clergyman IPPOLITO FORNASARI. Diplomats, statesmen and other noblemen were generally elected for their potential as correspondents rather than their scientific achievements (Hall 1982: 64; Gomez Lopez 1997: 35).[74] However, it is not possible to group Fellows into specific disciplinary or social groups since, as was typical of the time, many of them had multiple interests and occupations, and the concepts of science and sciences had not yet developed to what they are today. The count LUIGI FERDINANDO MARSILI, for instance, was an emissary, a soldier, a natural historian and the founder of the Academy of Sciences of the Institute of Bologna (1712). MARCANTONIO BORGHESE was Prince of Sulmona and Rossano but he was also "keen on naturalist curiosities" (Gomez-Lopez 1997: 38). Part of the group were also two humanists, the Catholic abbot and historiographer GIOVANNI BATTISTA PACICHELLI and the polemicist and historian GREGORIO LETI. Finally, three Fellows were engaged in the mathematical and physical sciences: the mathematician VINCENZO VIVIANI; the astronomer GIOVANNI DOMENICO CASSINI; and DOMENICO GUGLIELMINI, who had studied both medicine and mathematics and mainly carried out studies in the fields of astronomy and physics.

However, the list of contributors to the "desseins of the Royal Society", as OLDENBURG used to summarise, cannot be limited to those Italians who were

---

73 Middleton (1979: 158) translates a comment made by LORENZO MAGALOTTI, which displays a rather suspicious opinion regarding the person of UBALDINI; it was therefore thought worthwhile to include it here: "I [(MAGALOTTI)] remember [...] a certain Italian personage who calls himself one of the Ubaldini [a noble family of Montefeltro]. Of his birth I know nothing, but I know very well what he has done since. At present he professes the protestant religion, and face to face he confesses he is an atheist. There are those who say that he has been a friar and that, leaving his order, he went to Costantinople and there, after having himself circumcised, tried his fortune with little success. After having been in various trades in different parts of Europe and in several of the courts of heretical princes, he came to London where, spreading stories about his birth and displaying some superficial learning, he obtained some grant from the king and some sort of pension from the bishops, on which he is now living. He is young, perhaps 32 years old, but extraordinarily fat, so that in time he will be entirely unable to move. I did not admire any very particular uncommon talent in him [...]".
74 See also Crosland (1983).

formally made Fellows. Indeed, the papers, letters and records of the Society provide a great deal of names of both known and lesser-knownh Italians who kept the Society informed on the Peninsula's goings-on and sent books and writings which often appeared in the journal.

A group of "contributors" has therefore also been included in this study. This group can by no means be considered exhaustive in that the more one looks into the Society's archives the more one can find, but it is hoped that at least most of those non-Fellow Italians who were highly influential for the Society have been inserted. Among them we find the physicians FRANCESCO REDI, CARLO FRACASSATI and GIOVANNI BATTISTA GORNIA; the botanist PAOLO BOCCONE, who also enriched the Royal Society's repositories with a number of natural curiosities, to which a paper was dedicated (*Phil. Trans.* 1673: 6158–6161);[75] the mathematicians, physicists and astronomers GIOVANNI ALFONSO BORELLI, GEMINIANO MONTANARI, FRANCESCO LANA, GIUSEPPE CAMPANI, GIOVANNI BATTISTA RICCIOLI, STEFANO DEGLI ANGELI; the Milanese Canon and collector MANFREDO SETTALA;[76] the physician and discoverer of the scabies mite GIOVANNI COSIMO BONOMO; the clergyman with natural-philosophical interests GIOVANNI GIUSTINO CIAMPINI who also visited England and attended three Royal Society meetings (Cook 2002);[77] and LORENZO MAGALOTTI, the secretary of the Cimento Accademy, who on behalf of the PRINCE LEOPOLD OF TUSCANY, travelled to London in 1667/8, with the nobleman PAOLO FALCONIERI, to personally deliver to the Royal Society the famous account of experiments made by the Academy, *The Saggi di Naturali Esperienze* (Middleton 1969: 283).[78]

---

75 These natural curiosities were "un-common" red and white pieces of coral; "a certain stony substance that is fossil and has the scent of bitumen", the uses of which, according to the writer, may be worth of study; "a not ordinary *sangui suga* or leech"; *sal armoniac* (ammonium chloride) from Sicily; and "many figur'd stones, shells, glosso-petras, fishes, plants, mineral bezoards of Sicily &c." (*Phil. Trans.* 1673: 6158–6161).
76 SETTALA probably thought to have been elected a Fellow of the Society as in one of his letters to OLDENBURG he writes as if he were one of them. In a meeting where this letter was read, the misunderstanding was noted and it was suggested to elect SETTALA in the following meeting; the matter, however, was not taken up again and SETTALA was never elected (Hall & Hall 1986: 496).
77 His full manuscript account, entitled *Iter in Britannicum,* is preserved at the Biblioteca Vallicelliana in Rome (Cook 2002).
78 The book was an account of the experiments made by the Accademia del Cimento between the years 1657 and 1667 (Middleton 1969: 283). Although the book was well received, it would seem that its contents did not, at least initially, arouse particular interest among the Fellows. The experiments within the book were seen as no novelty in

The lack of Fellowship does not mean that these men were of less interest to the Society. On the contrary, letters and papers show that there was strong interest in them. OLDENBURG for example frequently asked his informants for news on CAMPANI and his advances in lens making.

Hence, these contributors were never given Fellowship but were in many cases more valuable to the Society – and to the same Italian philosophers, since they helped spread the research of other Italian natural philosophers rather than just their own – than the actual Fellows, who in some cases did not seem to make any contributions at all. Names of contributors with brief biographical notes are thus included in Tab. 6.1. The table also shows the number of papers appearing in the *PTRS* per every 17th-century Fellow/contributor.

Of course, in some cases there were Italians who strived to make their own contributions, such as PAOLO MATTIA DORIA, who wrote several letters to PAOLO ROLLI and CROMWELL MORTIMER in the early 1700s asking for the Fellows' opinion on his work, yet never gained great visibility (see also Section 4.2.3). Moreover, articles that were sent to the Royal Society were not necessarily always published. In other cases, the name(s) of the original author(s) of a study did not appear within the publication.

Finally, there were also Italian men who were neither Fellows nor article or letter contributors, but who were nonetheless very well known and appreciated by the Royal Society. An example is the mathematician GIOVANNI ANTONIO DAVIA, who does not appear to have corresponded with the Society, yet travelled in 1681 to London and participated in a meeting held in his honour;[79] another example is LEONARDO DI CAPUA, a member of the academy of the Investiganti, whose book on the nature of damps was reviewed in the *PTRS* (*Phil. Trans.* 1694: 33–40); again, ANTON FELICE MARSILI's *Relazione del ritrovamento dell'uova di chiocciole* (Bologna 1683) indirectly reached the Society through M. MALPIGHI;[80] and GIACINTO CESTONI's findings on the generation of fleas are

---

British science. However, in 1683 it was decided at a Council Meeting that a translation by RICHARD WALLER (FRS) should be printed (Boschiero 2010). The translated title was *Essays of Natural Experiments, made in the Academie del Cimento, &c.* (see Section 1.3 for further information on the translation of the title of the book).

79 Brizzi, G. (1987). Davia, Giovanni Antonio. *Dizionario biografico degli italiani*. Vol. 33. http://www.treccani.it/enciclopedia/giovanni-antonio-davia_(Dizionario-Biografico)/.

80 *Phil. Trans.* 1683: 365–359 contains an account of MARSILI's book.

reported in *Phil. Trans.* 1699: 42–43.[81] Another example is that of EUSTACHIO DIVINI, an optician very well known in Europe for his production of telescopes, of which also CASSINI and the Cimento Accademy made use. He is referenced in various papers on optical matters,[82] and information about his work was sent to the Society in various letters.[83] His and CAMPANI's telescopes moreover appear to have been sent to the society by JOHN DODINGTON.[84] Other Italians of or from whom the Society received news were: CARLO BARTOLI, FILIPPO BONANNI, P. CAPUANI, and ANTON MARIA VALSALVA.[85] Henderson (2013) also mentions translations being made corporately by the Society, or privately by individual Fellows, of the following authors' books: FRANCESCO STELLUTI, *Trattato del legno fossile minerale nuovamente scoperto* (Rome 1637); JACOPO BAROZZI DA VIGNOLA, *Regola delli cinque ordini d'architettura* (Rome 1562); GIOVANNI VENTURA ROSETO, *Plictho de l'arte tentori* (Venice, 1540); ANTONIO NERI, *L'arte vetraria* (Florence, 1612); NICOLAUS STENO, *De solido intra solidum naturaliter contento dissertationis prodromus* (Florence, 1669). As a result, observations on the Italian contribution to the Royal Society's programme cannot just be made on what appears from the election of Fellows and the papers selected for publication but must be based on all activity relating to the Italians' excahnges with the Society.

In sum, the first major result that appears by looking at the list of Fellows of the Royal Society is that the Society elected several Italians in the 17th century but only some of them were helpful contributors to the plan of a universal natural philosophy. Moreover, Italians who were not made Fellows did not play any lesser role than formally elected Fellows and were very helpful correspondents

---

81 Cavazza, M. (2008). Marsili, Anton Felice. *Dizionario biografico degli italiani*. Vol 70. Treccani online http://www.treccani.it/enciclopedia/antonio-felice-marsili_%28Diz ionario-Biografico%29/
82 *Phil. Trans.* 1665: 131–132, 209–210, 362, *Phil. Trans.* 1677: 1005, *Phil. Trans.* 1668: 840–842.
83 EL/A/12, EL/D1/18, EL/I1/42, EL/O1/74 (in this particular letter addressed to ADRIEN AUZOUT, OLDENBURG expresses his wish to receive information about a controversy between DIVINI and CAMPANI concerning their lenses), EL/O2/18, LBO/1/112 and LBO/31/29.
84 In one of his letters, DODINGTON informs the Society that he is sending them some telescopes and microscopes of DIVINI and CAMPANI, and some books from MALPIGHI. (EL/D1/18). More on JOHN DODINGTON (1628–1673) can be read in the section on Fellows and Englishmen in Italy below. Much of his correspondence has been published in Hall & Hall 1965–1986.
85 On VALSALVA see Cavazza (1980: 112).

and/or communicators of knowledge and newsworthy material. Finally, by considering Italian contributors and Fellows together, it becomes apparent that the three major areas of interest in the 17th century were medicine, astronomy and physics. More detail on the topics of the papers will be provided below.

3.1.2 Creating a discourse community: Correspondence and information exchange

The plans of the Royal Society were especially promoted by its secretary HENRY OLDENBURG, who from the first years of the Society's existence sought contacts with scientists and noblemen from all over the world. The Society's intentions are well expressed in the following passage, in which OLDENBURG encourages the Italian LEOPOLD DE MEDICI to start cooperation:

> OLDENBURG to LEOPOLD DE MEDICI, 1667
> As the designs for the cultivation of a sound and useful philosophy that have been framed both by our august King [Charles II] and your highness [Prince Leopold of Tuscany] are so well matched, it seems entirely fitting that there should be a close and friendly relationship between our two societies [the Royal Society and the Academia del Cimento]. And indeed, when one examines the matter deeply, it seems that the successful execution of so arduous and burdensome a plan requires rather the joint labours of the industrious and wise men of the whole world in mutual co-operation, than those of this region or that alone.
> For here it is not a question of expounding some text of Aristotle or Plato, nor of unravelling the causes of this or that phenomenon, but rather of investigating and explaining the Book of Nature in the Universe, and of duly searching into the hidden purpose of its most wise Author, so far as human intellect can do so. For the task will not be completed by sectarian zeal and oratory, nor by warmth in disputation, nor by the arts of speaking, nor by the precipitate stitching together of systems; what is absolutely necessary is a choice and lasting association of those throughout the world who are skilful experts in Nature, and a diligent and unremitting examination into Nature through observation and experiment, carefully and frequently performed. (Letter 706, Hall & Hall 1966: 620)[86]

Hence, in order to cultivate an objective and universal philosophy of nature, the latter needs to be investigated by "diligent and unremitting examination" and not simply by drawing onto what was said by great men of the past. Further, it is

---

86 Letters exchanged with Italians were usually written in Latin or French. This and the following extracts in English are the translations made by Marie Boas Hall ad Richard Hall.

not sufficient to study nature in solitude; a thorough study of nature requires the cooperation of the best wits throughout the world.

OLDENBURG worked hard to collect information and create as many contacts as possible in Italy. To this objective, he exploited various means: English residents in Italy, travellers, merchants, and foreign acquaintances. In 1670, for instance, he sent JOHN DODINGTON, an English resident in Italy, a list of Italians to get in contact with.[87] By having letters and books delivered through travellers, he also got in direct contact with Italians whose names and works had come to his knowledge. The letters, written to encourage correspondence, tend to be more formal and encomiastic in tone than letters exchanged with closer friends and Fellows but always make clear what his intentions are, see for example the following extracts:

OLDENBURG to MANFRED SETTALA, 1667
Both my duty and inclination impel me to endeavour to further the object[ive]s of this Society in every way: now I understand that you are engaged upon successful study of the secrets of nature and of art and make great strides in both; moreover, it is agreed on all sides that cooperation and friendliness between learned men greatly foster and illuminate the liberal arts and our knowledge of things. Hence, by this letter, I wish to entice you among others into wholehearted co-operation in thought and study, to the end that through the experimental observation of things in nature the foundations of philosophy may be rendered more solid – the old ones being subjected to criticism and equally fruitful new ones be made publicly known.
Therefore **it is my earnest request that if anything out of the way, philosophically speaking, becomes known to you or to other learned men in your noble city [Milan] and in its neighbourhood, you will be so kind as to impart it to us.** To deserve well of you in return I shall gladly, so far as I can, recompense you for your philosophical generosity. Please take a chance on this. You will find me most zealous on your behalf. Farewell. (Letter 647, Hall & Hall 1966: 440–441)

OLDENBURG to MALPIGHI, 1667
With the utmost earnestness, I beg you to be so kind as to impart to us whatever in your later work appears to be philosophically notable, or whatever occurs to other skilled and learned men in Sicily that helps to promote philosophy. To the best of my ability I will gladly furnish you with everything that will in turn cause you and your countrymen to think well of me and recompense your philosophical liberality from our store. Put me to the test please. You will find me most zealous towards you. (Letter 740, Hall & Hall 1967: 92)

---

87 EL/O2/18. The Italians mentioned in this letter were TRAVAGINO, DEGLI ANGELI, MALPIGHI, MAGALOTTI, SETTALLA, DIVINI, CAMPANI, CORNELIO.

17th century 81

OLDENBURG to FRANCESCO TRAVAGINO, 1667
Do you, Sir, have the goodness to impart whatever occurs in Venice or the rest of Italy that is out-of-the-way, philosophically speaking. I will gladly add whatever I can, in exchange, to deserve well of you. Grasp the nettle, and try my zeal for you. Farewell.
(Letter 640, Hall & Hall 1966: 416)[88]

OLDENBURG to LEOPOLD DE MEDICI, 1667 (Continuation of the above cited letter)
Come then, noble Prince, **let us begin by combining our studies so that our good sense may persuade the philosophers of other nations to join with us. Let us all put together our intellects, labours, and resources** and omit nothing whatever which may enable us to restore the pristine dignity of humanity and recover our lawful right to its primitive dominion over subject creatures. It may hitherto have been enough to have sworn by the authority of others; but from now on we ourselves will rely upon our own minds, hands and eyes. (Letter 706, Hall & Hall 1966: 621)

As can be seen from the first lengthier extract, OLDENBURG tended to make his cooperation request after going through an explanation of the Society's principles and aims. This included praise towards the work and interests of his addressee. Finally, in a rather polite and standardised manner, he would ask his addressees to send him any information available on their own work or that of others (in bold).[89] In exchange for the information sent by the Italians OLDENBURG would offer to do the same (underlined in the extracts). In some cases, as an example of his good intentions, he promised or directly sent with the letter (as in the case of the letter to the PRINCE LEOPOLD OF TUSCANY)[90] a copy of Sprat's *History of the Royal Society*. Several of these letters encouraging cooperation were sent to learned men throughout Italy (Milan, Sicily, Venice and Florence in the examples). A copy of letter 647 to SETTALA, for instance, was also sent to THOMAS CORNELIO at Naples (letter 648, Hall & Hall 1966: 441). OLDENBURG

---

88 In this case it was FRANCESCO TRAVAGINO who had got in contact with OLDENBURG seeking for judgement on his work (letter 591, Hall & Hall 1966: 302–303). OLDENBURG then exploits this reply letter (640, Hall & Hall 1966: 414–416) to ask TRAVAGINO to send material from Venice.
89 If there were any notable natural philosophers temporarily or permanently staying in Italy, OLDENBURG would also ask his contacts to send him information about them. For example, through LORENZO MAGALOTTI OLDENBURG learned about the researches of the Danish NICHOLAUS STENO, who carried out fossil and blood studies while moving about Tuscany.
90 "P.S. The logic behind the Royal Society's design, its studies, progress, and endeavour, will be disclosed to your highness in its history lately published here. We modestly beg your highness to accept kindly the copy offered by the hand of Sir John Finch, His Majesty's Resident in the Court of Florence." (letter 706, Hall & Hall 1966: 619–622).

was of course not the only Fellow who corresponded with Italians, other Fellows did the same, such as BOYLE, SLOANE and others.

Contacts with individual Italian scholars were probably not enough for the Society, which also relied on third parties to collect information on Italy. *PTRS* articles, for instance, were sometimes extracted out of the French *Journal des Sçavans*.[91] And to further search into the work of the Italians, the Society exploited the connections of French Fellows such as the astronomer ADRIEN AUZOUT and the Huguenot counsellor HENRI JUSTEL. The exploitation of a varied number of resources was most likely due to the problems and risks related to postal services and the shipment of material through merchants, as can, for instance, be seen the letter extract below. Hence, searching for the same material from a number of different sources would give the Society more chances of being satisfied, as well as receiving different perspectives on a same subject.

> AUZOUT to OLDENBURG, 1666
> I have found nothing new here to send you. I should have liked to be able to send you the letters of Cassini and Campani, but I have only partial copies and I cannot easily secure others; for these gentlemen ordinarily send only one copy, on account of the postage. I am surprised that they do not send you any copies direct, or that the Englishmen in Rome do not do so, but perhaps you have now received some. (Letter 517, Hall & Hall 1966: 115)

The passage ends with AUZOUT's expression of surprise about OLDENBURG not receiving letters directly from Italians, which in fact he did. News on CASSINI's work, for instance, was also directly sent by CASSINI. OLDENBURG had also asked

---

91 E.g. "An Observation of Optick Glasses, Made of Rock-Crystal" *Phil. Trans.* 1665: 362; "Some Observations of Vipers" *Phil. Trans.* 1665; 160–162; "Observations of Some Animals, and of a Strange Plant, Made in a Voyage into the Kingdom of Congo: by Michael Angelo De Guattini and Dionysius of Placenza, Missionaries Thither. Extracted out of the Journal des Scavans" *Phil. Trans.* 1677: 977–978; "An Extract of a Letter Written by Signor Cassini to the Author of the Journal des Scavans, Containing Some Advertisements to Astronomers about the Configurations, by Him Given of the Satellites of Jupiter, for the Years 1676, and 1677, for the Verification of Their Hypotheses" *Phil. Trans.* 1676: 681–683; "Some New Observations Made by Sig. Cassini and Deliver'd in the Journal Des Scavans., Concerning the Two Planets about Saturn, Formerly Discover'd by the Same, as Appears in N. 92. of the se Tracts" *Phil. Trans.* 1677: 831–833; "An Extract of the Journal Des Scavans. of April 22 st. N. 1686. Giving an Account of Two New Satellites of Saturn, Discovered Lately by Mr. Cassini at the Royal Observatory at Paris" *Phil. Trans.* 1686: 79–85. *Phil. Trans.* 1695: 467–471, instead, was extracted out of the "Journal of Brunets *Progres de la Medecine*".

about CASSINI's and CAMPANI's letters to the English ambassador in Italy, SIR JOHN FINCH:

> OLDENBURG to FINCH, 1666
> Sir, I put you in mind, by ye Society's order, [...] yt you would please to communicate to us what is from time to time transacted among the excellent Virtuosi of Italy, especially those of Florence (yt have so eminent a Patron, as is Prince Leopold) of Rome, Naples and Bononaea. I desired further, to be fully informed by you, of the new way of Grinding Optick glasses, invented by Campani and its progress, ye length, aperture and charge of ye Glasses; as also of the new discoveries, made thereby; concerning wch we would gladly see those letters, wch we heare, passed between Cassini and Campani, and are in print, but not yet come hither. (letter 507, Hall & Hall 1966: 86)

This extract is also relevant to introduce the important role played by Englishmen residing in and travelling through Italy. Before asking more specific information about CASSINI and CAMPANI, OLDENBURG tells FINCH that the Society would like to receive news on what is transacted among all the Virtuosi of Italy, from north to south. Similar requests were also made to other Fellows in Italy.

Requests for news, of course, came from Italy too. The following extract from the end of a letter sent from Italy seems relevant to quote here in that it further shows how news on scientific activity was treated as a good to be bartered and to encourage the "production" of further knowledge:

> If I may be so happy as to receive sometimes from you an account of the Curious performances of your famous Royal Society, I shall make use of that favour, to animate the Virtuosi here to do something that may not be unworthy of your knowledge [...]. (*Phil. Trans.* 1672: 5066)

Moreover, this request was published in the *Transactions* together with the rest of the letter, and, since frequently salutations were cut from published letters, the keeping them in this letter meant that the publisher probably wanted readers to view it. The publisher was therefore forwarding the request of knowledge to all readers of the *PTRS*.

3.1.3 Fellows and Englishmen in Italy

Hence, throughout the 17th and 18th centuries another very important source of information for the Society were Englishmen in Italy. In some cases, they were trusted merchants, such as a MR. JOSEPH MAY, who lived in Venice, and took care of OLDENBURG's correspondence there, (Letter 885, Hall & Hall 1967: 460) and ROBERT BALLE, an English merchant working in Leghorn between 1662 and

1698, who then returned to England and was made Fellow in 1708.[92] A more detailed study on BALLE as a merchant Fellow has been carried out by Fisher (2001) who shows that BALLE was more than a mere messenger of knowledge, rather he acted as a "patron, contributing time, money and contacts to the Society" (2001:352). Among other things, BALLE made his own contributions to Italy-Royal Society relations by donating an Italian mathematical manuscript to the Society, translating an Italian letter (and possibly more), and promoting the election of the Italian ANTONIO MARIA SALVINI (FRS 1716). In his correspondence with his Fellow friends he also provided information on what were the Italian customs on the subjects being discussed, such as the plague and the use of smoke from tar, coal, tobacco and other materials to prevent infection (Fisher 2001: 361), and, related to his strong interests in botany and horticulture, news on the botanic gardens of Italy and the possibility of making plant exchanges between the two countries (Fisher 2001: 364). In the *PTRS* we find an account on the ice and snow preservation methods used in Leghorn (*Phil. Trans.* 1665: 139–140), communicated by ROBERT's brother WILLIAM BALLE and based on ROBERT's account.

In other cases, the contributing Englishmen were actual Fellows who were either travelling about Italy or had taken stable residence there. SIR JOHN FINCH (FRS 1663), stayed in Florence for several years as his Majesty's representative. He took an MD at Padua (1657), becoming pro-rector and syndic of the same University (1656) and later became professor of anatomy at the University of Pisa (1659–1665).[93] FINCH, other than collecting news on Italy, had been entrusted with handing SPRAT's *History* to Prince LEOPOLD. However, the book was eventually delivered by his close friend THOMAS BAINES (FRS 1663), who, like FINCH, had taken an MD at the University of Padua and followed his friend along his travels. It would seem that the reason why it was BAINES and not FINCH to deliver the book to the Prince was a religious one. In a 1668 letter, FINCH explains that he could not make any "Personall applications to his H[igh]nesse, since he is become an Ecclesiastical Prince [i.e. Cardinal de Medici]" (letter 918, Hall & Hall 1967: 541). BAINES instead appears to have (still) been a member of the Cimento Academy and was able to perform FINCH's duty.

---

92 He returned to Italy again in 1721 and continued to support the Society (Fisher 2001: 352). While in England he was elected to the Society's Council (1710–1721).
93 Hutton, S. (2004-09-23). Finch, Sir John (1626–1682), physician and diplomat. *Oxford Dictionary of National Biography*.

The 7th Duke of Norfolk, HENRY HOWARD (FRS 1672), instead, probably stayed in Italy for a shorter period of time, but he was still asked to "obtaine some Philosophicall Correspondents in ye chief Citty's of Italy, and particularly at Florence, Pisa, Bologna, Milan, Venice, Naples, Rome" (Letter 552, Hall & Hall 1966: 200). And within these cities there were individual Italians in which the Society was specifically interested. In this letter, for instance, we read the names of MICHELANGELO RICCI, GIOVANNI DOMENICO CASSINI, GILLES FRANCOIS GOTTIGNIES, HONORÉ FABRI, FRANCESCO REDI, GIOVANNI ALFONSO BORELLI, MANFREDO SETTALA AND GIUSEPPE CAMPANI.[94]

The diplomat and naturalist SIR ROBERT SOUTHWELL (FRS 1662) went on the *Grand Tour* of Italy in 1659 residing first in Rome and later in Florence. He sent information to the Royal Society and in particular on the activities of the Academia del Cimento (Gomez-Lopez 1997: 37). ROBERT BALLE's brother, CHARLES BALLE, also helped the Society with its Italian relations by acting as an agent between DOMENICO BOTTONI (FRS 1695) and SIR HANS SLOANE. Worth mentioning again is also RICHARD WALLER (FRS 1681), who translated the *Saggi di Naturali Esperienze* in 1683 (see Section 1.3). WALTER POPE (FRS 1663) spent two years in Italy and wrote a paper on the mines in the Julian Alps, which will be briefly summarised in Section 3.1.5.[95]

Finally, intelligence on Italy was also sent by non-Fellow Englishmen. An example is JOHN DODINGTON, who resided in the republic of Venice and corresponded extensively with the Royal Society between 1670 and 1673 transmitting news gained through correspondence with his Italian contacts (see *Phil. Trans.* 1672: 4066–4067 and 4067–4068). WILLIAM BADILY, an English captain sailing in Italy, wrote a letter about raining ashes following an eruption of Mount Vesuvius (*Phil. Trans.* 1665: 377). And JAMES CRAWFORD, residing in Venice, created various connections between Italians and the Royal Society (see Hall & Hall 1986).[96]

---

94 MICHELANGELO RICCI (1619–82, mathematician, Rome), GIOVANNI DOMENICO CASSINI (1625–1712, astronomer, Bologna, Paris), GILLES FRANCOIS GOTTIGNIES (1630–89, Belgian mathematician and astronomer, lived in Rome), HONORÉ FABRI (c. 1607–1688, writer on medical and mathematical topics, Rome), FRANCESCO REDI (1626–97, physician, Tuscany), GIOVANNI ALFONSO BORELLI (1608–79, mathematician and physiologist, Rome), MANFREDO SETTALA (1600–1680, Canon in Milan), and GIUSEPPE CAMPANI (c. 1620–1695, lens maker, Rome).
95 See also Clerke, A., & McConnell, A. (2008). Pope, Walter (bap. 1628, d. 1714), astronomer and writer. *Oxford Dictionary of National Biography*.
96 Other Fellows and non-Fellow Englishmen who contributed *PTRS* papers on Italy were: THOMAS PLATT (*Phil. Trans.* 1672: 5060–5066), FRANCIS VERNON (*Phil. Trans.*

Hence, it was thanks to the hard work of individual Fellows, travellers and merchants that links were established between the Royal Society and Italy starting from the very first years of the Society's existence and continued into the 18th century.

To conclude, it may be here reminded that several Italians visited or lived in England too. Italian travellers in England generally wrote accounts about their travels and their impressions on England, such as LORENZO MAGALOTTI and his *Relazione d'Inghilterra*.[97] MAGALOTTI moreover learnt English and was one of the first writers to spread a knowledge of British literature in Italy (Waller 2012). Italian residents in England, instead, helped the Society with its foreign relations and acted as intermediaries between the Society and Italians, such as the already mentioned BERNARDO GUASCONI (see Section 3.1.1) and GIOVANNI AMBROGIO SAROTTI, who kept correspondence with Italians, proposed their letters in meetings and assisted Italian visitors to the Royal Society.[98]

3.1.4 Topics

After having created the spreadsheet listing all of the *PTRS* papers collected for the corpus, a quick glance of the papers immediately revealed that there is material coming from Italian sources nearly in every single 17th-century volume of the *Transactions*.[99] Moreover, the first year of the *PTRS* alone contains 14 Italian-research-based papers.

It ought to be reminded that science as we mean it today did not exist in the first centuries of the Society's existence; until the 19th century the study of science was not considered a professional activity but rather a recreational one, and gentlemen pursuing scientific activity did not limit themselves to one specific subject area. This was also due to the fact that there was no clear-cut distinction between different scientific disciplines, which gradually developed throughout the centuries. Hence, among the sampled papers it is not uncommon to find papers covering topics that would today pertain to very different disciplines.

1676: 575–582), MARTIN HARTOP (*Phil. Trans.* 1693: 827), OCTAVIAN PULLEYN (*Phil. Trans.* 1695: 537–539), the brother of Dr. ROBERT ST. CLAIR (*Phil. Trans.* 1698: 378–381), WILLIAM AGLIONBY (*Phil. Trans* 1699: 183–186), and *Phil. Trans.* 1669: 1028–1034 was communicated by "some inquisitive English merchants now residing in Italy".
97 See Waller (2012).
98 For a list of the Italian Fellows who either visited England or lived there see Hall (1982).
99 Except for the years: 1681, 1682, 1687, and 1692.

A book of PAOLO BOCCONE's,[100] for instance, which is described in the *PTRS* (*Phil. Trans.* 1694: 33–40), deals with miscellaneous natural philosophical matters such as "the cures and preservatives from the Plague"; "a certain man that after his wife's death suckled his child at his own breasts"; "an account of the several museums or repositories of curiosities to be seen in Italy"; "four reasons why plants are green all year" etc.

That said, it is possible to identify some topics of preference. The most numerous papers were those dealing with astronomical matters (about 30 papers), the majority of which are based on GIOVANNI DOMENICO CASSINI's work.[101] The principal topics of this period were the observation of comets, the discovery of the satellites of Saturn made by CASSINI, spots on the Sun and on Jupiter, lunar and solar eclipses, the satellites of Jupiter and a paper on the motion of the earth. Related to the astronomical interests of the Fellows are a series of papers on optic glasses and telescopes (about nine papers, some papers contain information on advances both in optics and astronomy). The key Italian specialists here were GIUSEPPE CAMPANI, EUSTACHIO DIVINI and GIOVANNI ALFONSO BORELLI.

The second large group of papers covers topics of medical interest (17 papers).[102] These papers reflect the scientists' desire to learn about human nature – still very much unknown – and its afflictions. Indeed, among these papers various anatomical studies by MARCELLO MALPIGHI, LORENZO BELLINI, GIACOMO GRANDI and GIOVANNI MARIA LANCISI are present.[103] Noteworthy among the medical papers is the

---

100 *Osservazioni Naturali, ove si contengono materie Medico-Fisiche, &c.* (Bologna 1684) (Natural Observations containing several Medico-Physiscal and Botanical Matters).
101 GIOVANNI DOMENICO CASSINI (1625 –1712) was born in the Savoyard State and lived and worked around Bologna until 1669, when he permanently moved to France. He helped set up and directed for the rest of his life the Paris Observatory. His son, JAQUES CASSINI, was also a member of the Royal Society; however, since he was born and lived all of his life in France and wrote in French, he was not inserted in the list of Italian Fellows.
102 Berti (2019) analyses 17th- and early 18th-century Italian-research-based medical papers extracted from the corpus.
103 Such as "An Account of Some Discoveries Concerning the Brain, and the Tongue, Made by Signior Malpighi, Professor of Physick in Sicily" (*Phil. Trans.* 1666: 491–492); "An Extract Out of a Lately Printed Epistolary Address, Made to the G. Duke of Toscany Touching Some Anatomical Engagements, of Laur. Bellini, Ord. Anat. Prof. at Pisa" (*Phil. Trans.* 1670: 2093–2095); "An Extract of an Italian Letter Written from Venice by Signor Jacomo Grandi, to an Acquaintance of His in London, Concerning Some Anatomical Observations, and Two Odd Births" (*Phil. Trans.* 1670: 1188–1189);

Extract of a Letter from Jean Marie Lancisi, Prof. Anat. Rom. To Mr. Bourdelot, Giving an Account of Mr. Malpighi, the Circumstances of His Death, and What Was Found Remarkable at the Opening of His Body. Being Art. I. of the 3d. Journal of Brunets Progres de la Medecine. (*Phil. Trans.* 1695: 467–471)

MALPIGHI, whose portrait still hangs in the Royal Society's halls, seems to have been of particular interest to the Society for his advances in medicine and as a correspondent. In the *Transactions* there are two papers written by him and the above account of the circumstances of his death and his autopsy. MALPIGHI, too, seems to have held the Society in high esteem as he dedicated and sent his works to them during his lifetime and "signed with his hand three days before his death, his posthumous works, which he had ordered to be deliver'd to his collegues of the Royal Society at London" (*Phil. Trans.* 1695: 469). Indeed, all of his works were sent to the Society and published by their official printers (Cavazza 1980: 109–111).[104]

Following the studies on animal and human anatomy are a number of papers investigating the nature of blood. These are reports of experiments carried out mainly by the physician CARLO FRACASSATI, such as "An Account of Some Experiments of Injecting Liquors into the Veins of Animals" (*Phil. Trans.* 1666: 490–491).[105] Although there are only two papers on this subject in this century, worth mentioning are also the studies on the nature and effects of viper venom. The key figure of these studies is the physician FRANCESCO REDI, who started studying viper venom upon request of FERDINAND II in the 1660s (Schickore 2010: 575).[106] REDI's work was very influential for contemporaneous

---

"An Extract of a Latin Letter, Written by the Learned Signior Malpighi to the Publisher, Concerning Some Anatomical Observations, about the Structure of the Lungs of Froggs, Tortoises, & c. and Perfecter Animals; As Also the Texture of the Spleen, & c." (*Phil. Trans.* 1671: 2149–2150); "Extract of a Letter from Jean Marie Lancisi, Prof. Anat. Rom. To Mr. Bourdelot, Giving an Account of Mr. Malpighi, the Circumstances of His Death, and What Was Found Remarkable at the Opening of His Body" (*Phil. Trans.* 1695: 467–471).

104 *Dissertatio Epistolica de Bombyce* (1669); *Dissertatio Epistolica de Formatione Pulli in ovo* (1673); *Anatome Plantarum* (1675); *Opera Omnia* (1686); *Opera Posthuma* (1697).
105 See also "An Experiment of Signior Fracassati upon Bloud Grown Cold" (*Phil. Trans.* 1666: 492) and "Two Extracts out of the Italian Giornale de Letterati; The One, about Two Experiments of the Transfusion of Blood, made in Italy" (*Phil. Trans.* 1668: 840–842).
106 The first of the papers dealing with REDI's researches (*Phil. Trans.* 1665: 160–162) is based on a paper in the French *Journal de Savants*, which reports about REDI's *Observations on Vipers* (1664). The English *PTRS* paper reports, among others, REDI's following findings: (1) that viper venom is produced in "two *vesicles* or *bladders*, which cover the teeth", and that these vesicles release the venom when compressed; (2) that

and future studies in this field. An example is here seen with a 1672 paper by THOMAS PLATT,[107] who writes a letter to the Society reporting about a series of experiments carried out on pigeons and other animals by thrusting vipers' teeth into the animals' bodies. The purpose of these experiments was to view whether REDI's observations on vipers were reliable. Although the author seemed already convinced of the truthfulness of REDI's study, the repetition of his experiments did appear to corroborate his work. The experiments were carried out in Italy, at LORENZO MAGALOTTI's house over several days. Other than THOMAS PLATT, various other Italian and English men arrived at the house to witness the running of these experiments (see the section on witnessing below).

There are four papers on Italian mines, their minerals and the way of extracting them: one is on mines in Friuli from which mercury was extracted (*Phil. Trans.* 1665: 21–26); another one focuses on experiments made by MARCO ANTONIO CASTAGNA, "superintendent of some mines in Italy", with *amianthus* (asbestos) (*Phil. Trans.* 1671: 2167–2169); and still CASTAGNA claimed to have discovered a mine near Bergamo of stones with cavities that contained a particularly fragranced balsam, which appeared to have healing properties (*Phil. Trans.* 1671: 3059). And finally, a paper originally written by FRANCESCO LANA comments on CASTAGNA's work, expressing his belief that a mine of crystal near Brescia, under CASTAGNA's superintendence, was in actual fact not a mine of crystal but

> a plenty of nitrous steams, which might withal hinder vegetation in those places, and coagulate the dew falling thereon. And that those exhalations were rather Nitrous, than of another kind [i.e. crystal], I was induced to believe, because Niter is not only the natural *coagulum* of water, as is manifest in artificial glaciations but also it ever retain the above said sex-angular figure. (*Phil. Trans.* 1672: 4068–69).[108]

---

the viper's venom is not deadly if ingested but it is if "rubbed into the wound" and that sucking the venom out of the wound will help save the life of the animal or person who has been bitten; (3) that the consumption of vipers' meat does not make thirsty as GALEN and "modern Physicians" of his time appeared to claim; (4) that the salt of vipers, which appears to have been esteemed by alchemists of the time for medicinal purposes, had no particular purging properties.

107 "An Extract of a Letter Written to the Publisher by Mr. Thomas Platt, from Florence, August 6. 1672. Concerning Some Experiments, There Made upon Vipers, Since Mons. Charas His Reply to the Letter Written by Signor Francesco Redi to Monsteur Bourdelet and Monsieur Morus" (*Phil. Trans.* 1672: 5060–5066).

108 The papers on CASTAGNA's work with mines and FRANCESCO LANA's paper, were all originally published in the Venetian *Giornale de Letterati* in Italian (Boella & Galli 2015: 59). The Royal Society extracted them from there, translated them into English and published them in the *PTRS*.

CASTAGNA's work and experiments were also of interest to the Fellows; it will be seen below (Section 3.4.1) that further papers related to his research are present in the *PTRS*.

About three papers deal with archaeology and antiquaries, two subjects that will considerably increase in the following century what with the popularity of the *Grand Tour*, which gradually extended its southern roots to the extreme south of Italy (see D'Amore 2015 and 2017). Related to the romantic fascination with the south of Italy and with wild nature are also six papers on earthquakes and volcanos in Sicily, Campania and one case of a "very odd eruption of fire out of a spot in the earth near Fierenzola" (*Phil. Trans.* 1698: 378–381). Ample space is given in the *Transactions* to the volcanic eruption of Mount Etna in 1669[109] and to the destruction caused by the earthquakes that occurred in Sicily in 1693, which destroyed many towns and cities.[110]

In the sampled corpus there are also about nineteen papers dealing with accounts, reviews, and/or simple lists of Italian books; however this figure will probably not be representative of the actual number of papers dealing with Italian books due to the method used in collecting the papers for the corpus, i.e. by searching the *PTRS* search engine inserting Italian keywords (see Section 2.3). Book accounts are often titled as such and therefore do not always contain words referring to the authors or to the book titles in their headings.[111] See, for instance, the following example:

Of some Philosophical and curious Books, that are shortly to come abroad.

1. Of the *Origine* of *forms* and *Qualities,* deduced from *Mechanical* Principles; by the Honorable *Robert Boyle* Esq.
2. *Hydrostatical Paradoxes,* by the same Both in *English.*

---

109 "An answer to some inquiries concerning the eruptions of Mount Ætna, an. 1669. Communicated by some inquisitive English merchants, now residing in Sicily" (*Phil. Trans.* 1669: 1028–1034).
110 "An Account of the Earthquakes in Sicilia, on the Ninth and Eleuenth of January, 1692/3 Translated from an Italian Letter Wrote from Sicily by the Noble Vincentius Bonajutus, and Communicated to the Royal Society by the Learned Marcellus Malpighius" (*Phil. Trans.* 1694: 2–10); A Letter from Mr. Martin Hartop at Naples, to the Publisher. Together with an Account of the Late Earthquake in Sicily (*Phil. Trans.* 1693: 827–830); "An Extract of the Account Mentioned in the Foregoing Letter, Taken out of an Italian Paper. Written by P. Alessandro Burgos. Printed First at Palermo, and Afterwards at Naples" (*Phil. Trans.* 1693: 830–838).
111 On the structural and linguistic features of book reviews in the *PT*, see Gotti (2011).

3. A Tract of the *Origine* of the *Nile*, by Monsieur *Isaac Vossius*, opposed to that of Monsieur *de la Chambre*, who is maintaining, That *Niter* is the principal cause of the Inundation of that River.
4. A Dissertation of *Vipers*, by *Signor Redi*, an *Italian*.
5. A Discourse of the *Anatomy* of *a Lyon*, by the same.
6. Another, *De Figuris Salium*, by the same.
7. A Narration of the Establishment of the *Lyncei*, an *Italian* Academy, and of their Design and Statues: the Prince *Cesi* being the Head of them, who did also intend to establish such Philosophical Societies in all parts of the World, and particularly in *Africa* and *America*, to be by that means well informed of what considerable productions of Nature were to be found in those parts. [...]. (*Phil. Trans.* 1665: 145–146)

This is a list of books that were about to arrive in England. Points 4–7 refer to Italian books, three of which are by FRANCESCO REDI and an anonymous one on the history of the Lincei academy. Half of the books advertised in this paper came thus from Italy, and another booklist in the corpus is dedicated *only* to Italian books.[112] News on newly printed books arrived from the Royal Society's correspondents, but also directly from Italian booksellers, as is shown in the following extracts, in which the writers comment about the scarcity of book trading relations with Italy:

> of the whole work, we think it will not be unacceptable to the Reader, to find it here entirely *Englished*, especially since the Book it self is yet very scarce in *England*, the commerce between our and the Italian Stationers being very slow, if there be any at all. (*Phil. Trans.* 1673: 6195)

> Our Commerce with the *Italian* Booksellers having been for several years less constant, it may not be amiss to present the Reader with an Account of some of the most Curious relating chiefly to Natural History of Philosophy. (*Phil. Trans.* 1694: 33)

MALPIGHI moreover wrote to OLDENBURG that one of the Society's booksellers did not "get on well with the Italian booksellers" (MALPIGHI quoted in Rivington, 1984).

Booklists did not report any detail about the books but only included the references. Accounts and reviews of books, instead, reported more detail about the contents of the book(s). Reviews moreover provide useful insight into what were the opinions of the Fellows of the Italian works described.

Worth mentioning, given their informational relevance, are the papers of general updates. See for instance the following extract reporting "Some Communications from Rome and Paris" (*Phil. Trans.* 1675: 309):

---

112 "A Catalogue of Books Lately Printed in Italy" (*Phil. Trans.* 1698: 426–428).

Signor *Borelli*, who is now at *Rome*, pretends to have lighted upon a way of building *Gallies* with several Tires of Oars, of different heights, which he esteems to be more convenient, more speedy and stronger than those that are now in use. He thinks also, that he can give an account of the possibility of the Gallies of the Ancients to a determinat number of tires; and he promises a Treaty of it, with demonstrations.

There is also at *Rome* a Bowle, which is so counterpoised, that it can stop of it self upon an inclined plane like *Keplers* watch. It stops upon all sorts of matter, and even upon a Looking-glass.

Pater *Gottignies* hath undertaken at *Rome* to write an *Algebra* after a new manner. He gives it the Title of *Logistica universæ Mathesi inserviens*. It is to consist of four Books, whereof the first is already printed.

The author simply writes a few paragraphs on any news he has managed to collect from Rome and Paris. This was probably done to satisfy the Society's requests to receive intelligence from Italy (and other countries). Of the same kind but providing much more detail are two papers by PIERRE SILVESTRE,[113] which represent a good example of how information and, in this specific case, medical intelligence was collected and communicated to the Society. The papers are inserted in the *PTRS* with the following titles:

> A Letter from Dr P. Silvestre, of the Coll. of Phy. & F. R. S. to the Publisher, Giving an Account of Some New Books and Manuscripts in Italy. (*Phil. Trans.* 1700: 613–614)

> A Letter from Dr Peter Silvestre, F. R. S. to the Publisher, concerning the State of Learning, and Several Particulars Observed by Him Lately in Italy. (*Phil. Trans.* 1700: 627–634)

At the opening of the 18th century, SILVESTRE travelled through Italy and visited physicians, universities and academies collecting information for the Royal Society. He reported some information in a letter, and then, since the Fellows desired to be "more particularly informed of the virtuosi" he had seen in Italy and "of the state of learning there, chiefly as to natural philosophy and physick", he added the second more detailed supplementary letter.

In the first letter, he focuses on medical books that were being published in Italy in that period. He proceeds by mentioning the places he visited, the physicians he met, the research they were working on, and occasionally some further curiosities about the physicians or his conversations with them. The following extract provides an example of SILVESTRE's manner of proceeding:

---

113 Also anglicised as PETER SYLVESTER (1662–1718, FRS 1699). SILVESTRE was a French physician. He arrived in England as WILLIAM OF ORANGE's physician in 1689 (Source: Royal Society's *Fellows Directory*).

I saw at passing *Florence*, Monsieur Bellini, he is at present busie in writing the anatomy of the body of man, in the *Tuscan* language. He assured me this work was wrote so clearly, and that he had taken such pains to explain the functions, by examples from ordinary mechanicks, and the commonest things, that the most ignorant could understand them […]. At *Rome*, I saw some manuscripts of the late famous Borelli at the Scholæ Piæ, where he died. One of them was a discourse of his *de volatu hominum*, wherein by mechanicks he pretends to make up the natural defects a man has to fly. There are also many other academical discourses […]. Some others of these discourses were by him read in the Academy[114] of the Queen [Christina] of Sweden, and ready for the press. I had almost forgot to tell you that I saw at *Bononia*, a very fine preparation of the human organ of hearing, […] the author thereof Senior *Valsalva* told me he would speedily publish something, not being satisfy'd with what is already made publick upon that subject. (*Phil. Trans.* 1700 613–614)[115]

Letters such as SILVESTRE's were frequently sent in this period both by English and Italian correspondents allowing the Society to be relatively well informed on the state of Italian medical research. However, only SILVESTRE's letters (EL/S 2/26, EL/S 2/27 and EL/S 2/28) were translated (from French) and inserted in the *Transactions* in full.

In the second letter, SILVESTRE provides more detail about his travel and the people he met, sending over to England books and natural curiosities. He digresses in further, at times mundane, detail, which allowed the Fellows to gain a picture of the Italian cultural scene. For instance, he explains that in Padua "he enquir'd for the most eminent men of that University" but he unfortunately found that most of them were out of town since it was vacation time. He expresses his appreciation for GIANGIROLAMO SBARAGLI, but found that he was disliked because of his antagonism towards MALPIGHI. He visited the Collegio Romano and the Museum Kircherianum. In Naples, he was surprised to find "a great many persons applying themselves to the corpuscular philosophy and mathematicks". He also met

---

114 Queen CHRISTINA's court in Rome was a lively centre for natural philosophy and was frequently visited by learned travellers, including PIERRE SILVESTRE. CHRISTINA extended her patronage to the Accademia Fisico-matematica, founded in 1678 by GIOVANNI GIUSTINO CIAMPINI (Cook 2004: 4).
115 Italians mentioned in this paper: MARCELLO MALPIGHI, GIANGIROLAMO SBARAGLI, JOHN BAPTISTA TRIUMPHETTI (GIOVANNI BATTISTA TRIONFETTI), GIOVANNI MARIA LANCISI, [?] SANGUINETUS, ANTONIO DI MONFORTE, MONSIEUR GIMELLI (GIOVANNI FRANCESCO GEMELLI CARERI), LORENZO BELLINI, GIOVANNI ALFONSO BORELLI and ANTONIO MARIA VALSAVA.

Signior Joseph Valeta, a gentleman who has a very good library, and has learnt a little English, on purpose to understand English books, for which he has a very great value. He lent me a manuscript of his that he will speedily publish. His design is to commend and encourage the Experimental Philosophy. (*Phil. Trans.* 1700: 627–634, 629)[116]

He goes on to list the names of Italian physicians and other men of learning, some already known and some new, providing detail as to their lives and careers. For instance, he says that BELLINI had become Professor Emeritus and physician to the Grand Duke of Tuscany and that DEL PAPA had become physician to the Cardinal DE MEDICI. He then goes back to more specifically medical curiosities describing some wax carvings of the muscles and internal viscera that he had been shown in Genoa.[117] He praises them saying that he could hardly distinguish them from the parts of a real corpse and emphasises the utility that such material could have in the study of medicine:

> If there was half a dozen of these wax carvings, in several views, to shew at any time the structure of humane bodies, it would not only shorten the study of anatomy, but besides make it a great deal less nauseous to the beginners. (*Phil. Trans.* 1700: 627–634, 630)

He closes off with some of his own observations on a distemper that was frequent in Lombardy and Savoy and with a list of natural curiosities that he sent with the letter.

It is thanks to letters such as the above that the Society was informed on the state of learning in Italy and all over the world. Learned men and their work would eventually come to the knowledge of the Fellows, and real samples of natural curiosities added a further sense of truthfulness to the reports.

From a cultural point of view, another paper worth mentioning is "An extract of a letter, written March 5, 1672 by Dr. THOMAS CORNELIO, a Neapolitan Philosopher and Physician, to JOHN DODINGTON Esquire, his Majesties Resident at Venice; concerning some observations made of persons pretending to be stung by tarantula's". CORNELIO had evidently promised DODINGTON to send

---

116 Italians mentioned in this paper: POMPEIO SACCHI, FRANCESCO SPOLETI, CAVALIER SORANZO, DOMENICO GUGLIELMINI, GIANGIROLAMO SBARAGLI, MARCELLO MALPIGHI, MONSIGNOR LUCA TOZZI, [?] SINIBALDI, GIORGIO BAGLIVI, RAFFAELE FABRETTI, FILIPPO BONANNI, PAOLO BOCCONE, TOMMASO CORNELIO, LEONARDO DI CAPUA, GIUSEPPE VALLETTA, TOMMASO DONZELLI, ANELLO DI NAPOLI, OTTAVIO SANDORO, GIOVANNI BATTISTA GARNIERI, NICOLA PARTENIO GIANNETASIO, LORENZO BELLINI, GIUSEPPE DEL PAPA, GIUSEPPE ZAMBECCARI, PASCASIO GIANNETI, ANTONIO MAGLIABECHI, VINCENZO VIVIANI and [?] COLECHIANI.
117 For more on this subject see Taddia (2009).

him some tarantulas, which were due to arrive; meanwhile he decides to keep his friend in Venice entertained by sending him this account of people convinced of having been stung by tarantulas. The piece reveals some popular beliefs of 17th century southern Italy:

> The same person affirm'd to me, that all those that think themselves bitten by *Tarantula's*, (except such, as for some ends fain themselves to be so,) are for the most part young wanton girles, (whom the Italian writer calls *Dolci di Sale*,) who by some particular indisposition falling into this melancholly madness, perswade themselves according to the vulgar prejudice, to have been stung by a *Tarantula*. And I remember to have observed in *Calabria* some women, who seised on by some such accidents were counted to be possess'd with the Divel; it being the common belief in that Province, that the greatest part of the evils, which afflict man-kind, proceeds from evil Spirits. (*Phil. Trans.* 1672: 4066–4067, 4067)

The tarantula of Puglia was evidently a topic of interest and it is also mentioned in an account of PAOLO BOCCONE's *Museo di fisica ed esperienze* (Venice 1697). JOHN RAY, the author of the paper, comments about how this was "a beaten Subject, and of which more hath been said than it's true" (*Phil. Trans.* 1699: 53–67, 57). He does not refrain from expressing his disbelief in what BOCCONE reports about people stung by tarantulas – who BOCCONE calls the "tarantolati" (misspelt as "tarantati" in the *Transactions*) –; namely, that if the *tarantolato* does not kill the spider immediately after having been bitten, he will be affected by the spider's venom for years. According to BOCCONE, among the various symptoms that the venom circulating in the body causes – such as nausea, loss of appetite and extraordinary strength – the most interesting one appears to be the desire of the *tarantolato* to dance. This particular symptom is said to recur annually during the summer season, since the spider's venom appears to ferment in the body (Boccone 1697: 101–103). In the same book, BOCCONE also writes about the tarantulas of Corsica (also called "malmignatto" by the locals) and of Sardinia.

Finally, apart from a small group of four papers on natural history, the remaining are all individual papers on a variety of topics.

3.1.5 Landscapes of northern and central Italy

As far as Puglia is concerned then, the main topic of interest in the *Transactions* appears to be the tarantula. Much more attention was given to the remaining parts of the Italian south. The Bourbon kingdom of Naples and Sicily with their natural and archaeological beauties – earthquakes, volcanos, and, later, the discoveries of the ancient cities of Pompei and Herculaneum – have been amply described by D'Amore (2017). Not much, instead, is found in the literature about the Fellows' impressions of the northern parts of the Italian peninsula. It is therefore the purpose of this section to report what natural, artistic and architectural features drew the Fellows' curiosity to the northern and central parts of Italy.

An unmissable stop for travellers in the north was SETTALA's museum in Milan (Preianó 2016). SETTALA was an avid collector of all sorts of natural and artificial objects from all over the world – he had travelled himself to the East and to Africa to learn about those cultures and bring memorabilia back to Italy. He kept his memorabilia and scientific instruments in his own museum in palazzo Settala in Milan, which is frequently defined as a *Wunderkammer* but was more than that. SETTALA did not limit himself to collecting and exposing his treasures in his museum; he was himself a scientist and used his collections to study and make experiments with them. He made his own creations too; a well-known example is his devil automaton: a head and bust of the devil, carved in wood, which could, by turning a leva, roll its eyes and move its tongue, make a sound and spit smoke from the mouth. He moreover regularly corresponded with the Society and provided them, for instance, with the geographical dimensions of the city of Milan (EL/I1/52). Further, he communicated that among the mountains of Piedmont, in the valley of *Lancy* (Lanzo), grew a plant called *Doronicum*, around which mercury could be found, and that in the mountains near Genoa great deposits of cockleshells were discovered (*Phil. Trans.* 1666: 493, more detail is provided in the original letter, see Hall & Hall 1986: 454–457). From the areas of Bergamo and Brescia there are reports about their mines (*Phil. Trans.* 1671: 3059 and 1672: 4068). From Padua, an account of its "Apponensian baths" (baths of Abano Terme), the waters of which were "hot", "stinking" and "yield a great deal of very fine salt; of which the natives serve themselves in their ordinary occasions" (*Phil. Trans.* 1672: 4067–4068, 4067). Many papers came from the Tuscan cities; however, these consist primarily of book accounts, astronomy and medical papers. From Venice and Padua, which at the time were part of the Republic of Venice, and from Bologna (Papal States) there are also several papers concerning medical topics. The papers from Rome cover a variety of topics, including descriptions of its architectural features, an example of which will be provided here. Overall, in the 17th century *Transactions*, physical descriptions of the northern and central parts of Italy are only a few, however some information by FRANCIS VERNON, TANKRED ROBINSON, GIOVANNI AMBROGIO SAROTTI, GIOVANNI BATTISTA DONI, and WALTER POPE was published.

WALTER POPE wrote an account of the mercury mines in Friuli (*Phil. Trans.* 1665: 21–26). He provides a detailed description of the territory, explaining that Friuli was part of the Venetian Republic and that the mines were situated at *Idria*

(Idrija), a valley in the Julian Alps (today part of Slovenia). The territory was part of the Emperor LEOPOLD's I possessions and the inhabitants spoke Slavonic. POPE was clearly struck by the area and shows his appreciation throughout the description:

> In going thither we travell'd several hours in the best Wood I ever saw before or since, being very full of *Firrs, Okes,* and *Beeches,* of an extraordinary thickness, straitness, and height. The Town is built, as usually Towns in the Alps are, all of wood, the Church onely excepted, and another House wherein the Overseer liveth. When I was there, in *August* last, the Valley, and the Mountains too, out of which the *Mercury* was dugg, were of as pleasant a verdure, as if it had been in the midst of Spring, which they there attribute to the moistness of the *Mercury;* how truly, I dispute not. That Mine, which we went into, the best and greatest of them all, was dedicated to Saint *Barbara,* as the other Mines are to other Saints. (*Phil. Trans.* 1665: 21)

Having himself experienced it, he explains in detail how the workmen would climb down into the mines by means of ladders and having to duck at various points due to the lack of space. He describes what minerals were found and how mercury was extracted. He praises the engines employed in the mines "the Wheels, the greatest I ever saw in my life; one would think as great as the matter would bear: all moved by the dead force of the water" (1665: 23). He details the workers' pay (six or seven pence), working hours (six) and explains that all of them in time become paralytic.

VERNON's paper (*Phil. Trans.* 1676: 575–582) derives from a letter of his travels "from Venice through Istria, Dalmatia,[118] Greece, and the Archipelago, to Smyrna". He describes the places he saw and informs the reader on what, according to him, is worth visiting, for instance:

> That which is most worth seeing in *Dalmatia,* is *Spalatro; where* is Dioclesian's Palace, a vast and stupendous fabrick, in which he made his residence, when he retreated from the Empire. It is as big as the whole town; for the whole town indeed is patch't up out of its ruines, and is said by some to take its name from it. The building is massive; there is within it an entire Temple of *Jupiter,* eight-square, with noble Porphyrie pillars, and Cornice, worth any bodies admiration.
> There is a Court before it, adorned with *Ægyptian* pillars of that stone called *Pyropoicilos,* and a Temple under it, now dedicated to St^a *Lucia;* and up and down the Town several fragments of Antiquity, with Inscriptions and other things, worth taking notice of. (*Phil. Trans.* 1676: 576)

---

118 Istria and Dalmatia, today mostly Croatian, were part of the Republic of Venice in the 1600s.

His comments reflect British cultural and literary interests of the time: for example the strong interest in classic antiquities is shown with his numerous descriptions of Roman and Greek monuments; he moreover reports about the inscriptions from the monuments. The narrative starts from Venice, from which, aboard of a galley, they set out on their travel and reached Istria and Dalmatia. He briefly describes the ancient Republic of Pula and its amphitheatre built with two orders of Tuscan pillars, standing one over the other, and the lower pillars unusually standing upon pedestals. He then continues with the Roman temple and triumphal arch. From Dalmatia he provides some information on the city of *Zahara* (Zara/Zadar), described as a very well-fortified metropolis. The journey then proceeds passing by *Zabenico* (Šibenik), St. Nicholas and St. John's fortresses and *Fortezza Vecchia* (in Dubrovnik). Then he describes *Spalatro* (Split), *Salona* and the fortress of *Clissa* (Klis), part of the Venetian Republic and used for defence against the Turks. He then turns to the west coast of the Adriatic mentioning Lesina (Puglia) "where is nothing very remarkable" and back to the east coast with *Ragusi* (Dubrovnik), *Antivari* (Bar), Valona (Vlorë, Albania), to finally reach the Island of Corfu and Greece, on which the rest of the narrative focuses.

As far as central Italy is concerned, the letter by TANCRED ROBINSON (*Phil. Trans.* 1684: 712–713) provides an account of the bridges on the Tiber, which were subject to "great inundations". According to ROBINSON, the numerous bridges of the Tiber are built with "magnificence and art" and, although they are "more pompous" and rich in rare stones and statues than the French bridges, yet they are very solid. He provides the examples of the *Ponte Molle*, the *Pons Æmilius*, which had just been restored, the *Pons Fabritius* and the *Cestius*, "that leads over to the *Insula Tiberina*; in all which there are still very fair marks of the old Roman structure and design" (1684: 713).[119]

Two years later, following a flooding of the Tevere, GIOVANNI AMBROSIO SAROTTI sent an account of "a discovery made upon the inundation" (*Phil. Trans.* 1686: 227). The flooding caused severe damages to houses and aqueducts; yet it uncovered some vaults containing urns and sepulchres. Apparently, most of these artefacts were of little value; however, an urn found in a vault outside of the city, once opened

> there came out such a strong Smoake, that it made the man that was by it almost giddy, the smell was like Bitumen, but being quickly dispersed they found in the bottom of the said Urne an earthen Pot made up as a Lamp, full of a *Materia Oleosa*, which by degrees,

---

119 See also Section 3.2.5 for another account by ROBINSON.

as the cold Air got into it, grew hard. Several persons suppose this to be one of those perpetual Lamps the Antients mention [...].

The account by GIOVANNI BATTISTA DONI "Concerning a Way of Restoring the Salubrity of the Country about Rome" (*Phil. Trans.* 1670: 2017–2019) provides instead some information regarding the *Campagna* of Rome

> (which is that Tract of Land, that is destitute of Inhabitants and Trees, and extends it self for many miles, taking in *Latium,* and part of the ancient *Sabins,* and of *Tuscany,*) would be of great use to the State, and of subsistence to the people, if it could be Inhabited without that great danger to health, which now 'tis so much noted and fear'd for [...].(*Phil. Trans.* 1670: 2017)

This area is described as being "unhealthfull", and the air "bad and noxious" due to the southern winds, the stagnant waters, the lowness of the shore and the inconsistency of the weather, especially in the summer heat. However, he believes that the salubrity of the air could be restored if the stagnant waters were drained, the land tilled and the area inhabited again.

Finally, very important to the Society were also travel accounts written by travelling Fellows. JOHN RAY, for instance, wrote about his travels through Italy; some of his writings were published, while others only remain in manuscript form (see Hunter 2014). As far as the north is concerned, he wrote about Padua, Milan, Ferrara and Bologna. In his *Observations*, published by the Society in 1673, he provides detailed descriptions of the places he visited, including a description of SETTALA's museum. He was fascinated with SETTALA's glasswork, in particular with a series of "looking-glasses so disposed as by mutual reflexion to multiply the object many times, so that one could see no end of them: the best in this kind that I have anywhere seen" (RAY 1673: 245).

3.2 18th century

The 18th century, like the 17th, abounds with papers on Italy and Italian research. Once again, almost every issue of the century contains Italian subject matter, with the exception of the years 1710, 1716, 1760, 1790, 1792, 1794, 1796–1799. It should be noted moreover, that in most of the years where no Italian-research-based papers were found only single issues of the journal were published.[120] In

---

120 There were no issues for the years: 1707, 1709, 1711, 1715, 1718, 1719, 1725, 1726, 1732, 1734, 1736, 1737, 1741, 1742, 1745, 1747, 1751, 1754, 1755, 1756, 1758 and 1762. Further, it should be noted that the Royal Society have modified their cataloguing system for the *Transactions* while this research was being carried out, and many of the

fact, the Royal Society and the journal faced a crisis around 1752 due to MARTIN FOLKES' (PRS) illness, the death of the editor-secretary, and to the harsh satire that was targeted at the Society and the *Philosophical Transactions* for the triviality of its contents (see Section 1.2.1). As a consequence, the Royal Society took over the journal – which was formerly run privately by the Society's secretaries – and under their control publication slowed down and the journal's contents became more strictly related to what went on in the Society's Thursday meetings. The meetings too gradually changed becoming more focused on the reading of papers and less on the presentation of experiments (Hall 1991).[121] Moreover, new procedures of collective editorship were put in place; a 21-person committee was established to determine what was suitable for publication,[122] while the secretary was responsible for seeing the journal through the press.[123]

A total of 185 papers were collected for this century (1700–1799), 103 Italian Fellows and 22 Italian contributors were counted, and several English promoters of Italian science were identified. This section is arranged mostly like the previous starting with an introduction to the Italian Fellows and contributors (Section 3.2.1); a description of the English collectors of Italian scientific knowledge

---

articles in the corpus are now dated with the following year in the *PTRS* database. As a consequence, articles dated, for instance, 1749 in the corpus, are now dated 1750 in the *Philosophical Transactions* online database. This is probably due to the adaptation of dates to the Gregorian calendar, which was adopted in Britain starting from 1752. Previously Britain followed the Julian calendar, which began the new year on the 25th of March and was eleven days behind the Gregorian calendar, which started being used throughout Europe around this time. Due to the use of the Julian calendar, papers and letters written in January February and part of March carried the dates of the previous year.

121 It became an official requirement that all papers proposed for publication be first read at meetings. This remained the rule until 1892, when due to the limited time available in meetings it was decided that papers could be considered for publication even if all that was read of them in the meeting was the title (Fyfe & Moxham 2016: 2).

122 The Committee of Papers had editorial responsibility and mostly based its judgement on 300 to 500-word abstracts of papers read to the Society, although they could consult the original paper in full if needed (Fyfe et al. 2015: 13). The stamp of individual editors (secretaries and presidents) however did not disappear. The Committee of Papers was abolished in 1989 and the Fellows were given editorial responsibility instead, supported both by editorial advisory boards and by the Society's staff of professional publishers and editorial assistants (Fyfe et al. 2015: 4). By the mid-nineteenth century, papers considered for printing in the *Transactions* were usually sent to two referees for comment before the final decision was made by the Committee of Papers.

123 https://royalsocietypublishing.org/rstl/about

(Section 3.2.2); Section 3.2.3 provides an example of the processes behind the publication of a foreign paper; Section 3.2.4 reports on the main topics of the papers collected for the corpus; and Section 3.2.5 provides some further insights as to how Italy is portrayed in the *Transactions*.

3.2.1 Fellows and contributors

Indeed the 18th century brought with it a reform wave which also influenced election procedures. In 1761 it was agreed that foreign candidates should have their certificates signed by at least three Foreign Fellows as well as by at least three Fellows named in the Home List (Crossland 1983: 177). These restrictions were set primarily for the laxity of admission, which led to the election of a large number of men with no real interests in science.[124] The lack of proper evaluation of a candidate's qualifications and interests, moreover, could lead to political incidents. For instance, Count ALGAROTTI of Venice was elected in 1736 sponsored by the newly-elected PHILIP, EARL STANHOPE, who passed him off as "a gentleman of great knowledge in all parts of philosophical and mathematical learning" and "exceedingly well qualified" (Crossland 1983: 177), yet he appeared to trivialise NEWTON's theory of gravitation in his *Il Newtonianismo per le dame* (1737).[125] As a result, the Fellows needed to be sure of a candidate's worthiness by receiving assessment both from the candidate's countrymen as well as the Fellows. In 1765 it was moreover decided that no more than two foreigners should be elected a year, excluding royalties.[126] Instead, foreigners resident in England for at least six months could be elected as Home Fellows (Hall 1982: 69–70).

Despite these restrictions, which took a long time to become effective, the number of Italians elected to Fellowship in the 18th century amounts to 103, to which another 22 non-Fellows should be added as active correspondents and scientific contributors.[127] Given the incredibly high number of Fellows,

---

124 Opposition towards the admission criteria had been manifested much earlier; in the 1730s it was decided that new candidates should have their certificates signed by at least three Fellows, to which the foreign members rule was added in 1761.
125 Mazzotti (2013) however points out how ALGAROTTI's book, which in fact defended Newtonianism and presented with clarity and simplicity NEWTON's optic experiments and gravitational theory, played a key role in the diffusion and success of Newtonianism in continental Europe. See also Vicentini (2019).
126 While the numbers were still controlled, this restriction was relaxed after 1776 (Hall 1982: 70).
127 Hall (1982) includes 94 Fellows in her list; in some cases, this is because she does not include Fellows who had moved abroad. Missing from Hall's list are the

this section will be more schematic than the corresponding one in the previous 17th-century section and biographical information will only be provided for a selection of Fellows and contributors. The reader may however consult Tab. 6.2 in the appendix for more information on Fellows that are here only mentioned.

Very often natural philosophers had interests in different scientific fields and it is therefore not possible to group them into clear-cut research areas. Most of them however had interests in related fields, which made it possible to roughly divide 18th-century Italian Fellows into four main groups. The most numerous group includes physicians, biologists and botanists with 31 Fellows: ANTONIO VALLISNERI, EMANUELE TIMONE, GIOVANNI MARIA LANCISI, MICHELANGELO TILLI, RINALDO DULIOLO, FRANCESCO TORTI, PIETRO ANTONIO MICHELOTTI, LUDOVICO RIPA, GIOVANNI BATTISTA MORGAGNI, NICOLA CIRILLO, IACOPO BARTOLOMEO BECCARI, JERONIMO GIUNTINI, ANTONIO LEPROTTI, JACOBUS JATTICA, JOSEPH CERVI, ANTONIO COCCHI, GIUSEPPE LORENZO BRUNI, PIER PAOLO MOLINELLI, MARSILIO VENTURI, SAVERIO MANETTI, CARLO ALLIONI, ANTONIO MATANI, GIOVANNI FRANCESCO CIGNA, GIOVANNI BATTISTA CARBURI, LAZZARO SPALLANZANI, LEOPOLDO MARCO ANTONIO CALDANI, GIUSEPPE SAVERIO POLI, ANTONIO SCARPA, BRUNO TOZZI, VITALIANO DONATI and GIOVANNI MARSILI. Among this group worthy of notice are the physician EMANUELE TIMONE, who reported to the Society about the inoculation against smallpox in the Ottoman Empire;[128] MICHELANGELO TILLI, a botanist whose most notable contribution to science was the use of greenhouses to cultivate tropical plants in Europe; GIOVANNI BATTISTA MORGAGNI, who is today regarded as the father of modern anatomical pathology and who pursued his studies according to the experimental philosophy, corresponding regularly with the Society and sending books and information about his studies. IACOPO

---

following: JOSEPH CERVI, EMANUELE TIMONE, GIOVANNI GIACOMO MARINONI, [?] CARRON DI TOMMASO COUNT OF BRIANCON, COUNT LUDOVICO OF BELGIOIOSO, GIOVANNI BATTISTA CARBONE, GIUSEPPE LUDOVICO LAGRANGIA, ALBERTO FORTIS and MARSILIO VENTURI.

128 It is not certain whether TIMONE and the below listed GIACOMO PILARINI actually had Italian origins; however, it was decided to keep them in the list of Italian Fellows and contributors for their connections with Italy; both of them, in fact, graduated at the University of Padua (and Oxford) and PILARINI had served as a diplomat for the Venetian Republic and had his treatise on inoculation, *Nova & Tuta Variolas Excitandi per Transplantationem Methodus, Nuper Inventa & in Usum Tracta: Per Jacobum Pylarinum, Venetum, M. D. & Reipublicae Venetae Apud Smyrnenses Nuper Consulem*, published in Venice in 1715 (see Eriksen 2020; Berti 2021).

BARTOLOMEO BECCARI, a professor of medicine and chemistry known today for having discovered gluten, also regularly corresponded with the Society sending them updates on his own and others' research. The physician GIUSEPPE LORENZO BRUNI informed the Society not just about medical topics but also about various curiosities that occurred in Piedmont and which were reported about in five *PTRS* papers.

The second broad group includes physicists, mathematicians and astronomers with 19 Fellows: GUIDO GRANDI, GIOVANNI POLENI, FRANCESCO BIANCHINI, ANTONIO SCHINELLA CONTI, GIULIO CARLO FAGNANO DEI TOSCHI, GIOVANNI BATTISTA CARBONE, EUSTACHIO MANFREDI, GIOVANNI FRANCESCO CRIVELLI, GIOVANNI GIACOMO MARINONI, EUSTACHIO ZANOTTI, GIOVANNI FRANCESCO MAURO MELCHIORRE SALVEMINI DI CASTIGLIONE, GIOVANNI BATTISTA BECCARIA, PAOLO FRISI, SIMONE STRATICO, TIBERIO CAVALLO, GIUSEPPE LODOVICO LAGRANGIA, ALESSANDRO VOLTA, GREGORIO FONTANA, and BARNABA ORIANI. EUSTACHIO MANFREDI, mathematician, astronomer and minor poet, was a regular correspondent and contributed six papers to the *PTRS*. TIBERIO CAVALLO was a Neapolitan who moved to England where he began cultivating his scientific interests; he carried out studies in electricity and contributed 16 papers to the journal. He moreover acted as an intermediary between Italians and the Society. GIOVANNI POLENI was an eclectic professor of mathematics and astronomy who actively corresponded with the Society's Fellows and got involved in the century-long "vis viva" dispute, which centred on how to define the quantity of motion and force (Rusnock 1996). Five papers of his were published in the *PTRS*. ANTONIO SCHINELLA CONTI was also involved in an international scientific dispute, the LEIBNIZ-NEWTON calculus controversy, where he acted as an intermediary. He moreover spoke English and translated English literary works into Italian. The mathematician and astronomer GIUSEPPE LODOVICO LAGRANGIA, better known as LAGRANGE, is well known today as a contributor to various scientific fields (analysis, number theory, mechanics and astronomy). Other than the Royal Society, he was a member of several academies – the Berlin academy (1756), the Royal Society of Edinburgh (1790), the Royal Swedish academy of Sciences (1806) – and together with GIUSEPPE SALUZZO and GIANFRANCESCO CIGNA he founded the Royal Academy of Sciences of Turin (1783). Needless of introduction is ALESSANDRO VOLTA, the inventor of the Voltaic pile, but worth highlighting, instead, is that the Royal Society was the first he chose to communicate his findings to. He was also a member of several other institutions and one of the founding members of the Milanese Istituto Lombardo Accademia di Scienze e Lettere (1797). Six papers on VOLTA's work have been published in the *Transactions*.

The third broad group includes diplomats, statesmen and other noblemen who, notwithstanding the calls for changes in the election procedures, were still elected in great numbers. The group includes 26 Fellows: CARRON DI TOMMASO COUNT OF BRIANCON, FRANCESCO CORNARO, VENDRAMINO BIANCHI, PIETRO GRIMANI, COUNT GIOVANNI ANTONIO BALDINI, NICOLÒ TRONI, NICOLO ALERBO D'ARAGONA PRINCE OF CASSANO, FRANCESCO MARIA D'ESTE PRINCE OF MODENA, ANTONIO NICOLINI, FOLCO RINUCCINI, CAVALIERE OSORIO, CAVALIER DE BAILLOU, PIETRO PAOLO CELESIA, GIOVANNI CARAFA, GIAMBATTISTA ALBERTINI, GUISEPPE ANGELO SALUZZO, FRANCESCO LORENZO MOROSINI, DOMENICO CARACCIOLI, ABONDIO REZZONICO, LUIGI CARLO MARIA COUNT OF BARBIANO DI BELGIOIOSO, FRANCESCO MARIA VENANZIO D'AQUINO, LUIGI MALASPINA DI SANNAZZARO, ANTONIO MARIA LORGNA, GASPARE CERATI, PIETRO ANDREA CAPELLO and GIUSEPPE TOALDO. While of most of them very little is known and, like their predecessors of the 17th century, they made no contributions to the Society's scientific goals, a few of these Fellows proved to be useful members of the Society. An example is NICOLÒ ALERBO D'ARAGONA, Prince of Cassano, who contributed two papers on volcanology and natural phenomena and was interested in receiving "accounts of any new Inventions or Discoveries relating to Geography, Navigations, Astronomy etc." (1738, LBO/25/38). Worthy of notice is also ANTONIO MARIA LORGNA, an army official who studied mathematics, physics and astronomy and founded the Italian National Academy of Sciences in 1782.[129] His election certificate (EC) is one of the first that follows the 1761 rule of having a foreign candidate elected by three Fellows as well as three Foreign Members; it was written in Italian:

> Attisto io sottoscritto che il Sig Colonello Antonio Mario Lorgna al servizio della Serenissima Repubblica di Venezia nel dipartimento delle fortificazioni, e Professore di Matematica nel Colleggio militare stabilito in Verona e stimato si in questo Stato, che per tutta l'Italia, ed anco fuori peruno de piu distinti Geometri del nostro tempo, avendo dato piu sagi de suoi studi per via di opere da lui stampate, e per premi riportati nelle l'Academie estere con molto applauso; Onde l'addozione di si distinto Soggetto non puo reccare se non onore e vantaggio a qualunque Accademia, o Societa Letteraria che gliela accordi; in fede della qual verità ho considerato, come atto di giustizia, il fare al preffato Sig.re Colonello il presente certificato. Da Verona li cinque novembre mille settecento Settanta sci munito del sigillo delle mie Armi
> Paolo Frisi [...]; Giovanni Marsilli [...]; Ottaviano Conte abbe Guasco; [Charles] Blagden; John Paradise; Charles Peter Layard; W[illia]m Parsons. (EC/1787/20)

---

129 It was first based in Verona as the Società Italiana; today the Academy is based in Rome.

There is also an earlier Latin proposal with two different Italian proposers, EUSTACHIO ZANOTTI and FRANCESCO MARIA ZANOTTI, who signed the certificate together with FRISI (EC/1769/11).[130]

The final broad group includes humanists and antiquarians with 17 Fellows: the Marquis GIOVANNI GIUSEPPE D'ORSI, LORENZO MAGALOTTI (who had been an active contributor since the 17th century), ANTONIO MARIA SALVINI, LUDOVICO ANTONIO MURATORI, PAOLO ANTONIO ROLLI, CARLO TAGLINI, FRANCESCO SCIPIONE Marquis of Maffei, Count FRANCESCO ALGAROTTI, ANTONIO FRANCESCO GORI, FRANCESCO MARIA ZANOTTI, CAMILLO PADERNI, GIOVANNI BATTISTA PASSERI, OTTAVIO DE GUASCO, GIUSEPPE-MARIA PANCRAZI, RIDOLFINO VENUTI, MARCO FOSCARINI and FILIPPO VENUTI. Noticeable in this group are the philosopher and literate FRANCESCO MARIA ZANOTTI, who corresponded with the Society informing them on the goings-on of the Academy of Sciences and Arts of Bologna; CAMILLO PADERNI, a painter and art restorer, who contributed eight papers on antiquarian and archaeological topics; and PAOLO ROLLI, a London-based Italian teacher and librettist, who is today known not only as a man of letters but also for translating and spreading English literature in Italy and Italian literature in England. The question may arise as to why a man of letters such as ROLLI would have an interest in being elected to the Society. Dorris (1967: 61) suggests that "his membership would seem to be justified by a real interest in the scientific problems which were then perplexing his contemporaries, and in particular the nature and properties of heat", which is probably based on the paper that ROLLI contributed to the *PTRS* (see Section 3.2.3); however his scientific interests seem to start and end here. The second probable reason for ROLLI's interest may be the prestige that came with this role and, indeed, ROLLI did append his FRS title to his name on various publications following his election (Dorris 1967: 190). From the opposite perspective, what was the Society's interest in a man who prior to his election had made no contributions whatsoever to the study of nature? Their interest in ROLLI was very likely related to the role ROLLI could serve as translator and intermediary between the Italians and the Royal Society. His little but not unimportant contribution to international science will be examined in Section 4.1.3.[131]

---

130 The English signatories of the earlier certificate were NEVIL MASKELYNE, MATT RAPER, HENRY CAVENDISH and CHARLES MORTON.
131 A few Fellows have been left out of the above four groups; these are: the natural historians GIUSEPPE AVERANI and ALBERTO FORTIS, an unknown GATUCCI, FILIPPO DI PAOLI, GIAMBATTISTA RECANATI, DOMENICO FERRARI, CELESTINO GALIANI, MICHELANGELO GIACOMELLI, GIULIO SACCHETTI and ANTONIO OTTAVIO BAIARDI.

Several Fellows promised to become regular correspondents but did not however keep their promise and the only letters which the Society received from them were letters thanking for their election and making further promises (e.g. JATTICA, CRIVELLI, GALIANI). On the contrary, the non-Fellow Italians who were included in the study as contributors all appear in the *Transactions* with at least one paper. The majority of the contributors were physicians, i.e.: GIOVANNI COSIMO BONOMO, FORTUNATO BIANCHINI, DOMENICO CIRILLO, GIACOMO PILARINI, ANTONIO BENEVOLI, CARLO CRUSIO, GIOVANNI BATTISTA PAITONI, GIUSEPPE DEL PAPA, FLAMINIO PINELLI and ROCCO BOVI. There were also a few physicists: FELICE FONTANA, who contributed three studies on poison and gasses to the journal; JOSEPH PIAZZI, who became Fellow towards the end of his life in the early 19th century; Father GIOVANNI MARIA DELLA TORRE, whose microscopes were of interest to the Fellows; and the well-known physician and physicist LUIGI GALVANI, who carried out influential researches in what became known as the field of animal electricity, a subject of strong interest between the late 18th and early 19th centuries. The contributors further included the humanists ANTONIO MAGLIABECHI, GIUSEPPE VALLETTA, PASQUALE PEDINI, the diplomat FRANCESCO IPPOLITO, the geologist and naturalist LAZZARO MORO, the missionary GIUSEPPE DA ROVATO, the geographer GIOVANNI RIZZI ZANNONI and a V PUCCI, secretary to the Grand Duke of Tuscany, who held correspondence with SLOANE, JURIN and DEREHAM sending papers and books of Italian scholars.

In the 18th century, evidence of a few rejections of candidates was found. One concerned the Count CARLO LUIGI MOROZZO, first Major in the Susa Regiment in the Service of the King of Sardinia and member of the Academy of Sciences of Turin. MOROZZO was proposed in 1787 but his election certificate has a note reporting "Ballotted for & rejected April 3d 1788" (EC/1787/35). The same note is reported on the election certificate of GIUSEPPE ANTONIO TESTA, professor of medicine and member of several Italian academies, who was rejected on the same day (EC/1787/28).[132] While the reasons behind MOROZZO's and TESTA's rejection remain unclear, the rejection of another Italian, Count PAOLO ANDREANI – a friendly correspondent of CHARLES BLAGDEN, secretary of the

---

132 Neither of them reached three Italian signatures on their certificates, which could be one of the reasons for their rejection, especially considering that the other Italian Fellow elected on this occasion, LORGNA, had been proposed according to the new regulations (see above); however, no evidence for this was found and other Italian candidates elected after 1788 did not have three Italian proposers.

Society at the time (1784–1797) – was found to be a political one. ANDREANI was proposed in 1793 by various English Fellows and TIBERIO CAVALLO (EC/1792/18). In a letter to BLAGDEN, JOSEPH BANKS explains that "Andreani had more black than white balls, he is said to have had very inappropriate conversation relative to the government of England and to have decanted publicly in Praise of Republicanism [… and] that he had been […] noticed by several"; hence, "Count Andreani was blackballed, which I am sorry for as I voted for him" (CB/1/1/116). Another Italian rejected in 1756, on the grounds that he was insufficiently known (Hall 1982: 69), was the Count MAFFEO D'ALBINI, a nobleman of Udine and Friuli, who was visiting London at the time of his candidature (EC/1756/12).

Frequently occurring in the Society's 18th-century archival material were also unpublished and rejected papers, several of which have been referenced in Tab. 6.2 in the appendix.[133] One interesting example is a paper by the Marquis LUIGI MALASPINA, who cultivated interests in architecture. The paper, which he had originally written in Italian and had translated into English for the Society, was criticised because the author used "French terms of Art which are not at all intelligible in English. It has been translated by a Person, not at all versed in the Subject, and not much in the English Language. The terms of Art are not Proper and very often are defective […]" (AP/3/16).

Finally, there were always Italians who did not have any direct or indirect links with the Society but whose work was nonetheless well known to the Fellows. An example is the famous physician ANTONIO MARIA VALSALVA, who had been known to the Society since the 17th century, and whose work is referenced and discussed in several *PTRS* papers. Also the studies of the physician STEFANO LORENZINI with electric rays and eels were known and referenced in the *Transactions*.[134] Another interesting example, since the Society did not admit

---

133 The high presence of rejected papers (Archived Papers, APs) is related to the establishment of the Committee of Papers. Fyfe et al. (2015: 13) report that the APs were either rejected or even filed without being read, and that they comprised a striking proportion of foreign papers.

134 Other Italians referenced in the *PTRS* are: BERNARDINO RAMAZZINI; PIER ANTONIO MICHELI; GIOVANNI FRANCESSCO PIVATI; PIO FANTONI; PAOLO POSI; GIOVANNI BATTISTA PIRANESI; GIUSEPPE BENVENUTI; ABBATE ANNIBALE OLIVIERI; GURNACCI; FRANCESCO REDI; SCIPIO FERREUS and GEROLAMO CARDANO (15th-century mathematicians); COUNT DE GIOENI; BRUGNATELLI, LUIGI VALENTINO; GIOVANNI DOMENICO MARALDI; ABBOT LERCARI, ABBOT CUZZONI; PAOLO RICOLVI; ANTONIO RIVANTELLA; CANONICO BRIGOLI; GIUSEPPE VERATTI; CANONICO MAZZOCCHI; SIGNOR F. CORSINI; MARQUIS CARNAVALLIA; SIGNOR MATHENCI; SIGNOR MARINI; PROFESSOR CASALI; PROFESSOR CANTERZANI; MR. COLTELLINI; DR. LAMI;

female Fellows until the 20th century, is MARIA GAETANA AGNESI, an Italian female mathematician, whose book *Instituzioni analitiche ad uso della gioventu' Italiana* was treasured in the Society's library and translated into English in 1801.[135]

3.2.2 Fellows and Englishmen in close contact

For the previous century, it was seen how merchants, Englishmen and Fellows travelling or resident in Italy all contributed in their own ways to creating links between the Society and Italian scholars; and HENRY OLDENBURG was provided as the example *par excellence* of the secretary's foreign relations (Section 3.1.2). But what happened after OLDENBURG's death in 1677? The Society's foreign relations continued to be kept by its secretaries – and foreign secretaries after 1723 (Poliakoff 2015). Contacts were still established by means of correspondence and travelling friends, Fellows, merchants and diplomats. Not all of OLDENBURG's successors succeeded in keeping up with the vast and complex network of exchanges that OLDENBURG himself had built, but all of them made their efforts with some being more successful than others. Names of 18th-century secretaries and presidents that more frequently recur in the Italian-English correspondence are those of HANS SLOANE, secretary from 1693 to 1713 and president from 1727 to 1741; JAMES JURIN, secretary between 1721 and 1727; WILLIAM RUTTY, who became second secretary in 1727 until his death in 1730; CROMWELL MORTIMER, SLOANE's assistant between 1729 and 1740; MATTHEW MATY, foreign secretary between 1762 and 1766; JOSEPH BANKS, president between 1778 and 1820; and CHARLES BLAGDEN (1784–1797).

From a textual point of view, the correspondence of JAMES JURIN provides a good example of OLDENBURG's diplomatic legacy. After a period of decline under the office of EDMOND HALLEY, JURIN extended the Society's connections with continental natural philosophers by relying on expatriate Englishmen (Rusnock

---

CANONICO GIUSEPPE RECUPERO; TARGIONI TOZZETTI; MERCATI; FERRANTE IMPERATO; PRINCE OF TORREMUZZA; ANTONIO FRANCESCO GORI; ALBERTO FORTIS; SIGNOR TEMANZA; ABBOT BOTTIS; BERNOULLI; DE CESARIS, REGGIO; ANTONIO SANTORELLI; DR. SERAO; FATHER ANTONIO DI PETRIZZI; ABBE TATA; EMMANUEL SCOTTI; XAVERIO [SAVERIO] GATTA; PIETRO FABRIS.

135 *Instituzioni analitiche ad uso della gioventu' Italiana* (Milano, 1748) and its translation, *Analytical institutions, in four books, by Maria Gaetana Agnesi, translated into English (with an introduction and an addition) by the late John COLSON; now first printed, from the translator's MS., under the inspection of John HELLINS* (London, 1801).

1996: 18). He invited the English resident in Italy THOMAS DEREHAM to initiate correspondence with the following:

Sir,
Our Illustrious President, who embraces every opportunity of promoting ye ends of ye Institution of ye Royal Society, has lately been pleased to inform me of ye kind offer you made to him, when in England, of Corresponding with ye Secretaries of ye Society.
He did me ye honour at ye same time of laying his Commands upon me, to enter into a Correspondence with you which with your permission. I shall begin by ye Letter I have now ye honour of writing to you. I am perfectly sensible, that I can in no particular do better, or more acceptable Service to ye Royal Society, than on opening a way for their being better informed, as I hope, Sir, by your means they will be, of what passes among that Learned & Inquisitive Nation, with which you reside. The Royal Society has so just a regard & Veneration for ye memory of ye Galilei, the Borelli, Malpighi, and Bellini, y^t she can never be incurious of what is doing in a Country, y^t produced those Great & Excellent Genii. I flatter myself, Sir , that it will be no little pleasure to your self, as well as to me, to be a means of opening a Philosophical Communication between two Nations, among both which there are so many generous Spirits united in ye same noble design for ye common benefit & information of Mankind. I shall on my part take due care by ye punctuality & exactness of my correspondence to shew with how great Esteem and Respect I am, Sir.
your most obed. & most humble Serv .
James Jurin
R.S Secr.
P.S. [...] It will perhaps be a pleasure to some of your Italian Friends, and particularly to the Illustrious Signor Morgagni, who is in great Esteem here, to know, y^t ye Myotomia of ye late Mr. Cowper is now very forward in ye Press. It is printed in Folio, in a very pompous manner, wth 66 Copper Plates, wch Mr. Cowper left behind him, most of them etched with his own hand. (JURIN to DEREHAM, 1722, in Rusnock 1996: 90–91)

The letter displays the same features of OLDENBURG's correspondence: formal and reverential tone; praise towards Italy and its most famous scientists; request to receive Italian knowledge; and promise to return the service highlighting the benefits that the exchange could provide both sides.

Another Fellow who collected information on Italy was ROBERT MORE, who had travelled through the country in 1750. The account of his Italian travel was published in the *PTRS* (1749: 464–467),[136] together with another paper on "the

---

136 See also PIERRE SILVESTRE's *PTRS* accounts, written at the opening of the 18th century, described in Section 3.1.4. It should not be forgotten moreover that a great deal of books were published for travellers through the Italian Peninsula and *Grand Tourists*, such as WILLIAM BROMLEY's *Remarks made in Travels through France and Italy* (1693) and JOHN RAY's *Observations topographical, moral, & physiological;: made in a journey*

method of gathering manna near Naples" (1749: 470–471).[137] The physician TANCRED ROBISON too provided an account of the botanical and zoological curiosities he saw in his travel throughout the peninsula (see Section 3.2.5). Other receivers and communicators of several Italy-related papers were WILLIAM WATSON (FRS 1741) who was the addressee of several letters in his fields of study, i.e. medicine, animal electricity and botany; the naturalist HENRY BAKER, who provided papers on natural phenomena, physics and electricity (FRS 1744); JOHN SWINTON (FRS 1729) a British priest, writer and academic, who lived in Livorno and Florence in the 1730s, became a member of the Academy Degli Apatisti at Florence and of the Etruscan Academy of Cortona in Tuscany and contributed eight papers on the subject of numismatics.[138] THOMAS HOLLIS (FRS 1757) corresponded with CAMILLO PADERNI and provided information about the discovery of Herculaneum in the *PTRS*. Sir WILLIAM HAMILTON, envoy to the King of Naples (1764–1800), wrote about the volcanos Etna and Vesuvius and the nature of the soil around Naples.[139] His residence in Naples, Palazzo Sessa, "was an exclusive *salon* and *the* reference point for all 'respectable' foreigners in the south of Italy" (D'Amore 2017: 9). His guests could moreover make use of HAMILTON's Italian connections and be escorted on their travels.

---

through part of the Low-countries, Germany, Italy, and France (1673). However, the focus of this research is on the articles published in the *Transactions* and the related sources found in the Society's archives. Due to the considerable size of the corpus and related epistolary and bureaucratic sources, published books have only been considered in very few cases.

137 *Manna* is the name attributed to a variety of plants that produce a sweetish substance or resin. The substance itself is commonly called *manna*. ROBERT MORE in the paper believes that the trees from which the manna is extracted are the Flowering Ash or Manna Ash (Fraxinus ornus), which is the source of a sugar-alcohol, mannitol, which has been used medicinally (MANNA. *Encyclopædia Britannica*. Encyclopædia Britannica, inc. (2016) https://www.britannica.com/topic/manna-plant-product. Last accessed Jan 2020). On the extraction of Manna from Italian trees see also TANCRED ROBINSON's travel account (*Phil. Trans.* 1714: 473–483) and CIRILLO's paper *Phil. Trans.* 1770: 233–238. Also MORE's travel account talks about the extraction of Manna from the Flowering Ashes. The several descriptions of this process suggest that it was something quite remarkable and clearly attracted the Fellows' interest.

138 Several Fellows were made members of Italian academies. The Institute of Science of Bologna, for instance, elected CHARLES WALMESLEY, JAMES BRADLEY, STEPHEN HALES, JOHN STRANGE and EDWARD WARING in the second half of the century (Cavazza: 2002: 16).

139 See D'Amore (2017) for more on the English-Italian exchanges with regard to the discovery of Herculaneum and volcanology.

HAMILTON himself climbed Vesuvius over fifty times, toured the islands in the Neapolitan area, crossed the Strait of Messina and visited Sicily, which was to become the next big step in the *Grand Tour* of Italy (D'Amore 2017: 10). ABBE NOLLET (FRS 1735) contributed papers on electricity and geology. DIDACUS DE REVILLAS (FRS 1734) worked as an intermediary between Italian astronomers and the Royal Society and contributed papers on astronomy and medicine. JOHN STRANGE (FRS 1766), British minister resident in Venice from 1773, had interests in archaeology and geology and published four *PTRS* papers in these and related subjects (see Section 3.2.5, see also Ciancio 1995).

Finally, worthy of special notice is Sir THOMAS DEREHAM (1678–1739, FRS 1720), baronet of West Dereham in Norfolk and a Roman Catholic, who took residence in Florence in 1718 and later in Rome, at the court of JAMES III, the Old Pretender. DEREHAM held a large amount of correspondence with both Italian scholars and the Society's Fellows (most importantly JURIN, SLOANE, and RUTTY). He acted as an intermediary of science between Italians and the Society, forwarding both ways letters, books and papers and promoting projects undertaken by the Society (see for instance the meteorology project in Section 3.2.4). He moreover acted as translator, translating his correspondence, issues of the Society's *Transactions* into Italian, and of the *Giornale de' Letterati* into English. In a letter to SLOANE, dated 1734, he wrote that the (translated) *Transactions* "had so good an effect in these parts, that many curious Virtuosi here have learned our language". DEREHAM was passionate about science and took his role as a scientific diplomat enthusiastically; this can be seen in his reply to JURIN's letter (quoted above):

Sir
About a fortnight ago I received in Rome your favour of ye 19th March, & was very glad to find that our Illustrious President has remembered the offer I had made of corresponding with the Secretaries of the Royall Society by the actuall notice you are pleased to give me & should have sooner acknowledged the honour that by your means is done me, only I would not return an empty letter , therefore without loss of time I wrote unto a friend at Pisa, where Father Grandi, Dr Averani ,& Dr Tilli our learned Fellows are Professors, that they might be informed with the design of ye Society of opening a Philosophical Communication between two Nations, among both which have been, & are so many generous spirits , as you say, united in the same noble design, for the common benefit, & information of mankind . Father Grandi answered me that he will be very proud of obeying the R.S. in whatever commands shall be laid upon him about mathematicall matters, as he has done upon other occasions. D^r. Averani replied that he has some Philosophicall observations, & experiments lately made, that he will very soon offer unto the R. Society. D^r. Tilli is about printing in a very pompous manner the catalogue of the several herbs of the Phisick garden, with many fine copper plates,

& remarks about natural History, which as soone as published he will present by my means unto the R.S.
& when I shall be at Florence , as I reckon about a fortnight hence I shall stir up there, & at Pisa by as much as shall lay in my power an emulation to contribute to so great an end. [...] (DEREHAM to JURIN, 1722, in Rusnock 1996)

DEREHAM eagerly accepted to become the scientific intermediary between Italians and the Society and, further, he purposely waited to send a reply in order to already furnish his first letter to JURIN with Italian scientific news and to inform him of the enthusiasm displayed by his notable Italian friends upon the prospect of transmitting theirs and others' research to the Royal Society "for the common benefit, & information of mankind". As far as the *PTRS* is concerned, DEREHAM provided material for many articles, in that he forwarded to the Society several Italian treatises; yet his contribution is acknowledged in only five papers. The fields he was mainly involved in were medicine, natural philosophy and mathematics.[140]

3.2.3 A complex network of exchanges

Indeed, foreign papers reached the Society and final publication in the journal through the Society's correspondence network. This was by no means a simple process, in that a paper or letter of a foreign scholar could pass from its author, to an intermediary, into the hands of a merchant or diplomat, and finally to the Englishman or Fellow who would present it to the Society. The same paper could be translated either by one of the Society's translators or secretaries or by the intermediary. Alternatively, as was frequently the case, rather than translating the whole paper, the Fellows would only translate an extract of it, generally a summary containing the essence of the original paper. In this section, a *PTRS* paper is provided as an example of this process.

In 1744 the following paper appeared in the *Transactions*:

> An Extract, by Mr. Paul Rolli, F. R. S. of an Italian Treatise, Written by the Reverend Joseph Bianchini, a Prebend in the City of Verona; Upon the Death of the Countess Cornelia Zangari & Bandi, of Cesena.

---

140 See also Rusnock (1996 and 1999), Findlen (2009), Cavazza (2002) and Fontes Da Costa (2002a). I am also working on DEREHAM's correspondence and translations for a future publication. Among DEREHAM's Italian contacts were: POLENI, CARBONE, MAFFEI, TILLI, GRANDI, AVERANI, CIRILLO, LORENZINI, VALLISNERI, MANFREDI, BIANCHINI, RAMAZZINI, BECCARI, and TAGLINI.

To Which are Subjoined Accounts of the Death of Jo. Hitchell, Who was Burned to Death by Lightning; And of Grace Pett at Ipswich, Whose Body was Consumed to a Coal. (*Phil. Trans.* 1744: 447–465)

Focusing on the first part of the title, we learn that the paper is about the death of an Italian Countess and that the account of her death was written by GIUSEPPE BIANCHINI.[141] The name of the literate PAOLO ROLLI (FRS 1729) also appears, but his role in the matter has not been made explicit. By looking into the Society's letters, it was discovered that ROLLI himself had asked the Society to become a Fellow explaining that he had already translated from English to Italian[142] and offering the same service to the Society if required (LBO/19/ 140 and EL/R1/ 71). And indeed, ROLLI did translate Italian letters for the Fellows, including the letter on the lady of Cesena. Moreover, from the letter exchanges behind this publication it appears that BIANCHINI was not the sender of this paper but that this had been put together and sent by another Italian, the Marquis FRANCESCO SCIPIONE MAFFEI, who was hoping to become a Fellow – he was eventually elected in 1736. Hence, this paper shows how complex the network of the Society's foreign relations was and how important a role was played by the intermediaries – which were two in this case (MAFFEI and ROLLI). Without the help of intermediaries many writings would have never reached the Society. Further, another informant on the subject was the earlier mentioned THOMAS DEREHAM, who forwarded information on MAFFEI's account to HANS SLOANE (LBO/19/191).

Finally, the paper of the lady of Cesena does not only report MAFFEI's piece, but also two further accounts "of the death of Jo. Hitchell, who was burned to death by lightning" (1744: 461) and of "Grace Pett at Ipswich, whose body was consumed to a Coal" (1744: 463). Both these cases occurred in England. What is relevant here is not the content of the papers, but the fact that three different

---

141 GIUSEPPE BIANCHINI (1704–1764) was an erudite of Verona and secretary to the Academy of Church History. His scientific interests led him to write his opinion on the death of the Countess ZANGARI. The account was published in a booklet in 1731 and was reprinted four times in less than two decades (*Parere sopra la cagione della morte della Sig. Contessa Cornelia Zangari ne' Bandi*, Verona 1731 and 1733, Rome 1743 and 1758). In his account, BIANCHINI supported the possibility of spontaneous human combustion. The book had a minor international resonance: it was not only reported about in the *Transactions* but also in the Dutch journal *Bibliothèque raisonnée des ouvrages des savans de l'Europe* (Amsterdam, XXXIX) (Rotta, S. (1968). Bianchini,Giuseppe. In *Dizionario Biografico degli Italiani*, vol. 10. Treccani).
142 MILTON's *Paradise Lost*.

accounts on a similar phenomenon – death by apparent human combustion – were grouped in one paper. This reflects the Society's objectives of gathering experimental data from multiple sources and perspectives in order to attain as truthful and accurate a study of nature as possible (see also Section 3.1.2).

3.2.4 Topics

Although it is still not possible to distinguish clear-cut scientific disciplines as early as the 18th century, an attempt to divide the papers in broad subject areas was made. The grouping of the papers according to their field of study allowed us not just to give some order to an otherwise very numerous and varied group of papers, but also to compare the topics of preference with those of the previous and following centuries.

A very copious group concerned the field of astronomy with 27 papers spread throughout the century and a concentration of papers in the 1720s and 30s. While the name of CASSINI is still remembered, the main contributor in astronomy for this century was GIOVANNI BATTISTA CARBONE (seven papers), a Neapolitan Jesuit who had established an observatory in Lisbon (Udías 2014).[143] Following were EUSTACHIO MANFREDI (five papers), GIOVANNI POLENI (two papers), EUSTACHIO ZANOTTI (two papers) and FRANCESCO BIANCHINI (two papers). The topics concerned the observation of comets, lunar and solar eclipses, spots observed on Pluto, Jupiter's satellites and the transit of Mercury. While the astronomical papers provide a good example of how the Society collected observations of a same event from different perspectives and for the attainment of a more precise knowledge of the firmament, these papers tend to go straight to the point and are therefore less insightful for the study of Anglo-Italian relations.

Six papers in biological subjects were counted, including two studies on fungi and three studies on marine organisms (corals, spongiae and pearls of oyster shells). A key figure for the Society in this area was VITALIANO DONATI, a correspondent and a contributor of three papers. Another subject that was much sought after by the Fellows was fireflies, or *ignes fatui*. The Fellows were

---

143 Possibly because CARBONE moved to Lisbon, not much Italian research about him was found. CARBONE however, contributed a considerable number of observations and up until at least 1719 he was still working in Italy (Giovanni Battista Carbone (1694–1750). In: Jesuit Science Network, online version, accessed 03/11/2018. http://jesuitscience.net/p/114/). Very recently CARBONE has started receiving more attention from the Italian scholarly community and more focused research about this astromer can be found in De Simone et al. (2022).

fascinated by the light emitted by these insects and wanted to understand its cause; they therefore requested and collected accounts on them from various Italians, as it was known to them that Italy abounded with fireflies.

A total of 28 papers were counted concerning the medical field. While in the previous century the topics of major interest were anatomical observations, blood studies and extraordinary medical cases, these seem to play a minor role in 18th-century Italian *PTRS* papers. Interest in these topics continued in the first years of the new century but were then overshadowed by several studies on infectious diseases and their cures. The most important example is the smallpox and the practice of inoculation (variolation) – i.e. the extraction of the virus from the infected pustules and its introduction into the skin of a non-infected individual in order to procure a less severe infection than naturally-contracted smallpox, which then leads to the patient's immunity to the virus.[144] This practice, which originated in Asia, was introduced and developed in Europe in the 18th century with contributions from physicians from various parts of the world. Key investigators of smallpox inoculation from the outset were EMANUELE TIMONE (three papers), who reported to the society about smallpox inoculation in the Ottoman Empire, and GIACOMO PILARINI, who was the first to practice inoculation outside Asia. It was indeed their papers, published in the second decade of the 1700s, that sparked off the interest and future researches in this practice and soon various members of the English upper classes were being inoculated against the disease.[145] Inoculation had also been attempted in Italy with some success, as THOMAS DEREHAM told JURIN in 1725. DEREHAM moreover translated into Italian a treatise of JURIN's on inoculation[146] and CHARLES MAITLAND's pamphlet *An Account of Inoculating Smallpox* (London, 1722).[147] Further topics of interest included the studies of MALPIGHI's pupil and successor, ANTON MARIA VALSALVA, and in particular his observations "upon the Cataract of the Eye" (*Phil. Trans.* 1722: 194) and the discovery of an "Excretory

---

144 Other diseases which are reported about in the Italian papers are "a dreadful contagious Distemper, seizing the Black Cattle in the Venetian Territories" and in other parts of Italy (*Phil. Trans.* 1714: 46–49 and 1721: 83–86); a study of PINELLI's on the causes of Gout (*Phil. Trans.* 1727: 491–494); and case reports about the curing of various ailments.
145 The first case being Lady MARY WORTLEY MONTAGU's daughter, who was inoculated in 1721. MONTAGU was the first to introduce the actual practice of inoculation in England, after having lived and witnessed many cases of inoculation in Turkey.
146 See Berti (2021).
147 *Relazione…dell'innestare il vajioulo* (Florence, 1725).

Duct of the Glandula Renalis" (*Phil. Trans.* 1724: 190 and 270). Differently from his mentor, it was not VALSALVA himself who sent information about these researches, but they were collected and sent by DEREHAM. In 1705 they had moreover received and reviewed his famous book on the anatomy of the ear (Cavazza 2002: 8). The Society knew about VALSALVA by repute and, although these represented only minor examples of VALSALVA's research, they would have certainly published more material had they managed to get their hands on it.

Towards the end of the century, another important medical topic was given ample space in the *Transactions*; "animal electricity" – the term itself was coined by the physician and physicist LUIGI GALVANI, who through his experiments with frogs discovered that muscle contractions are powered by electrical impulses conducted through the nerves. Five Italian papers are dedicated to this subject and many more were inspired by GALVANI's discoveries.[148] Other Italians involved in this area of study and referenced in these papers are STEFANO LORENZINI, who carried out studies on electric rays and eels; GIOVANNI FRANCESCO PIVATI, who used electricity for medical treatment and whose experiments were rapidly heard about throughout Europe; FORTUNATO BIANCHINI (one paper); GIUSEPPE VERATTI (referenced);[149] and GIUSEPPE LORENZO BRUNI, a physician from Turin who corresponded with the Society and informed them on the Italian advances in animal electricity. The following extract from a paper based on BRUNI's correspondence with HENRY BAKER is worth quoting here. The paper represents a letter from BAKER to the Society's president and its aim is not only that of providing updates on the state of foreign researches in the medical uses of electricity, but also to encourage English scholars to carry out more research on the subject, in that, according to BAKER, the English "have not hitherto attended to the effects that may be thereby produced, any further than to assure ourselves they may be killed thereby" (*Phil. Trans.* 1748: 270). BAKER basically explains that, since the English have seen the deadly effects of electricity, they are refraining from investigating further into this subject and that "very few Trials have been made, to ascertain what, in distemper'd Cases, it can or cannot Perform". He then moves on to the description of foreign research on the subject and dedicates a lengthy section to Italian research:

---

148 Indeed, by searching for the term "galvanism" in the Society's journals more than 700 results come up; and more than 2000 results by searching "animal electricity".
149 Also GIUSEPPE VERATTI's wife, LAURA BASSI, was a specialist in the study of electricity; no mentions of her name however were found. For more on VERATTI's and BASSI's activity in electricity see Cavazza (2009).

The Philosophers in *Italy* and *Germany* have applied their Industry to discover by Experiment, how far Electricity may, simply and in itself, be of Service in several Diseases, and likewise how far it may conduce towards conveying the more subtile and active Effluvia of useful Medicines, either into the whole Body, or into some distempered Part. [...] Several Experiments to this Purpose, made at Venice by M. Pivati, and repeated afterwards by himself at *Leipsic* with the same Success. He gives Instances of saturating, by Electrification, with the *Effluvia* of Balsam of *Peru,* and of Sulphur, so as to produce very remarkable Effects; and of taking a Fit of the Gout away intirely, by conveying into the Part afflicted the sanative *Effluvia* of warm and discutient Drugs.
My ingenious Friend Dr. *Joseph Bruni,* one of the principal Physicians at *Turin,* and Fellow of our *Royal Society,* has likewise sent to me an Account, lately received by him, of Experiments made at *Rome,* and at *Bologna;* which I now, Sir, lay before you, in order to shew what Attempts to the same Purpose have been made in different Countries, and by different People. – The Doctor informs me, that at *Turin* they have repeated, with great Success, the electrical Experiments made in *England,* whereof I had sent him printed Accounts; that People all over *Italy* are busily at Work making electrical Experiments; and that, at *Bologna,* the electrical Power has been applied to the Cure of Diseases. (*Phil. Trans.* 1748: 272–273)

BAKER then reports a few cases from Bologna, Venice and Rome in which the use of electricity was claimed to have cured various kinds of distempers. Although the truthfulness of these accounts could neither be confirmed by BRUNI nor by BAKER, who were only the intermediaries of these reports, both of them encouraged the Fellows to carry out further research that could either corroborate or refute the claimed beneficial effects of electricity.[150] The same studies mentioned by BAKER were to be later criticised by the ABBE NOLLET (see Section 4.2.4). Finally, the medical group also includes a paper by GIOVANNI COSIMO BONOMO "Containing Some Observations concerning the Worms of Humane Bodies" (*Phil. Trans.* 1702: 1296–1299), which was one of the first studies on the scabies mite.

Another numerous group of papers consists of studies in electricity and magnetism (15 papers). Key Italians in the study of electricity *per se* were the physicist and literate GIOVANNI BATTISTA BECCARIA, who contributed six papers; ALESSANDRO VOLTA (six papers); and TIBERIO CAVALLO, the Italian resident in England (16 papers) (see Section 3.2.1). CAVALLO was moreover chosen to present the Bakerian Lecture for thirteen consecutive years (1780–1792).[151] The studies in electricity consisted of experiments; information on new instruments;

---

150 For more on the Italian studies in animal electricity and their transmission to the Society see Cavazza (2002 and 2009).
151 See CAVALLO in Table 6.2 in the appendix.

observations on natural electric phenomena; and methods of ascertaining the presence of electricity.

Eight papers focused on chemical topics. Worth mentioning here is FELICE FONTANA (three papers), who studied "inflammable Air". His studies led him to the discovery of the water-gas shift reaction in 1780.[152] The properties of airs, i.e. gasses, was another topic of interest in the 18th century; and contributions on this subject were also made by CAVALLO. Returning to FONTANA, he is moreover considered the father of modern toxicology having made many important studies on poisons (viper venom especially); one paper on this subject, a study on a poison called *Ticunas*, from the Americas, is present in the *PTRS* (*Phil. Trans.* 1780: 163).

A group of 16 papers have been classified as pertaining to the earth sciences. These studies include a broad range of topics such as a study on grottos; a mapping of the Adriatic Sea and another one on its tides; a study on the presence of petrifactions within the strata of the earth; descriptions of natural disasters such as avalanches; spectacular natural phenomena such as the appearance of red lights in the sky; giant causeways in the Euganean hills and more. The papers came from all parts of Italy and were sent both by Italians and Englishmen in Italy (HAMILTON and STRANGE especially). Adding to this group are 21 papers in volcanology and seismology,[153] nearly half of which were sent by WILLIAM HAMILTON. Thanks to HAMILTON's letters, from the 1760s to the end of the

---

152 Carbon monoxide reacts with water vapour to form carbon dioxide and hydrogen. The water-gas shift reaction is exploited for the production of hydrogen in manufacturing processes.

153 Eruptions of Vesuvius: 1730, reported by CIRILLO (*Phil. Trans.* 1731: 336–338); 1737, reported by D'ARAGONA (*Phil. Trans.* 1739: 237–252); 1751, reported by SUPPLE and PARKER (*Phil. Trans.* 315–317 and 474–475); 1757, minor eruption and earthquake reported by PADERNI (*Phil. Trans.* 1757: 619–623); 1761 by MACKINLAY (*Phil. Trans.* 1761: 44–45); 1764 and 1767 by HAMILTON (*Phil. Trans.* 1767: 192–200 and 1768: 1–14); 1779 by HAMILTON (*Phil. Trans.* 1780: 42–44).

Eruptions of Mount Ætna: 1755 in *Phil. Trans* 1755: 209–210. After the Sicilian earthquakes of 1693, another earthquake of 1717 was reported by DOMENICO BOTTONI. His treatise on the subject was reviewed in *Phil. Trans.* 1724: 151–158. Another earthquake that took place in Puglia in 1731 is also reported about by CIRILLO (*Phil. Trans.* 1733: 79–84). A violent earthquake that caused severe damage in Calabria is reported by HAMILTON and FRANCESCO IPPOLITO (*Phil. Trans.* 1783: 169–208 and 209–vii, see also Placanica 1985). There are also accounts of two earthquakes felt in Livorno in 1742 (written by PEDINI, *Phil. Trans.* 1742: 77–90) and in Turin in 1755 (DONATI, *Phil. Trans.* 1755: 612–616).

century, the Society had a full record of Vesuvius' activity and the state of the Kingdom of Naples from a geological point of view. After years of sending detailed accounts, HAMILTON himself defined the bulk of his letters as his "Vesuvian diary" and hoped "one day to have the honour of presenting these curious manuscripts [...] to the Royal Society, if it should think them worthy of a place in [its] Library" (*Phil. Trans.* 1786: 367). HAMILTON was fascinated by Vesuvius and its territory and developed a true love for this volcano; he believed that "volcanoes should be considered in a creative rather than a destructive light" (1786: 378) – given the variety of effects and physical changes that a volcanic eruption brings about when it occurs – and was disappointed when in accounts of earthquakes "nature is taxed with being malevolent, and bent upon destruction" (*Phil. Trans.* 1795: 73).

Many botanical observations are also found in various papers, only one of which was specifically dedicated to the subject. A topic that was clearly of interest to the Fellows, as it is described and discussed in various papers, was the Manna tree (*Fraxinus Ornus*, Flowering Ash) and the extraction of a sweetish resin commonly referred to as "manna".[154] The manna was extracted in the warm seasons by making an incision into the bark of the tree and letting the substance flow out, nearly like water out of a tap, with the difference that the manna hardens once it reaches the outer surface. There were differences in the type and colour of the manna according to the area where the trees grew – Naples, Calabria and Gargano most importantly. The process could also differ; for instance, a cannula could be employed in the extraction. Manna had various medicinal uses and was traded as an article of merchandise.

Also separately were considered the studies in meteorology (eight papers). Worth mentioning here are THOMAS DEREHAM; WILLIAM DEREHAM, a Fellow who worked on meteorology himself and collected foreign studies in the field forwarding them to the Society; GIOVANNI POLENI (two papers); and JAMES JURIN. Rusnock (1996) reports on a meteorological project in which Italy was involved. By the 1720s meteorology had become a well-established field of research among the Fellows, and improvements in instrumentation and theory encouraged the systematic recording of meteorological observations. In the 17th century, FERDINAND II of the Cimento Academy had already encouraged the creation of a network of correspondence in meteorology, which was taken on by Frenchmen, Germans and Englishmen. The keeping of weather journals became a common hobby among educated Europeans, and the physician and Secretary

---

154 See Section 3.2.2 for more on the topic.

JAMES JURIN launched a new meteorological project in 1723.[155] The proposal was advanced in the *PTRS* and in a separate print which he distributed throughout Europe, in Latin, to ensure response from the international community of men of learning. He moreover provided specific guidelines for the collection and recording of observations. JURIN managed to recruit observers from all over Europe and America. In Italy, it was THOMAS DEREHAM who recruited observers for him – from Pisa, Padua, Bologna and Naples. But the absence of standardised instruments and measurements, added to the difficulties of managing such a widespread group of researchers and maintaining their interest in time, made the management of the project rather strenuous. JURIN talked about these difficulties to THOMAS DEREHAM, the representative of the Italian researchers, who gave JURIN his advice:

> puff him [NICOLA CIRILLO] up a little in order to encourage him the more to proceed in his work, for he is of a Nation, that loves excessively to be flatter'd, & as he will certainly spread all over Italy your answer, it may serve others of his temper to contribute the sooner to the accomplishment of the great undertaking (DEREHAM to JURIN 1725, in Rusnock 1996:29)

The accomplishment was the compilation of a "Natural History of the Air" made up by all of the collected observations. JURIN did not reach his final objective as he left his office before he could complete it; the project did however have a long-lasting effect on the recording of weather observations. Moreover, WILLIAM DEREHAM, published several abstracts of the meteorological diaries, written between 1723 and 1728, in the *PTRS*, which included the observations made by NICOLA CIRILLO. POLENI and IACOPO BARTOLOMEO BECCARI also made contributions to the project.

Other hot topics in the 18th century were archaeology, antiquarianism and numismatics (26 papers). MARTIN FOLKES, who was President of the Royal Society between 1741 and 1752 and vice-president before this time, was an antiquarian and numismatist, his influence may thus justify the high presence of papers in these non-scientific subjects around the mid of the century. Main informants in this area were the painter CAMILLO PADERNI (eight papers), who minutely informed the Society about the excavations at the site of Herculaneum; the VENUTI brothers – NICOLÒ MARCELLO, RIDOLFINO and FILIPPO – who contributed papers on Greek epigrams and numismatics; and English residents

---

155 His interests in meteorology were related to his profession (physician) as he believed in the existence of a correlation between weather and diseases. An earlier meteorological project had been launched by ROBERT HOOKE in 1663 (Rusnock 1996).

and travellers in Italy, such as the above mentioned JOHN SWINTON and others who contributed individual papers on the subject.[156] These papers encouraged learned travellers to visit the Southern parts of the Italian peninsula and, together with the papers on the eruptions of Mount Etna and Vesuvius, they extended the routes of the *Grand Tour* of Italy to the wildest and remotest areas of the South (see D'Amore 2017).

The remaining papers include a small group of six mathematical papers; six travel accounts; three papers in optics; two in mechanics; and individual papers on a variety of subjects classified as "other". One interesting paper from the latter group is a letter from OLIVER JOHN, which reports about the Italian use of a device called "arcuccio" in childcare (*Phil. Trans.* 1731: 256). This was a semi-cylinder in wood or iron that was placed above the new-born either in the bed or in the cot and was then covered with a blanket. Its purpose was to keep the child warm, avoiding suffocation.[157] The paper itself is very brief but furnished with a drawing of the device. The arcuccio here represented also included a gap through which the mother could insert her breasts to feed the baby and a wooden bar to lean on for her comfort. The author explains that the use of the arcuccio was widespread and compulsory in Florentine midwifery and suggested that the English adopt this practice too, given the high number of children declared "overlaid in the Bills of Mortality".

To conclude, the subject areas of the papers seem to proportionally reflect the number of Italian Fellows per field of interest. Indeed, the papers in medicine and astronomy are much more copious than others, which reflects the higher presence of physicians and astronomers among the Italian Fellows.[158] Instead, differences in specific topics of preference seem to be related to the interests of individual Italians, in the case of medical and electricity papers; while in the case of the papers in meteorology, volcanology, seismology, archaeology and antiquaries, the Italian contributions were determined by the specific interests of the Society who searched and asked for information in those areas. The subject areas moreover would seem to be in line with the overall presence of papers in these fields in the *Transactions*; i.e. not only Italian-research-based papers but

---

156 JOHN MONRO, ROGER GALE, MR RAMSAY, DAVID ERSKINE BAKER, MR HOARE, JOHN WARD, JOSEPH WILCOX, and CHARLES MORTON.
157 Other purposes were to protect the child from insects and light (ARCUCCIO o ARCHETTO. In *Dizionario ostetrico, ad uso delle levatrici, del Dottor L. P.*, 1803).
158 The group of statesmen, diplomats and noblemen was more numerous than that of the astronomers; however, as it has been seen, very often these Fellows were not contributors of scientific knowledge.

all *PTRS* publication trends. Periods in which one research area was particularly present are often due to the influence of the secretary and/or president in office.[159]

3.2.5 Beauties of northern, central and southern Italy

> *I cannot have a greater pleasure than to employ my leisure hours in what may be of some little use to mankind; and my lot has carried me into a country, which affords an ample field for observation. (WILLIAM HAMILTON, Phil. Trans. 1769: 21)*

As for the previous century (Section 3.1.5), this section focuses on the textual imagery of Italy that reached the Fellows in the 18th century. A simple eleven-page account by the English physician and naturalist TANCRED ROBINSON (FRS 1784) provides a perfect example of the curiosities that a natural philosopher could find of interest in the different parts of the Italian peninsula. (*Phil. Trans.* 1714: 473–483). Indeed, ROBINSON provides a miscellaneous travel account from a natural philosophical perspective, focusing especially on botanical, zoological and geological features of Italy. He reports on different curiosities that attracted him in the various places he visited. His narrative goes up and down, from north to south and south to north, in an unorderly manner. He starts from the south, discussing the wild colourful plants of prickly pears and the great quantity of insects that fed on them and that were physiologically endowed with devices to ease their nourishment and breeding in that environment. He then moves to the mountains of the north where he saw "great varieties of perfect Shells, that never occurred to me on the Italian Shores" (474). He continues with the different plant species he found along the *Via Appia*, with a particular interest in the southern Ash trees, from which Manna was extracted, and the cork trees with the *locals* working on them:

> Before I enter'd the beautiful *Campania* of *Naples* large Woods of *Cork Trees* grew on each side the Road, where the Inhabitants were decorticating them. I ask'd if the Trees did not perish: they answer'd, some did, but the Acorns return'd annual Supplies. The Women and Children wore Shoes made of the Bark. (*Phil. Trans.* 1714: 474)

He continues with observations on Vesuvius, Solfatara, the Monte di Cinere and the "hot streams" that pervaded the country of Naples. Then back into botanical observations with a listing of the variety of plant species he found there, and then,

---

159 See Hall (1991) for more on the general topics discussed at meetings and published in the *Transactions* and their relation to the secretaries and presidents in office.

> When dark Night came on, I could see Multitudes of Luminous Flies thro' the *Campania of Naples*: perhaps our Male *Gloworm*, or flying *Cicindela*, may abound there; not but that many other Insects may carry such Lanthorns about them. The *Scorpions*, creep out about that time; and I have found them often in Bed, with the *Punaises*.
>
> The Hedges are full of *Lizards* of various Colours; and the *Cicada's* chirp and sing towards Evening. I observ'd several Species of stinging Spiders in the Corn Fields, some of which, in hot Harvests, may prove *Tarantula's*; The Poysons of Animals and Plants increasing with the approach of the Sun, and the Hearts of Climates. Abundance of Silk-Worms were spinning on the Trees and Shrubs; the Birds prey'd upon them, before they could change into *Papilio's*, as they do upon swarms of *Locusts*. (*Phil. Trans.* 1714:477)

After a description of frogs, tortoises and snails he ate "served up with Oyl and Pepper", he moves back to the central and northern parts of Italy. Here, ROBINSON gives ample space to the fish he saw and ate in Genoa and Civitavecchia, providing the common names given to the fish by the Italians – such as the "bocca in capo" or "prete" (*uranoscopus*, stargazer), the "mola" (*mola mola*, sunfish), the "pesce balestra" (*Balistes capriscus*, triggerfish), the "pesce porco" (*Oxynotus centrina*, angular roughshark) and many others. From the Alps he reports about the countless species of birds he saw with their different colours and variegation of feathers. On the Laguna of Venice:

> I saw several Species of *Mergi, Lari, Colymbi,* and other Water Fowls. Most of which Div'd. I was surpriz'd with the Variety of them, having not seen so many on other Coasts: perhaps the hard Winter had forc'd some unusual Birds thither. The Monks and Fryers told me, they eat some of those Sea Birds in *Lent* and on Fast Days, because they liv'd upon Fish, and had a piscose Taste. (*Phil. Trans.* 1714: 481)

He then moves from the wild mountain goats and marmots of the Alps to the buffalos of Lombardy and Naples – mentioning the leathers they produced from their skin, the snuffboxes and combs from their horns, and the cheese from their milk – and closes the letter with reflections on some luminous appearances he saw in the firmament "over *Vesuvius*, the *Strombulo* Islands, and towards *Ætna* in dark Nights" (483). Positive nomination strategies recur in the description of natural and artificial curiosities,[160] and a series of adjectives and nouns are repeated throughout the account, such as *beautiful, rich, fat, large, colourful, colours, variety, variegation, many* etc., which provide a positive, warm, surprising and attractive image of Italy.[161]

A shorter travel account published just over three decades later by ROBERT MORE (FRS 1730) (*Phil. Trans.* 1749: 464–467), confirms some of the Italian

---

160 The author, in fact, also briefly describes some of the antiquities of Rome.
161 On ROBINSON see also Section 3.1.5.

features that attracted the Fellows in their *Grand Tour* of the country, and contributed to increasing their interest and widen their prospective destinations. Indeed, Rome and the Kingdom of Naples were visited for their antiquities and volcanoes. Further, at Terni another point of interest was the Cascata delle Marmore, an 165m high waterfall created by the ancient Romans. From his visit to Bologna, MORE was impressed by the museum of the Specola, which had been improved since former accounts with collections "of the Count *Marsigli*, Marchese *Cospi*, *Aldrovandus*, and others, [and which] form the finest Sett of natural Curiosities I ever saw" (466). The Apennines had been recommended to him by MARTIN FOLKES (PRS), to whom the letter with the account was addressed, for their "continual Fires", which MORE was able to see too.

A less impressed view of Italy came from the French ABBE NOLLET (FRS 1734) who, due to his doubts as to the Italian research carried out in electricity, travelled to the country with a "Desire of knowing how far these Things were true" (*Phil. Trans.* 1749: 368). Following his travels he reports:

> I have made the whole Tour of *Italy*, which has enabled me to make many Observations relating to Natural Philosophy. I have made from Experiments at the *Grotto del Cani*, near *Naples*, which take off a good deal, in my Opinion, of the Marvellous of that famous Phænomenon. I propose to myself the Honour of transmitting them upon some future Occasion, as my Letter is already too long. The Eructations from *Vesuvio* were very great when I was there, and were the Prelude to three Earthquakes, which happened just after my Departure, and which I was fortunate enough not to be Witness of. (*Phil. Trans.* 1749: 369)

Something that did impress him however were

> The *Lagunes of Venice*, and the Waters of the *Mediterranean* Sea, appear luminous everywhere in Summer, in dark Nights; I have discovered, that this Light proceeds from a very small Insect, which multiplies prodigiously. I have heard all my Life, that the Water of the Ocean appears sometimes luminous: It may possibly proceed from the same Cause, and I should be very glad of a particular Inquiry into this Fact. (*Phil. Trans.* 1749: 369)

NOLLET's successive paper on the Cave of dogs (Grotta del Cane), near Pozzuoli (*Phil. Trans.* 1751: 48–61), offers a natural philosophical perspective on it. He carries out experiments to learn about the nature and effects of the "vapour" that is present in the grotto, rather than providing a description of the grotto itself.

From the meteorological papers, Pisa comes across as the "piss-pot of Italy" although this epithet was not intended in a critical way but was due to the high amount of rain that was measured there annually (*Phil. Trans.* 1708: 334–336 and 1742:77). The city was moreover said to be very windy, with three different kinds of winds crossing it daily.

Antiquities, as it has been seen, were a major topic of interest regarding the south and Rome; however, there are also some accounts of antiquities from the north. *Phil. Trans.* 1744: 540–549 for instance, is a report on various stone inscriptions dug out near Turin; the content of these inscriptions together with the further discovery of vestiges of an ancient fabric and various medals revealed, according to the author, "the great Antiquity" and power of the ancient city of Industria.

From a geological point of view, a very detailed description of the Giant's Causeways in the Venetian State is provided by JOHN STRANGE. His long account (*Phil. Trans.* 1775: 5–47) provides drawings and a minute description of the composition and shape of the prismatic basalt columns found on the Monterosso (near Abano Terme, Padua) and the Monte del Diavolo (San Giovanni Ilarione, Verona), reasoning on their origin and comparing them with similar natural phenomena in Ireland and France. Following the interest shown in this account, STRANGE sent a second account (with drawing) on the basalt columns discovered at Castelnuovo (Teolo, Padua) (*Phil. Trans.* 1775: 418–123).

On the region of Abruzzo, an impressive account is given by HAMILTON (*Phil. Trans.* 1786: 365–381):

> The whole country from Arpino, the native place of Marius, by Isola, Sora, Civitella, and Capistrello, to the lake of Celano, is, in my opinion, infinitely more beautiful and picturesque than any spot I have yet seen on the Alps, in Savoy, Switzerland, or the Tyrol. The road is not passable for carriages, and indeed is scarcely so, even in summer, for horses or mules, and is often infested with banditti; a party of which, consisting of twenty -two, had quartered themselves in a village which I passed through, and left it but a week before my arrival. There are many wolves and some bears in the adjacent mountains. The tyger-cat, *gatto pardo*, or lynx, is sometimes found in the wood of this part of Abruzzo. (*Phil. Trans.* 1786: 369)

Here HAMILTON provides a description of the Lago Celano (or lago Fucino). This account is precious in that the lake no longer exists as it was drained in 1878. HAMILTON states that it was the most beautiful lake he had ever seen, describing it as being surrounded by very high mountains covered in snow and many villages with "rich and well-cultivated farms". The lake furnished "an abundance of fish, but not of the best quality: a few large trout, but mostly tench, barbell and dace" and "thousands of water snakes, pursuing and preying upon a little fish like our thornback, but much better armed". He then describes the Emissary of Claudius, which was an under-ground drainage tunnel built in rock by the emperor Claudius[162] to allow the lake, which had no natural outflow, to

---

162 According to Suetonius, the tunnel, built thorough Monte Salviano, required 30,000 men and took 11 years to complete. Due to earthquakes and overgrowing vegetation

flow into the Garigliano river. Indeed, the water on the lake increased daily and was destroying "the rich and cultivated plains on its borders" (368). Claudius' Emissary, he goes on, "remains nearly entire, though filled up with rubbish and earth in many parts, and of course useless" (368).[163] Yet, he "would make no doubt, but that if [the drainage] was cleared and repaired, it would again answer its purpose. In its present state it is a most magnificent monument of antiquity" (369). In the same paper, HAMILTON also reports of his travels to the Islands of Ventotene, Santo Stefano, Ponza, Palmarola, and Zannone in the Tyrrhenian Sea focusing on their geological features.

Finally, Puglia is once again represented in the *Transactions* as the land of the tarantulas and of *tarantism* – an "illness" characterized by an extreme impulse to dance, which was widely believed by the southern Italians to be caused by the bite of the tarantula. Interest in this spider and the *tarantati* (those affected by the tarantula bite) had been shown since the 17th century (see Section 3.1.4). However, DOMENICO CIRILLO reports that this phenomenon was by the 1770s only believed by the inhabitants of Puglia, while in Sicily and in the Kingdom of Naples they no longer believed in such folkloristic tales. Indeed, the "cure of the bite of the Tarantula, by music, has not the least truth in it", "it is only an invention of the people, who want to get a little money, by dancing when they say that the tarantatism begins" (*Phil. Trans.* 1770: 237). The only other mentions of this region are made for the presence of manna tree woods and earthquakes. See also Section 4.2.5 for further notes on the research areas related to main Italian cities.

3.3 19th century

The 19th century started at the mid-point of JOSEPH BANKS' term as President of the Royal Society. He was elected President in 1778 and held this position until his death in 1820. BANKS was a person of both social status and scientific attainments, he was wealthy and influential and also a botanist and patron of botanists. He therefore perfectly represented the 17th- and 18th-century ideal of gentlemanly science. During his 42-year term, the Society prospered, although there was a growing conflict between non-genteel Newtonian scientists – who pushed for reform and a purging of the Society from non-scientific members – and the more conservative genteel scientists. Indeed, the Society, as an independent

---

      the drainage tunnel stopped functioning. A new drainage canal was built in the 19th century, which led to the complete draining of the lake.
163 It is here not said that Claudius' tunnel was extended and deepened by Hadrian. Hadrian's extension and the 19th-century works destroyed most of the original tunnel.

institution funded only by means of the Fellows' subscriptions, still needed to rely on the memberships of the rich and wealthy in order to survive (Hall 1984). During this period new specialised Societies were founded, such as the Linnean Society (1788), which focused on natural history; the Royal Institution (1800), focused on the practical application of scientific findings; the Horticultural Society (1804); the Geological Society (1807); the Society for the Promotion of Animal Chemistry (1809); and the Astronomical Society (1820). While some of these were affiliated to the Royal Society, the Geological and Astronomical Societies were seen (by BANKS) as a threat to the Royal (Hall 1984:5 –7; Atkinson 1999: 32; Gascoigne 2003: 7).

The struggle between professional and genteel scientists continued after BANKS' death, and through the rest of the 19th century both the Royal Society and its publications underwent considerable changes. The Society started moving towards an increasingly scientific orientation, and scientific interests and achievements were to become the primary qualities for the acceptance of new members. After a sharp period of criticism in the 1820s-1830s, followed a period of reform and revision. This included attempts to raise the standards of papers proposed for publication; a reorganisation of the Council with six Vice-Presidents covering the widest possible range of scientific subjects; the (at first informal) introduction of a refereeing system by two members of the Council; the establishment of Committees in each department of science; candidates to Fellowship were to be scrutinised more closely and limited in number; and there were attempts to economise the printing of the *Transactions*, which had always been a drain on the Society.[164] Abstracts of papers were to be prepared by the junior secretary to submit to the Committee of Papers and, in 1832, a new journal was created which initially only published paper abstracts and news reports. This was the *Proceedings of the Royal Society*. Not all of the papers published in abstract form in the *Proceedings* were then published in full form in the *Transactions*. The *Proceedings* eventually developed into a journal in its own right, extending its contents to obituaries, shorter articles and articles that were considered unfit for the *Transactions* (Atkinson 1999: 40). The gradual specialisation of the sciences eventually led to further major changes in the Society's publications: in 1887 the *PTRS* was divided into *Philosophical Transactions A* – focusing on the mathematical and physical sciences – and *Philosophical Transactions B* – focusing on the biological sciences.[165] Further, specialised sectional committees were

---

164 For more detail on the reform period see Hall (1984: 63–82).
165 The same division was then made at the beginning of the 20th century for the *Proceedings*.

appointed to referee the papers and, after 1898, the names of referees were no longer recorded in the journal books in order to ensure the anonymity of the reviewers (Atkinson 1999: 44). Another novelty at the end of the 19th century was the introduction of a Parliamentary Grant system, which allowed the Society to aid scientific development while remaining an independent body. Moreover, the Society advised the Government on a wide range of problems.[166] Hence, in the course of the 19th century the Society and its journals were slowly taking the form of the highly scientific institution and journal that are known today.

The corpus for this century includes 52 papers, both from the *Philosophical Transactions*, and the *Proceedings*. The Italians elected in this century were fewer, 12 Foreign Members in total, but this is not surprising considering the restrictions that were first tacitly and then officially established for Fellow admissions. To these at least 11 non-elected Italian contributors should be added.

Once again this section is divided like the previous; however, there will not be a section on the portrayal of Italy through the papers as the greater scientific orientation of the Society meant that there was no longer space for non-scientific topics in its journals.

3.3.1 Foreign Members and contributors

The Italians elected as Foreign Members (FMs)[167] in the 19th century reflect the changes that were slowly taking place at the Royal Society. Indeed, only 12 Italians were elected in this century and their contributions to the *Philosophical Transactions* were few – 17 papers contributed by only seven of the Fellows. However, their value as scientists was considerable and their work was not only known to the Society but in various cases also awarded with medals. The greater scientific orientation of the Society can be seen with the election of a majority of physicists, mathematicians and astronomers; these were: GIUSEPPE PIAZZI, GIOVANNI ANTONIO AMEDEO PLANA, FRANCESCO CARLINI, MACEDONIO MELLONI, PADRE ANGELO SECCHI, LUIGI CREMONA, PIETRO TACCHINI, and GIOVANNI SCHIAPPARELLI. PIAZZI was a mathematician and Astronomer Royal in Palermo, who made only one contribution in the 18th century before he was elected. PLANA, also a mathematician and astronomer in Turin, made no

---

166 For this reason, the Society, although independent, is regarded as the UK's *national scientific academy*.
167 From the 19th century foreigners started being elected as "Foreign Members" and no longer as "Fellows" and were inscribed on the "foreign list". Some 18th century Fellows were also qualified as "Foreign Members".

contributions to the journal but was awarded the Copley Medal in 1834 for his studies on lunar motion. CARLINI was the director of the observatory in Milan, he sent three papers to the Society, of which mention is made in the *Proceedings*; however, the papers do not appear to have been published in the *PTRS*. MELLONI was a physicist "celebrated for his discovery of the different scales of the diathermaneity of transparent & coloured media" (EC/1839/29) and was awarded the Rumford medal in physics in 1834. His work is referenced in several papers, but he made no contributions to the journal. However, he did collaborate with the Fellows in order to create a connection between the Royal Academy of Sciences of Naples and the Royal Society. The two societies exchanged several issues of their journals and portraits (MS/581/114 and 115). Another Rumford medal (1886) was awarded to TACCHINI, an astronomer of the Collegio Romano, who also collaborated with the Society. SECCHI instead, director of the Roman observatory, contributed three papers and corresponded with the Fellows. CREMONA, a mathematician, and the astronomer SCHIAPARELLI of Milan do not appear to have had any relevant contacts with the Society.

The remaining Fellows included the chemist DOMENICO PINI MORICHINI, who was known to the Society for his discoveries of the magnetic properties of the violet rays of the Solar Spectrum. His laboratory in Rome was visited by HUMPHRY DAVY and MICHAEL FARADAY in 1814. Although his claims to be able to magnetise a needle, by exposing it to the violet rays of the solar spectrum, failed miserably in front of the two visiting Fellows, his work continued to arouse interest in Britain and is acknowledged in a paper by MARY SOMERVILLE (Patterson 1983), which has been included in the corpus. SOMERVILLE repeated MORICHINI's experiment in England – without a precise knowledge of how MORICHINI had proceeded – and found his claim to be correct. Another chemist was STANISLAO CANNIZZARO, an Italian senator and artillery officer who carried out research in atomic and molecular weights. He was awarded the Copley Medal in 1891. Only one nobleman was elected in this century, and this was LEOPOLD II, the Grand Duke of Tuscany. As was always the case with Royalties, their election did not follow the rules to which other candidates had to subscribe, and LEOPOLD's certificate only reports "His Imperial and Royal Highness Leopold II Grand Duke of Tuscany" and the date of election (1838). In this century new Foreign Members also received a "diploma", which certified their election to the Society. The diploma was generally sent to them through the post accompanied with a kind reverential letter. The last Foreign Member, AUGUSTUS BOZZI GRANVILLE, an Italian patriot and MD with interests in midwifery, contributed three papers to the *PTRS*. He had travelled extensively and settled in England in 1813, where he became surgeon to the English fleet and Foreign Secretary to the Geological Society. In the 1820s-1830s the Society once again faced sharp criticism from the commercial papers and from some of its Fellows. One of these attacks came from GRANVILLE and his *Science without a Head, or the Royal Society Dissected by One of the 687 F.R.S. - - -sss* (1830). It was initially published anonymously and then later a second edition, *The Royal Society in the XIXth Century* (1836), was published reporting his name. Here, unlike his contemporaries, he denied that English science was in decline. On the contrary, he thought science in England, and in London especially, flourished; however, this did not include the Royal Society which, according to him, was in need of reform:

> He proposed two areas of reform: in the general management of the Society, and in the election of its officers. He depicted the Society as a kind of Venetian Republic or oligarchy, with the Council more or less continuous at the whim of the President; he deplored the dullness of the meetings, at which much formal business was read out, the paper was frequently interrupted by announcements of the results of voting for candidates, while the audience slumbered, aware that they would not be allowed to discuss the paper being read. He believed that the Royal Society should be divided into scientific classes, with a small 'Free class' for persons interested in science but not practicing it, the whole being limited in number. (Hall 1984: 56–57)

His views on the presidentship were more conservative;[168] he had admired BANKS' controlling office, and thought that a President of high rank, and thus above parties, would be a suitable choice for the Society.[169]

As in the previous centuries the contributions of non-elected Italians were no less important than those of the Foreign Members. In fact, this group had more papers published in the *Transactions* (20 papers), than the FMs. The contributors comprised eleven Italians, i.e.: CARLO MATTEUCCI, ANGELO RUFFINI, ANGELO MOSSO, LEOPOLDO NOBILI, COUNT VINCENT PICCOLOMINI, PASQUALE STANISLAO MANCINI, RAFFAELE ARGENTATI, GIOVANNI ALDINI, FRANCESCO BRIOSCHI, GIOVANNI BATTISTA GRASSI, and ANTONIO PANIZZI. The physiologist CARLO MATTEUCCI was not only awarded the Copley Medal in 1844 for his researches in animal electricity, but also published 17 *PTRS* papers.[170]

---

168 This stands in opposition to the attacks directed at the Society by his contemporaries, who believed that the Society's members, and therefore also its presidents, should only be practicing scientists.
169 The candidate he suggested was AUGUSTUS FREDERICK THE DUKE OF SUSSEX, (FRS 1828, PRS 1830–1838) who was eventually elected winning by only eight votes over the mathematician and astronomer JOHN HERSHEL (FRS 1813, Sec. 1824–1827).
170 Only twelve have been included in the corpus.

The Society held him in great esteem as is shown by JOHN FREDERIC DANIELL's communication of the award:

> Dear Sir,
> I never sat down with greater pleasure to the performance of my official duty [For. Sec. 1839–1845], than I do this day to announce to you that the President and Council of the Royal Society of London for the promotion of Natural Knowledge, have unanimously awarded to you the Copley Medal for your Researches in Animal Electricity. This medal, though neither remarkable for its beauty, nor its pecuniary value, is the most ancient medal of the Society, and regarded as its most honourable distinction; and I most heartily rejoice that your name will stand so worthily in the long list of eminent Philosophers, by whom it has been rendered illustrious.
> I sincerely wish that we could anticipate the pleasure of seeing you amongst us at the Anniversary on the 30th [Nov.] to receive this distinction personally from the hands of our President [...].
> I truly rejoice that I was one of those who had the high gratification of seeing the unambiguous experiments by which you have established one of the most important discoveries of the age; and I trust that your life and energies will long be spared to enable you to continue your successful researches in a field of such promise. [...]. (DANIELL to MATTEUCCI, 15 Nov. 1844, MS/581/107)

MATTEUCCI in fact continued GALVANI's experiments in animal electricity; he proved the electrical properties of animal tissues and was the first to detect an electrical current in the heart. LEOPOLDO NOBILI also had interests in electricity as physicist; he contributed one paper to the journal. ANGELO RUFFINI instead was a low-profile histologist and embryologist whose studies focused on sensory endings. Due to financial difficulties he abandoned his post as Director of the histology laboratory of the University Hospital of Bologna to become a municipal doctor in Lucignano (Tuscany) where he was cut off almost entirely from the centre of scientific life (Eccles 1975: 70). During this period however he kept correspondence with the physiologist CHARLES SHERRINGTON (FRS 1893), whose influence in the Society enabled the funding for the publication of RUFFINI's illustrated monograph *Sulla presenza di nuove forme di terminazioni nervose nello strato papillare e subpapillare della cute dell'uomo* (Siena, 1898) and of a paper published in the *Journal of Physiology* ("On the Minute Anatomy of the Neuromuscular Spindles of the Cat, and on their Physiological Significance" vol. 23, 1898) (Eccles 1975 and NLB/17/38). His discoveries in the fields of physiology and embryology were several including that of the neuromuscular spindles, highly specialised receptor organs in mammals and humans. RUFFINI's work was eventually applauded, and his struggles were rewarded with the chair of Histology and General Physiology at the University of Bologna (1912–1929). ANGELO MOSSO was also a physiologist whose studies focused on the digestive and circulatory systems, and the improvement of medical instruments in the field. In 1892 he was invited by the Royal Society to hold the Croonian Lecture in physiology (see appendix Tab. 6.3), which was consequently published in its original French in the *Transactions* and an English abstract in the *Proceedings*.[171] Mosso moreover founded in 1876 the laboratory Col d'Olen, at the foot of Mount Rosa, for the purpose of investigating human physiopathology. At the founding of the laboratory he asked the Royal Society whether they could appoint two of their Fellows for a position there; the request was granted (NLB/35/419 and NLB/35/738). Little is known about Count VINCENT PICCOLOMINI, who sent the Royal Society a paper on the "Geographical position of the principal points of the triangulations of the Californias and of the Mexican coasts" (*Proceedings* 1843: 196–197). PASQUALE STANISLAO MANCINI instead was more of a would-be contributor in that the paper he sent to the Society in 1837, on the application of electricity to the study of earthquakes, was refused publication. Initially, the Society probably lacked a person who could translate it from Italian; it was then translated a year after and, although MANCINI had requested "a mild judgement of his paper [...] in consideration of him being only twenty years of age", it was refused publication on the grounds that the style of the paper was "somewhat redundant and pompous" (RR/1/160). A similar fate concerned two papers by GIOVANNI ALDINI, a well-known physicist who continued and extended GALVANI's research in animal electricity. Although they had been read at the Society's meetings, his proposals do not appear to have been published in the journals. The papers concerned galvanism and gas-light illumination (see ALDINI in appendix Tab. 6.3). ALDINI could also speak and write English and French and had visited England in 1803 where he publicly performed one of his most famous experiments, i.e., the electro-stimulation of the limbs of an executed criminal at Newgate. ALDINI was moreover the one to send the Society a copy of GALVANI's *De Viribus electricitatis* in 1796 or 1797, together with his own *De animali electricitate dissertatio* (Bologna, 1794) (Hall 1982: 72). RAFFAELE ARGENTATI was an inventor interested in flight. He created a locomotive apparatus to give direction to aerostatic globes (hot-air balloons) and gave news of it to various academies in Europe including the Royal Society. The paper he proposed for publication was not however published. FRANCESCO BRIOSCHI was a Milanese mathematician, patriot and Dean of the University of Pavia. One paper by him was published in the journal, although it remains

---

171 These two papers were not included in the analysis as they were found after the analysis was carried out (*Phil. Trans.* B 1892: 299–309; *Proceedings* 1892: 83–85).

unclear whether it had been sent by BRIOSCHI or others; he does not appear to have had any relevant correspondence with the Society. GIOVANNI BATTISTA GRASSI was a zoologist and physician who made important contributions to the field of parasitology, among which was the first description of the life cycle of the malarial parasite. He moreover carried out studies on eels and termites, which he communicated to the Society and for which he was awarded the Darwin Medal in 1896. This award provoked a minor controversy at the Royal Society as SALVATORE CALANDRUCCIO, who had collaborated with GRASSI on the award-winning studies, claimed that he was "at least equally entitled with Professor Grassi to the credit of the discoveries" (NLB/23/1/818). Consequently, the Royal Society appointed a special Committee to examine the claim and eventually replied to CALANDRUCCIO that they "made a most careful examination of the various documents he sent. President and Council, while fully recognizing the fact of his collaboration with Professor Grassi in the researches relating to Leptocephali, have concluded that there is no valid reason for making any change in the award of the Darwin Medal of 1896 to Grassi" (NLB/23/2/212).

Finally, a different kind of contribution to the Royal Society was made by ANTONIO PANIZZI, an Italian patriot and refugee in London, who is mostly known for becoming Italian Professor at the University of London and librarian (and later Principal Librarian, 1856–1866) to the British Museum. Less known instead is that he was also hired by the Royal Society to compile a catalogue of their library, which in the early 19th century was in serious decay. PANIZZI started collaborating with a Library Committee set up for the purpose in 1832. Hall reports that PANIZZI

> was appalled to find old books in such a bad state of repair, that many were useless, especially the non-scientific ones (which he estimated in 1833 to be a third of the whole). By the end of 1833 a series of classes (ultimately eighteen in all) had been determined and the Committee had given Panizzi assistance on where to put 'difficult' subjects, ranging from acoustics to alchemy. (Hall 1984: 72)

The difficulties of the task were further enhanced by the deficiency in storage and shelving, which was later provided for with new rooms and a specially built gallery. In 1835 the first drafts of the catalogue were sent to the printer. These were revised by the Society's committee which in 1836 offended PANIZZI by requesting that all notes stating matters of opinion should be removed from the catalogue. Indeed, PANIZZI was a perfectionist and saw the catalogue as his own creation interspersing it with comments in first person (Hall 1984). Further, at the Anniversary address the President gave the credit of the catalogue to the members of the Committee, to which PANIZZI replied with a 56-page long pamphlet where he defended himself and stated that he had been prevented from completing his work and had not been paid all the monies due.[172] In this pamphlet he traces the various steps of the dispute by quoting extracts from the letters exchanged with the Society's secretary, MARK ROGET.[173] With the Anniversary address of the following year (1837) the President ended PANIZZI's collaboration. A long series of letters were then exchanged, with some of them being printed. PANIZZI was eventually paid most of the money due. The catalogue was published in 1836 deprived of the author's name.[174] Although PANIZZI was not given the deserved credit, the catalogue remains a valuable tool, which gave order to the Society's library. The division of the books moreover provides a good summary of the subject areas that interested the Society in the course of its existence and how they were viewed and grouped in the minds of the Fellows. These were:

> Mathematics; Astronomy; Mechanics, Hydrostatics, hydraulics and acoustics; Optics; Tables on various subjects; Chemistry, pneumatics and meteorology; Electricity, Galvanism and magnetism; Natural philosophy (general works on); Geology and mineralogy; Botany and agriculture; Zoology; Anatomy, physiology and medicine; Natural history (general works on); Transactions; Reports of the House of Commons; Journals; Voyages; Miscellaneous. (PANIZZI 1836: I)

The list also reminds us that the concept of science and its different branches were gradually becoming closer to the modern idea. Indeed, the papers now display an increased use of the word *science*, as opposed to the all-encompassing *natural philosophy* that was common in the 17th century. Moreover, the Fellows and Foreign Members now used highly specific terminology in defining both their field of interest and their profession.

3.3.2 Fellows and Englishmen in Italy

The Royal Society's foreign relations were kept by means of correspondence from its secretaries and foreign secretaries. While in the previous centuries some of them had more intense correspondence with Italians, in the 19th century secretaries appeared to have mainly formal exchanges with Italy.

---

172 *A letter to His Royal Highness* [the DUKE OF SUSSEX (PRS 1830–1838)] *the President of the Royal Society, on the new catalogue of the Library of that institution now in the press.* (In Royal Society library, TRACTS RS/6/4).
173 See also: Emblen (1969 and 1970: chapter 14) and Biagetti (2001: 11–129).
174 The library's record, however, provides PANIZZI's name as the author. In 1839 a second version "revised by members of the Society's Catalogue Committee" was also published.

Continued help was also provided by travelling Fellows and British and Italian diplomats, such as a Mr. Falconer, British consul at Livorno in the first part of the century, who was the recipient of various packages from the Society and was entrusted with forwarding them to the Italian Fellows in the northern and central parts of the Peninsula. Italian consuls in England were also invited to the Society's Anniversary meetings to receive medals on behalf of Italian scientists who could not attend personally.[175]

As far as the Fellows in Italy are concerned, an important publication that concerned the Neapolitan earthquake of 1857 was published by Robert Mallet (FRS 1854). Mallet was an engineer and chemist who became interested in earthquakes after learning that they could be accounted for by scientific laws. He therefore applied his knowledge of mechanics to the interpretation of earth movements and their propagation. He carried out many experiments to study the propagation, velocity and damage of elastic waves which, given the weak tremors that occurred in the British Isles, had to be conducted by means of artificial shocks, produced by controlled explosions (Ferrari & McConnel 2005). His studies were considerable advances compared to the seismological studies of the 18th century. In 1857, when Mallet read in the papers the announcement of a great earthquake that caused severe damage in Salerno and Potenza, he immediately started working to receive funding to go on a mission to study the Italian earthquakes. His request was backed up by the Royal Society, the Geological Society of London and the Royal Geological Society in Dublin. Mallet began a month-long trek in the affected areas of the Kingdom of Naples collecting a great deal of data and descriptions. His report also included photographs, which was one of the first cases of the use of photography for scientific purposes. Mallet submitted his report on his investigations to the Royal Society for publication in the *Philosophical Transactions* in May 1860. The publication however was delayed because the manuscript was excessively long for the *PTRS* (700 pages of text with a further 100 pages of appendices) and required the reviewing of three referees of different specialties. It was eventually published in book form in 1862 as *Great Neapolitan earthquake of 1857: the first principles of observational seismology*; while an abstract was published in the *Proceedings* (1859: 486–454). Also Sir Charles Lyell (FRS 1826) travelled to Italy in 1828 and 1858 to carry out studies in the field of geology. His observations are reported in an 88-page

---

175 For instance, his Excellency General Ferrero, Ambassador for Italy, attended the Anniversary Meeting on 30 November 1896 to receive the Darwin Medal on Grassi's behalf (NLB/13/835).

long paper "On the structure of lavas which have consolidated on steep slopes; with remarks on the mode of origin of Mount Etna, and on the theory of 'craters of elevation'" (*Phil. Trans.* 1858: 703–786).

Another Fellow who settled in Italy was ANTON DOHRN (FRS 1899), a German zoologist who founded the Stazione Zoologica in Naples in 1872. This was the first independent international institution for research in marine biology. The physician HENRY HOLLAND (FRS 1815) visited Italy in 1815 and wrote a report on the production of Sulphate of Magnesia at the Monte della Guardia, near Genoa (*Phil. Trans.* 1816: 294–300). Unlike the papers of the 18th-century Fellows visiting Italy, his report is entirely focused on the production process of this salt and free of opinions and personal comments.

Finally, a couple that travelled through Italy in the *Grand Tour* style, were WILLIAM (FRS 1817) and MARY SOMERVILLE. None of their travels is reported about in the PTRS, as there was no longer space for this kind of narrative in the journal; however, as "scientists" – WILLIAM was a physician[176] and MARY a science writer with interests in mathematics and astronomy – their trip of the Italian Peninsula also had scientific objectives among others. They stopped in Milan, Bassano, Venice, Padua, Bologna and Florence, although here, while they were "hospitably entertained", their hosts were rarely scientific (Patterson 1983: 28). They met the botanist ALBERTO PAROLINI who showed them his "very pretty botanical garden" in Bassano, and GIOVANNI BATTISTA AMICI,[177] astronomer and botanist, in Modena. AMICI showed them his instruments, among which was his new catatropic microscope, which later made him famous as an optician and microscopic botanist. Interestingly, SOMERVILLE also notes that AMICI's wife read and spoke English; and she herself studied Italian during her 10-month stay in the Peninsula. Their trip then continued to Rome and Naples. Patterson (1983) reports that their letters mention the observation of natural phenomena such as volcanic eruptions, meteorological oddities and new plants and animals, and that they visited galleries in Rome and made the tour of Pompeii, but reported little about technological processes or experimental science. Due to WILLIAM's poor health they returned to Italy in 1838[178] staying in Rome, Florence, Genoa, Lake Como and Siena. Here MARY continued to cultivate her scientific interests

---

176 Physicians were not considered scientists and one of the main points of criticism towards the Society in the 1820s–1830s was precisely the presence of too many medical men among its members.

177 AMICI's work was also known to the Society and he correspondend with JOHN HERSCHEL.

178 They would remain in Italy until their deaths in 1860 (WILLIAM) and 1872 (MARY).

and enjoyed the privilege that the Grand Duke of Tuscany, LEOPOLD II (FRS 1838) had granted her to borrow books from his private gallery. And when away from Florence, their life-long friend AMICI continued to furnish her with scientific materials and instruments. However, SOMERVILLE gradually felt that she was losing touch with the state of science in England for she and her husband had "completely adopted the dolce far niente of this country; life is a pleasure in this heavenly climate without seeking anything more" (SOMERVILLE quoted in Patterson 1983: 193).

3.3.3 Topics

Of the 52 papers collected for the 19th-century part of the corpus, eleven concerned medical topics.[179] Four of the medical papers were English translations of 18th-century papers published originally in Latin and concerned the inoculation of smallpox and a disease that affected cattle in various parts of Italy early in the 18th century. The remaining medical papers included anatomical observations on the prostate gland and on the Malpighian bodies; and various reports of single and multiple autopsies, one of which was focused on the study of cholera, which had reappeared in Italy in 1886.

Another thirteen papers concerned animal electricity, the majority of which were by MATTEUCCI. Most of MATTEUCCI's papers formed a series of "memoirs" on his "electro-physiological researches". His and others' papers in the field show that a whole new set of specialised terms related to GALVANI's name, but independent of him, had by now become part of the jargon of this field of study; these included names of instruments such as the *galvanoscopic frog* – a very sensitive voltmeter used to detect weak electric currents; it consisted of a frog's leg with electrical connections to its nerve –; *galvanometer*; and *galvanism*, which comes to be used in the papers as meaning treatment through electrocution. The term *galvanism* was accompanied by verbs such as *apply* (apply galvanism), *employ*, *use*, and *react to* (see for instance *Phil. Trans.* 1817: 22–31).

Notwithstanding the higher presence of astronomers in the list of Italian Foreign Members, only five papers concerned astronomy. Four of these papers were related to ANGELO SECCHI: one concerned his drawings of a spot on the moon called "Copernicus", which was highly appreciated by the Society, and was followed by another paper with SECCHI's explanations of the drawing; another was a comparison between SECCHI's and WARREN DE LA RUE's photographs of a

---

[179] In actual fact, there are fourteen, but three papers were published both in full form in the *PTRS* and in abstract form in the *Proceedings*.

lunar eclipse; and the last was a paragraph-long report of SECCHI's description of the Observatory of the Collegio Romano.

Seven papers concerned physics (three) and electricity (four). Only two of these papers were written by Italians, VOLTA and MATTEUCCI, while the remaining were intertextually related to Italian research (MORICHINI's, GALVANI's, and NOBILI's). The paper by VOLTA represented his famous communication of the invention of the voltaic pile (*Phil. Trans.* 1800: 403–431, see also Section 4.2.3).

Four "papers" concerned meteorology, although three of these turned out to be only the titles of papers by CARLINI that were read at the meetings and given notice of in the *Proceedings*, with no comments as to their reception.

Three papers concerned the fields of volcanology and seismology. These included the abstract of MALLET's report of the *Great Neapolitan Earthquake* of 1857; LYELL's paper on the structure of lavas; and a rediscovered 16th-century letter, which reported about an eruption of Etna in 1536 (*Proceedings* 1858 316). LYELL's and MALLET's papers show that this field of study was much more advanced in the 19th century; indeed, not only was the study of earthquakes and volcanos much more "scientific", but also Etna and Vesuvius had by now become a means to an end – namely the advancement of geology – rather than being only a matter of interest in themselves.

Finally, two papers concerned mathematics, one of which was written by FRANCESCO BRIOSCHI and the other was a report of an 18th-century paper by GRANDI. The remaining are individual papers in geography, chemistry and zoology.

3.4 Language of the papers and translation practice

Before moving on to the critical linguistic analysis of the papers (Chapter 4) the present section looks in detail at the languages used in the Fellows' correspondence and published papers, which are revealing of changing communicative practices both at the editorial level and among the international scientific community.

3.4.1 In the 17th century

Most of the 102 papers collected for the 17th century are written in English. Only fourteen are written in Latin, four partly in Latin and partly in English,[180]

---

180 *Phil. Trans.* 1668:693–698, for instance, is an English report of a controversy, but the arguments of one of the Italians involved are reported in the original Latin. *Phil. Trans.* 1670: 2093–2095 has introduction and conclusion in English while the central body of the paper has been left in the author's original Latin.

and one in French. Among the remaining 83 English papers however, some distinctions ought to be made. In this group we find:

- papers written in English directly, for example by English travellers in Italy;[181]
- papers that have been written in English based on some Italian's account or publication;[182]
- and finally papers originally written in Italian, Latin or French that have been translated into English.[183]

It was not always possible to determine with certainty whether a paper was a translation or not.[184] In some cases this kind of information was explicitly reported in the title, with expressions such as the following: "out of an Italian printed paper, English'd as follows", "English'd out of the *Journal de Scavans*", "Extracted Out of the Ninth Italian *Giornale de Letterati*; And English't as follows", "English'd out of the French" etc. In other cases, the translated nature of a paper was made apparent by the use of 1st-person pronouns, place deixis and action verbs that showed that what was being written took place in Italy and was written by the original Italian author. Finally, in a few cases, the writer reports the information but then seems to (implicitly) quote some parts of the original wording in translation. The number of papers that are explicitly marked as being translations or appear to be translations amount to about 26 papers; while the remaining are mostly reported.

The actual writers or translators of the sampled 17th-century *PTRS* papers are generally anonymous. The tendency is to put the Italian source of information and the addressee of their letter, but not the person who dealt with the translation or the writing of the paper for publication. In any case, at least until 1677, the translator of most papers was generally OLDENBURG himself.

---

181 See, for instance, "Mr. Francis Vernons Letter, Written to the Publisher Januar. 10th. 1675/6 giving a Short Account of Some of his Observations in His Travels from Venice Through Istria, Dalmatia, Greece, and the Archipelago, to Smyrna, Where This Letter Was Written" (*Phil. Trans.* 1676: 575–582).
182 E.g. "An Account of Some Discoveries Concerning the Brain, and the Tongue, Made by Signior Malpighi, Professor of Physick in Sicily" (*Phil. Trans.* 1666: 491–492).
183 E.g. "An Observation and Experiment Concerning a Mineral Balsom, Found in a Mine of Italy by Signior Marc-Antonio Castagna; Inserted in the 7th. Giornale Veneto de Letterati of June 22. 1671, and Thence English'd as Follows" (*Phil. Trans.* 1671: 3059).
184 Due to the high amount of papers collected and the limited time available, it was not possible to view all the original manuscripts from which each article came from. Only a selection of papers were compared with their source texts.

The papers in Latin are mostly astronomy papers[185] and about a third of all the astronomy papers are in this language. According to Henderson (2013: 108), while there is evidence in the bureaucratic archival material of requests for translations from foreign vernacular languages, there appear to be no such requests for material written in Latin, which would suggest that the Fellows felt comfortable with this language. Further, it is generally understood that Latin, as the international language of learning, started being purposely published in the *Transactions* to allow continental readers, who did not know English, to read *PTRS* papers. As stated then, some of the sampled astronomical papers, but not all, have been left in, or translated into, Latin.[186] One possible reason for the presence of these select Latin papers could be that most of these papers are the fruit of the combined work of scientists from different countries[187] or, put together, represent discussions between astronomers; and therefore needed to be in Latin in order to be understood by all the parties involved.[188] A possible corroboration of this could be the following statement made by OLDENBURG in the title of one of the Latin papers:

> Some Considerations [...] Here delivered in the Latine Tongue, Wherein they Were Written by the Author, as Chiefly Regarding the Learn'd in Astronomy [...] (*Phil. Trans.* 1670:1168–1175).

OLDENBURG here would seem to suggest that the paper was willingly left in Latin as it would have principally been of interest to astronomers. The statement thus presupposes that astronomers – generally educated gentlemen – were skilled in this language and it would seem to entail that people who were not astronomers might not have understood Latin, hence, possibly justifying the absence of an English translation. The publishing of papers in Latin, while it opened the Society to continental natural philosophers, was on the other hand limiting for English philosophers.[189]

---

185 They amount to nine papers. Of the remaining, three are medicine papers, one is on mathematics, and one is a list of books.
186 Indeed OLDENBURG did not only translate Latin papers into English but often translated papers into Latin for publication in the *Phil. Trans.* He moreover tried to discourage readers from reading unofficial Latin translations of the *Transactions*.
187 E.g.: "Mr. Flamsteads, Mr. Townlyes, Mr. Haltons, Signor Cassini's and Monsieur Hevelius's, Observations of the Late Eclipse of the Sun" (*Phil. Trans.* 1676: 662–667).
188 See Ortore 2022 for further observations on the continued use of Latin in the field of astronomy.
189 Henderson (2013: 108) makes a similar point when she writes:
> Oldenburg's prefacing of Latin letters in *Philosophical Transactions* with comments such as "This...we shall give the Reader in the same Language and Words, in which

3.4.1.1 English'd out of the Giornale de' Letterati

A number of papers published in the *Transactions* were extracted out of different Italian journals going by the name of *Giornale de' Letterati (GdL)*. The first of these journals was founded by FRANCESCO NAZZARI in Rome in 1668 and continued until 1679. It was based on the model of the *Philosophical Transactions* and the *Journal de Savants*, publishing mainly about natural philosophical matters. Moreover, NAZARI wrote to OLDENBURG asking him to correspond and regularly translated and published in the *GdL* papers from the *Transactions* (Gomez Lopez 1997). A second Roman *GdL* was published by GIOVANNI GIUSTINO CIAMPINI (1675–1683), then one in Parma (1680–90), in Modena (1692–95), in Venice – first as *Giornale Veneto de' Letterati* (1671–1680 and 1687–1690) and later with the title *Giornale de' Letterati d'Italia* (1710–1740) –, Florence (1742–45), and Pisa (1771–96).[190] The Royal Society took its Italian papers mainly from the Roman and Venetian journals.

3.4.2 In the 18th century

In the 18th century some novelties were observed regarding language and translation practices. Firstly, more evidence was found of Italians learning English inspired by the scientific achievements of Britain and the Royal Society. Further, an increasing number of British scientific works were being translated into Italian; for instance, NEWTON's *Principia* (1687) were translated into Italian by TOMMASO NARDUCCI (ca. 1722–1726) (Mazzotti 2013); STEPHEN HALES' *Vegetable Staticks* (1727) was translated by MARIA LUISA ARDINGHELLI in 1756 (Cavazza 2002); while CELESTINO GALIANI had had one of LOCKE's essays translated into Italian but it was banned by the inquisition (LBO/21/134).[191]

---

the Author of it desired, it might be inserted in this Tract" explains the impetus for his changed stance of Latin (allowing particular Fellows to carry on controversies with continental philosophers), but also suggests an unwillingness on his part to move away from his original intention of presenting the business of the Society for a wider English audience. At the same time he was aware of the need for an authorized Latin translation for the benefit of continental audiences, who were not generally able to read English.

190 On the *Giornale de' Letterati* see Generali (2016 and 2012), Sabba (2018), and Dooley (1991).
191 Hall (1991: 138) points out how, paradoxically, works on natural religion – such as DEREHAM's *Astro-theology* (1714 and 1728 in Italian) and GEORGE CHEYNE's *Philosophical Principals of Natural Religion* (1705 and 1729 in Italian) – were permitted to pass all censorship, while works on natural philosophy were still restricted.

News of the *PTRS* being received by Fellows in diverse parts of Italy was also more frequently found in the letter exchanges. For instance, the *PTRS* were sent to DEREHAM in Rome, to HAMILTON in the Kingdom of Naples, and to the Institute of Bologna. DEREHAM moreover translated into Italian and published a few issues of the *PTRS*, spreading them among his Italian acquaintances.[192] Fewer comments were also being made regarding post issues, although occasional losses of letters still occurred and natural philosophers would have to make sure not to lose the opportunity of sending material whenever a ship was leaving for England, as these were not very frequent.

As to the corpus of Italian-research based *PTRS* papers, in this century the names of translators start appearing in the publications. Some of the translators were the secretaries themselves, who were generally appointed – other than for their scientific achievements – for their linguistic skills. In other cases, the translators were Fellows who knew foreign languages. Hence, to provide some examples, ROBERT WATSON (FRS 1751) and WILLIAM WATSON (FRS 1741) translated papers from Italian and French; THOMAS STACK (FRS 1738) and MATTHEW MATY (FRS 1751, For. Sec. 1762–1766; Sec. 1765–1776) from French; and JOSEPH CASPAR SCHEUCHZER (FRS 1721 and For. Sec. 1728) from Italian.

There were no explicit statements made of papers being translated from Latin, although translated extracts and papers from this language are present in this century. Interestingly, one of MICHELOTTI's papers, an "Account of an uncommon cure for violent vomiting of blood" (*Phil. Trans.* 1731: 129–145), had been translated from Latin into English (LBO/19/171); however, the paper was then published in the *PTRS* in the original Latin without the English translation. The translation was read at one of the meetings, and this is probably the reason why the paper was translated into English, i.e. to allow all the Fellows present at the meeting to understand it.

Another point to be noticed is that there is a slightly higher number of papers in French or translated from this language, which reflects an increase in the use of French as a lingua franca in the epistolary exchanges. Also Italian letters appear more frequently; while Latin, although still present, was not as frequent in the letters as in the previous century. This shows how the use of vernacular languages was gradually replacing the use of Latin in international communication; yet, as of the 18th century there were still many supporters of Latin. An example is shown by the numismatist JOHN SWINTON, who in one of his *PTRS* papers

---

192 Dereham (1734); see also Gibelin (1793).

provides a cover letter in English – for the Fellows – but chooses to write his paper in Latin, based on the following reasons:

> With regard to the language of the paper in which my remarks are contained, I shall only beg leave to hint, that it is understood by all who are proper judges of the performance. For this therefore I shall offer no other apology, than that the letter from Sig. Abate Venuti to Mr. Nixon, which occasioned it, is penned in the Latin tongue; and that the famous F. Corsini, the removal of whose doubt or suspicion was one of the principal objects I had in view, writes for the most part at least in the same language. I might however add, that many learned foreigners, who are particularly pleased with such disquisitions, are much better acquainted with Latin than any other tongue , except their own; and that it were to be wished the use of this noble language, in the republic of letters, were more general than it at present seems to be. (*Phil. Trans.* 1759: 682)

Here SWINTON seems to have felt obliged to account for his use of Latin over English and explains that Latin was used in his paper: (1) because the papers with which his own was intertextually related were written in Latin;[193] and (2) because learned foreigners still preferred the use of Latin over vernacular languages other than their own. Finally, he comments that it would be desirable that the use of Latin within the Republic of letters were more widespread than it currently was.

Returning to the papers, Tab. 3.1 below provides a general picture of the 18th-century published papers from a linguistic and translational point of view. The third column reports the results of the 17th century for comparison. It should be reminded, that during the linguistic analysis a distinction was made between

---

193 The ones published in the *Transactions* had however been translated into English for publication but SWINTON provides an example from the German *Acta Eruditorum*, in which the papers related to his were left in Latin, rather than being translated into German.

Tab. 3.1: Languages used and translation choices in 18th- and 17th-century *PTRS* papers

		18th century	17th century
		Total number and percentage within the group of 185 papers	**Total number and percentage out of the group of 102 papers**
Papers written in English:		89 (48.1 %)	57 (55.8 %)
Papers in Latin:		38 (20.5 %)	14 (13.7 %)
Papers in French:		2 (1 %)	1 (0.9 %)
Papers in English and Latin:		10 (5.4 %)	4 (3.9 %)
Total of papers translated into English:		46 (24.8 %)	26 (25.4 %)
	Of which:		
	translated from Italian	11	8
	translated from French	5	3
	translated from Latin	2	1
	translated with no indication of source language	23	14
	published in Italian with an English translation in appendix	5	0

translated papers, i.e. papers representing the original author's use of language, and reported papers, i.e. papers representing the English (or other nationality) Fellows' language use and point of view.

While in the previous century it was often not stated whether a paper was a translation or not, and often papers consisted of a miscellany of translated-quoted sentences from the original writing intermixed with editorial summary and commentary, in this century such miscellany was never observed; it was always clear – even when it was not explicitly stated – when the paper was translated and when it was a report of an Italian piece of research.

One result that immediately stands out is that there is an increase in the percentage of papers published in Latin in the 18th century. This is neither surprising nor contradictory with what was previously stated as to the increase in the use of French and a reduction in the use of Latin in the epistolary exchanges. In fact, the initial plan was to publish the *Transactions* in English only; however, as was explained above, the publishing of papers in Latin would make them more widely accessible; hence, starting from the 17th century some papers were left in Latin or even translated into this language, which explains why there is a gradual increase in the presence of Latin.

The ten papers both in Latin and English consisted of one of the following: (1) papers written in English with Latin extracts; (2) papers written in Latin with an English introduction and conclusion; (3) Latin papers with a cover letter in English; or (4) a collection of letters with some written in English and others in Latin.

In this century there are also five papers which have been published in the original Italian but have been provided with an English translation in appendix. These papers were all published in the 1780s, and therefore the publication in both languages must have been desired by the same Committee of Papers.

As to the 89 papers written in English, like in the previous century, they were either reports of Italian scientific news and research, or accounts written by Englishmen in Italy; or experimental reports and observations made following previous Italian experiments and findings.

Finally, fewer papers compared to the previous century were taken from other journals: only two papers were taken out of the Venetian *Giornale de' Letterati*; one from the German *Acta Eruditorum*; and none appear to have been taken out of French journals. However, Italian periodicals, books and proceedings of Italian academies were still being sent to the Society.[194] For further notes on translation practice in the 18th-century *PTRS* see also the following chapter.

3.4.3 In the 19th century

Further noticeable changes occur in the 19th century regarding the language used in the *PTRS*, letter exchanges and the practice of translation. Firstly, by now, Latin seems to have been completely abandoned as a lingua franca in the Italian-English correspondence. Italians generally wrote to the Society in French, and the Fellows replied in the same language. Many papers too were written in French and then translated into English, although seven papers were left in the original language.

Secondly, Latin was no longer present in the papers apart from one, which was a reproduction of a 16th-century Latin letter and was framed by an English introduction.

---

194 For instance, in the archives there is a paper by PINELLI that was published in the *Giornale de' Letterati d'Italia* "'An account of an abnormally-situated foetus from Giornale de Litterati d'Italia' by Flaminio Pinelli" (L&P/1/408); this however does not appear to have been published.

Thirdly, unlike the previous century, titles neither report the name of the translator, nor the translated nature of the text. Translated papers amount to a total of fourteen (26.9 %), while the papers originally written in English are 30 (56.6 %).

In the previous centuries most of the papers originally written in English were reports of Italian findings; in the 19th century instead, most of these papers were included in the corpus because intertextually related to Italian research (twelve papers). Only seven papers were reports of Italian writings; five were

Tab. 3.2: Languages used and translation choices in 19th-, 18th- and 17th-century *PTRS* papers

	19th century	18th century	17th century
	Total number and percentage within the group of 52 papers	Total number and percentage within the group of 185 papers	Total number and percentage out of the group of 102 papers
Papers written in English:	30 (57.6 %)	89 (48.1 %)	57 (55.8 %)
Papers in Latin:	-	38 (20.5 %)	14 (13.7 %)
Papers in French:	7 (13.4 %)	2 (1 %)	1 (0.9 %)
Papers in English and Latin:	1 (1.9 %)	10 (5.4 %)	4 (3.9 %)
Total of papers translated into English:	14 (26.9 %)	46 (24.8 %)	26 (25.4 %)

based on the Fellows' personal experiences in Italy, and six were written directly in English by the Italian physician GRANVILLE. Tab. 3.2 below summarises the 19th-century language presence in the *PTRS* and provides the results from the previous centuries for comparison:

As can be seen from the table the percentages of translated papers and papers written directly in English do not vary considerably across the three centuries; there is an average of 53.8 % of papers written directly in English and an average of 25.7 % of Italian papers that were translated into English. Latin papers instead rose in the 18th century but suddenly dropped in the 19th; while French papers increased through the centuries with a considerable rise in the 19th century.

Language of the papers and translation practice   147

The numbers for this century also include 25 papers published in the *Proceedings*. This journal started out in 1800 as the *Abstracts of the Papers Printed in the Philosophical Transactions of the Royal Society of London*, which published paper abstracts starting retrospectively from the 18th century – hence the presence in the corpus of four reprinted (and translated) 18th-century papers.[195] In some cases, the *Proceedings* only reported the title of papers that were read at the Society, as was the case with CARLINI's papers, which do not appear to have then been published in the *PTRS*. The *Proceedings* later developed into a journal in its own right, extending its contents to obituaries and shorter articles. Article length eventually became the main criterion to decide whether to publish a paper in the *Proceedings* or in the *Transactions*. As a result, while some papers were published in both journals – e.g. GRANVILLE's papers were three but total six in the corpus because both the abstracts published in the *Proceedings* and the full papers published in *Transactions* were included – other shorter papers were only published in the *Proceedings*. Finally, no papers in the corpus for this century appear to have been taken out of foreign journals.

---

195 The Royal Society published four volumes, from 1800 to 1843. Volumes 5 and 6, which appeared from 1843 to 1854, were called *Abstracts of the Papers Communicated to the Royal Society of London*. Starting with volume 7, in 1854, the *Proceedings* first appeared under the name *Proceedings of the Royal Society of London*. Publication of the *Proceedings* in this form continued to volume 75 in 1905 (https://royalsocietypub lishing.org/rspl/about).

Chapter 4 Discourse features

4.1 17th century 150
    4.1.1 Textual dimension 151
        4.1.1.1 Macrostructural features 151
        4.1.1.2 Language use 152
    4.1.2 Discursive practice 158
        4.1.2.1 Discourse representation 158
        4.1.2.2 Meeting minutes and letter exchanges 161
        4.1.2.3 Evaluation in the discourse of the *PTRS* 163
    4.1.3 Reporting disputes and disagreements 168
    4.1.4 Witnessing 173
    4.1.5 Toponymy 174
    4.1.6 Interdiscursivity and intertextuality 176
        4.1.6.1 Dialogicity in the discourse on *amianthus* 177
4.2 18th century 181
    4.2.1 Textual dimension 182
        4.2.1.1 Macrostructural features 182
        4.2.1.2 Language use 184
    4.2.2 Discursive practice 188
        4.2.2.1 Discourse representation 188
        4.2.2.2 Evaluation in the discourse of the *PTRS* 190
        4.2.2.3 Original, translation and publication: A brief comparison 193
    4.2.3 Reporting disputes and disagreements 195
    4.2.4 Witnessing 202
    4.2.5 Toponymy 206
    4.2.6 Interdiscursivity and intertextuality 208
4.3 19th century 210
    4.3.1 Textual dimension 210
        4.3.1.1 Macrostructural features 210
        4.3.1.2 Language use 211
    4.3.2 Discursive practice 214
        4.3.2.1 Discourse representation 214
        4.3.2.2 Evaluation in the discourse of the *PTRS* 216
    4.3.3 Witnessing and toponymy 216
    4.3.4 Interdiscursivity and intertextuality 218

4.1 17th century

> *I earnestly beg you to be content in future that our correspondence should be conducted without preliminary compliments, because being unwilling to mar with them that other part of good philosophy which is sincerity of heart, I commonly eschew adorning it with forms of speech, which are so often associated with lies.*
> (MAGALOTTI to OLDENBURG, 1667, Letter 863, Hall & Hall 1967: 411)

In this chapter, results regarding the discourse analysis carried out on the sampled *PTRS* papers will be presented. The following section (Section 4.1.1) focuses on the *textual dimension* of the corpus for the 17th century; specifically on the macrostructure and on the linguistic features which characterise the papers. The section on *discursive practice* (Section 4.1.2) focuses on Italian discourse representation and discourse evaluation drawing on complementary sources for a critical comparison with the *PTRS* papers. Witnessing and the meticulous naming of place names are then discussed as a distinctive feature of the 17th-century papers (Sections 4.1.4 and 4.1.5). Finally, the intertextual features of the papers are described and exemplified in Section 4.1.6.

Before continuing onto the next section one point ought to be reminded; namely, that the discourse analysis distinguishes between the different types of texts present in the corpus, i.e. translated papers and reporting papers.[196] Translated papers represent papers that were originally written by Italians (in Latin, Italian or French), and were translated into English. Reporting papers are instead papers written directly in English, which report about a particular topic related to Italians and Italian research. Since the different nature of these papers can influence the writing style, the two text types have been treated separately. In the first case, the writing style may be that of the original Italian author, only translated into another language; while in the case of reporting papers, the style will be that of the English (or other nationality) writer. Further, the translated papers display the Italians' opinions (where present), while the reporting papers generally display those of the reporter.

---

[196] French and Latin papers were not included in the linguistic analysis.

4.1.1 Textual dimension

The present section analyses the textual dimension of the sampled corpus for the 17th century. Textual dimension corresponds to the first level of Fairclough's (1992) approach to text analysis and focuses on the linguistic components of the text. Features analysed were text type and structural components, at the macrostructural level; and author positioning, level of informativity, abstractedness and narrativity (Bazerman 1988; Biber 1988; Atkinson 1992 and 1999) at the microstructural level (see Sections 2.1.1 and 2.1.2).

4.1.1.1 Macrostructural features

Starting from the overall macrostructure of the texts, it was observed that 38 out of 102 (37.2 %) papers were published as letters or extracts of letters (29 out of 38). The letter form is representative of the *Philosophical Transactions*' inception, in that the founding of the journal came from the Royal Society and OLDENBURG's project to publish their doings and enquiries regularly. And, since OLDENBURG was the editor of the journal, the published material came from his extensive private (or semi-private) correspondence. Letters could either be prefaced with a short introduction by the editor or be directly published in their full or abridged version.

Most papers were written in the form of: observations; reports of phenomena, of experiments and of personal experiences; news reports; replies to other papers; and the above mentioned book reviews/accounts/lists. However, it is not possible at such an early stage to identify clear-cut genres characterised by specific regularly recurring discourse features.

Titles tended to be long, self-explanatory and often pointed out who the source of the information reported was, as in the following example:

> An extract of a letter, written March 5. 1672 by Dr. Thomas Cornelio, a Neapolitan Philosopher and Physician, to John Dodington Esquire, his Majesties Resident at Venice; concerning some observations made of persons pretending to be stung by tarantula's: English'd out of the Italian. (*Phil. Trans.* 1672: 4066–40–67)

This heading is quite representative of most titles of this period and, as can be seen, it indicates the epistolary nature of the piece of writing, the Italian source of information, the English addressee, the topic of the paper, and, less common, an indication of its translated status with the source language. However, not all headings were this specific; some were quite short, indicating only the content of the piece, such as "Some Observations on Vipers" or "A Way of Preserving Ice and Snow by Chaffe" (*Phil. Trans.* 1665: 160–162 and 139–140). In these cases, the Italian source of information, or the Italian person related to the researches being reported, was often mentioned within the paper (FRANCESCO REDI and ROBERT BALLE in the two examples).

Lengthwise papers tended to be short, that is not exceeding four pages in 65 cases (64.3 %). A few papers were just a paragraph long, as in the case of "An Observation of Optick Glasses, Made of Rock-Crystal" (*Phil. Trans.* 1665: 332) and 33 papers were between four and fourteen pages (32.6 %). Only three papers were longer than fourteen pages (2.9 %).

Finally, both translated and reporting papers often contain a brief introduction written by the publisher. Sometimes the introduction is followed by footnotes with references to other papers and/or short comments. Moreover, the publisher often intruded in the translated texts with quick reminders in brackets informing the reader that the person speaking was the original author.

4.1.1.2 Language use

As far as author positioning is concerned, the sampled 17th-century papers often display authorial presence. Out of 102 papers, 41 (40.1 %) were marked as being characterised by an involved author-centred style. Half of the author-centred papers are translations and are therefore actually displaying the Italian authors' presence within the discursive event. The most frequently occurring features showing the author's presence were first-person pronouns, active-tense verbs and private verbs. See for instance the following examples from two translated papers and a reported text; features marking authorial presence are highlighted in bold:

> As to the *Hypothesis* of *Georg. Domenico Cassini,* touching the motion of the *Comet* about the Great Dog in the Circle, whose Centre is in a streight line drawn from the Earth thorough the said Star, **I believe** it will shortly be publish'd in print, as **a thought I lighted upon** in **discoursing with one of my Friends,** who did maintain, that it turned about a Centre, because that its *Perigee* had been over against the *Great Dog,* as **I had noted** in **my** *Ephemerides.* This particular **I did** long since **declare** to many of **my** acquaintance, whereof some or other will certainly **do me** that right, as to let the world know it by the Press. (*Phil. Tran.* 1665: 18–20, translated)

This paper is most likely a translation from ADRIEN AUZOUT's French or Latin. It features first-person personal pronouns; the use of private verbs (*believe* and *light upon a thought*) showing the author's psychological processes; active tense; and displays of the author's personal relations. The paper is in letter form, which, by definition, is characterised by an involved style; however, author-centredness was also observed in articles that were not in this form:

I was told, that they were produced from the dews, because (**forsooth!**) being gather'd over night, the next morning there would be found others at such a time only, when it was a serene and dewy sky; and that upon the herbs of the meddow, and without the bounds of those bare and sterile places never any Crystals were to be found; besides, that the ground having been in some places bared of all greens, and reduced to the condition of those other naked places, yet no crystals were ever seen to have been form'd there. But I, when **I had examined**, that in the neighbour-hood of that hill there was no mark at all of any Mines, **did conclude**, that it **might** be a plenty of nitrous steames, which might withal hinder vegetation in those places, and coagulate the Dew falling thereon. (*Phil. Trans.* 1672: 4068–4069, translated)

This second extract is a translation of a paper that was published in the *Giornale de' Letterati* of Venice. Like the previous letter it makes use of first personal pronouns; active-tense verbs (*conclude, examine*); modal verbs (such as *might* in the extract), which show the author's (FRANCESCO LANA's) stance towards what is being presented; and the use of the adverb *forsooth* accompanied by exclamation mark, which displays irony on the author's part.[197] The following extract instead comes from a reporting paper, thus showing the English author's involvedness in what is being written:

Since **my** last concerning the *Bridge at Pont St. Esprit*, **I have found** something more in that kind for **your** further diversion; **I doe not doubt** but **you** still retain a fresh *Idea* of the stately ruins of the *Modern Bridge at Avignon*, which hath yielded in many places to the extreme rapidity, and violence of the *Rhosne*. Its fall **in my opinion may** be ascrib'd to three defects; first, it was not so *Multangular*, as that at *St. Esprit*: Secondly, it wanted in 3 or 4 places, the little Arches dividing the feet of the great ones, and in those parts it hath suffer'd most; for where those useful sluces are, there **I observ'd** the bridge to stand still the most entire. Thirdly, the *Pedestals* (or as **you** very properly **call** them, *Horizontal Arches*) were not so *Geometrically* and exactly laid, as those of *Pont St. Esprit*, their jettings out were few, and they not gradually contracted, for that the force of the steam must be the greater upon the Fabrick. (*Phil. Trans.* 1684: 712–713, reported)

Other than the above-mentioned features of involvement, in this extract the dialogic nature of the letter form is visible through the use of second-person personal pronouns. Moreover, the three extracts, written by three people of different nationalities, suggest that placing the author at the centre of what was being done, was common use throughout Europe and not just in England or Italy.

---

197 JOHNSON (1755): "Forsoo'th. adv. […] I. In truth; certainly; very well. It is used almost always in an ironical or contemptuous sense […]".

Displays of modesty and humility, which are further features generally associated with an involved writing style were also present in a small number of papers. See for instance the following examples:

> Having been honour'd here with the place of Publick Anatomist of Venice, **though** I have given as yet but a **very slender** accompt of my performances, **in comparison of the illustrious example** of Mundinus, Veslingius, Molineta, &c. **yet** I shall acquaint you with some particulars that have occurred to me. (*Phil. Trans.* 1670: 1188–1189, translated)

> The said worthy person saith, that **although** he **did not at first intend any more than** to present his Glasses to some of the most famous Astronomers; **yet being earnestly sollicited** by his Friends from many parts, he offers to rate the price of them, according to what the most known Artists, such as *Campani* and *Divini,* have done [...]. (*Phil. Trans.* 1677:1005, reported)

In these extracts the original Italian authors (GIACOMO GRANDI and GIOVANNI ALFONSO BORELLI) are cautiously putting forward their piece of information by understating themselves and their work through the use of contrastive subordination (*though, although, yet*) and words that have a connotation of smallness (*slender* in contrast with the connotation of greatness suggested *illustrious* used with reference to past notable physicians; and *not...any more*). In both these cases, although the second is a reporting paper, the display of modesty is that of the original Italian authors. These were two out of three cases found in the *PTRS* corpus for the 17th century; however, original letters sent to the Society provide many more examples of these displays of modesty, suggesting that it was common practice among Italian natural philosophers. Simply, in most of the published papers editors cut out unnecessary formal conventions. The third case is by an English physician, TANCRED ROBINSON, who, through the use of hedging (distancing verb *presume* and modal *may*) and negation (*not unfit*), humbly suggests that the contents of his letter be published in the journal:

> Having lately had an Observation communicated to me by my Brother, which he made when he travelled through *Italy* last, differing from any I have hitherto met with in Natural Histories, **I presume you may think it not unfit** to be communicated to the Publick, and so give it a Place in the next *Philosophical Transactions*. (*Phil. Trans.* 1698:378–381, reported)

Humble expressions like the above exemplified were also frequently used by the editor in presenting the papers to the public:

> This came but lately to hand from that knowing person, Mr. *Henry Robinson;* and **was thought fit** to be now inserted here, that it might not be lost, though it hath happened above 30 years ago. It was contained in a Letter. (*Phil. Trans.* 1665:377)

From this Observation, which **was thought fit to be publisht** here, <u>Mathematicians are invited from time to time to make the like in their Countries, to see, whether in this change there be any regularity</u>. (*Phil. Trans.* 1670:1184–1187)

So far this Inquisitive Anatomist, which **the Publisher** (who very much doubteth, whether any Copies of this intimation, Printed at *Pisa* this very year, besides that one, which lately came to his hands, be found in *England*) **thought fit to insert** in these Papers, <u>thereby to administer occasion to our dextrous Anatomists here, with all possible diligence and care to pursue, joyntly with that *Italian* Professor, those important Inquiries about such considerable Subjects, as have been above related; comparing with their Researches in this matter the many notable Experiments, lately published in N°. 63. and 64.</u> of these Tracts, made and communicated by the Honourable *Robert Boyle*. (*Phil. Trans.* 1670: 2093–2095)

The Journalist having been informed, that Signor *Gieronymo Barbato*, publick Professor of Practical Physick at *Padua*, and Physitian in *Venica*, had written a Book upon that subject, and illustrated it with new Anatomical Diagrams, all ready for the Press; did, it seems, obtain the perusal of the Original Manuscript, and permission withal, to make an Extract thereof, which in this Journal he presents the Curious with, to stay their desire whil'st the whole Dissertation in printing. **This Breviate we thought fit to** *English* **here** out of the *Italian,* as followeth […]. (*Phil. Trans.* 1671: 2224–2227)

Notice also, in the first and third extracts, how the publisher strengthens the interest factor of the proposed papers by praising his sources ("from that knowing person", "this inquisitive anatomist"). Moreover, he directly addresses readers of the *Transactions* encouraging them to carry out further studies on these matters (second and third extracts, underlined) making what are today known as "calls for further research".

Only one case instead was found of the tendency towards miscellany that Atkinson (1999: 77) reports to be frequent among early *PTRS* papers. The case was a two-paragraph long musical digression amidst an experimental report on viper poison (*Phil. Trans.* 1672: 5060–5066).

Involved papers represent a majority in the 17th century, however, a rather high number of papers (24 papers, 23.7 %) have been marked as tending towards a more informational and non-author-centred style, in that they lacked the features that typically denote an involved and author-centred approach (first personal pronouns, private verbs, personal evaluations, amplifiers, emphatics etc.) but still made use of active tenses.

Although it is still too soon to speak of a tendency towards an object-centred and abstract approach, a few cases of high abstractedness were found. The papers that seemed to move away from the highly involved writing style belonged mainly to the fields of astronomy, physics, chemistry, medicine, and a few book accounts. This is itself interesting in that, apart from the book accounts, these fields – the so called *hard sciences* – are still characterised today by the highest level of abstractedness.[198] However, it is not yet possible to speak of a regular tendency towards an object-centred approach in the sampled papers for the 17th century. Below are two examples; one taken out of a reported paper, marked as having a more informational and abstract writing, and the other from a translated paper displaying a high level of abstractedness.

> The *other* is, that the said P. *Lana*, **having extracted** out of a Metallic Substance a very white Salt, the same was, upon the **application** of the gentlest heat, resolved into a **Golden-colour'd** liquor; which **being removed** from that warmth, as soon as it felt the cool Air, and even **by opening** the Glass wherein it was inclosed, did in a moment shoot afresh into the same Salt; and that (which seem'd oddest) whilst he was pouring it out of one glass into another during its fluidity, it was dispersed all over the glass it was poured into, suddenly congealing into most fine threds, many of which were extended from one side of the glass to the other, and, hanging as 't were in the Air, formed just like the subtilest Cob-webs, not at all rigid, but, by reason of their exquisite subtlety, pliable, and scarce perceivable by the Eye. (*Phil. Trans.* 1671: 3060, reported)

> **There was taken** a Glass-cane, about 1⅔ of a Fiorentin *braccia* or Ell, open at one end, of which above one Ell and a quarter **was fill'd** with Quick-silver, the rest with common water. This open end **was shut** with a finger, and **inverted** into a vessel with stagnant Mercury; then removing the finger, the Mercury began to fall out, for that the aggregat of the Quick-silver and water falling, the water remain'd in the upper-part of the inverted cane, now free from Air. This **being done**, the Cane **was thus exposed** to the open Air in the Month of *January*, in frosty weather, and in one night the water in it was congealed into Ice of a very good consistence. (*Phil. Trans.* 1671:2169–2170, translated)

Features displaying a tendency to being informational in writing are a higher use of nouns, nominalisations, compound-forms, prepositions, attributive adjectives, and conjuncts. An informational use of language lacks overt displays of the authors' presence and stance and is characterised by an overall nominal rather than verbal production. The first extract, although it still presents the writer's thought at some points, avoids, to a certain extent, the use of active constructions, personal pronouns and verbs of emotion, substituting them with

---

198 Atkinson (1992) points out similar findings as far as medical papers in the *Edinburgh Medical Journal* are concerned; namely that there was always a discrete level of informational production. Texts developed from a moderate narrative level to a progressively less narrative one. And as far as abstract information is concerned, no statistically significant pattern of variation is recorded, in fact, this type of article has been generally abstract and technical throughout the journal's existence.

participles (*having extracted, being removed*), nominalisations (*application*), and attributive adjectives (*golden-coloured*). Papers marked as abstract, instead, were written especially through the use of passives, which allow the object of study to be put in subject position. Thus, the focus is on the object and on the action rather than on the author of the paper who becomes distanced or effaced. Linguistic features marking abstract production are highlighted in bold type in the second extract above.

The absence of features of involvement is probably related to another aspect that stands out among the group of informational papers; that is, most of the informational papers are reporting papers (19/24). Writers and editors therefore often reported what was being said or done in Italy trying to avoid expressing personal judgement and being as neutral as possible (see Section 4.1.2.1). It will be seen later however, that this was not always the case (Section 4.1.2).

Finally, 20 papers (19.6 %) were marked as being midway between an involved production of language and an informational one. These papers did not display features of author-centredness but did not tend towards an informational production either, making extensive use of verbal language. In other cases, "midway" papers had sections that tended to be involved and other sections that were more informational. See, for instance, the following extract:

> These Observations we shall summarily present the Curious in these parts with, as they were lately presented (by Letter from his Excellency the Ambassadour of *Venice*, now residing at the Court of *France*) to the *Royal Society*, in some printed sheets of Paper, entituled, *MARTIS, circa Axem proprium Revolubilis, Observationes, BONONIÆ à JO. DOMINICO CASSINO habita;* come to hand *June 3. 1666.*
> In these Papers the Excellent *Cassini* affirms;
>
> 1. That with a *Telescope* of 24. *Palmes,* or of about 16 *Feet,* wrought after S. *Campani's* way, he began to observe *February 6. 1666* (st-n.) in the morning, and saw two dark Spots in the *first* Face of *Mars.*
> 2. That with the same Glass he observ'd *Febr. [?],* in the Evening, in the *other* Face of the Planet, two other Spots, like those of the first, but bigger.
> 3. That afterwards continuing the Observations, he found the Spots of these two Faces to turn by little and little from *East* to *West,* and to return at last to the same situation, wherein he had seen them first.
> 4. That S. *Campani,* having also observ'd at *Rome* with Glasses of 50. *Palmes* or about 35 *Foot,* likewise of his own contrivance, had seen in the same Planet the same *Phenomena.* (*Phil. Trans.* 1665: 242–245, reported)

This is the beginning of a paper, which presents an involved introduction – with first-person plural personal pronouns, praise (*the excellent Cassini*), and active tenses – but then changes, in the bullet points, into a purely narrative and informational use of language free of features of involvement. This final point leads to another typical aspect of language use in the sampled 17th-century papers, narrativity.

More than half of the 17th-century sampled papers were characterised by a narrative style (55 papers, 54.4 %) featuring past or present tense, perfect aspect, public verbs and third-person personal pronouns. An example is provided in the following extract, where narrative features are in bold type:

> I. **He hath observed**, that the poyson of Vipers **is** neither in their *Teeth*, nor in their *Tayle*, nor in their *Gall*; but in the two *Vesicles* or *Bladders*, which **cover** their teeth, and which **coming to be** compressed, **when the Vipers bite, do emit** a certain yellowish Liquor, that **runs** along the teeth and poysons the wound. Whereof **he gives** this proof, that **he hath rub'd** the wounds of many Animals with the *Gall* of Vipers, and **pricked** them with their *Teeth*, and yet no considerable ill accident **follow'd** upon it, but that as often as **he rubbed** the wounds with the said yellow Liquor, not one of them **escaped**. (*Phil. Trans.* 1665:160–162, reported)

Features marking narrativity in the above example are the use of the third-person pronoun *he*, perfect aspect (*hath observed, hath rub'd*), historical present (*gives*) and past simple (*pricked, follow'd, rubbed, escaped*). Narrativity is generally found to be frequent in the early *Philosophical Transactions*, and this is especially true in the case of the many reporting papers sampled for the present study, which mostly deal with the work of a third Italian party, making a narrative and/or descriptive use of language necessary.

4.1.2 Discursive practice

The focus of this section is on discursive practice, the second level of Fairclough's approach to text analysis. Discursive practice concerns processes of text production, distribution and consumption. More specifically, this section analyses discourse representation and the intertextual relations between the 17th-century papers under study.

4.1.2.1 Discourse representation

Discourse representation refers to the explicit incorporating of other texts focusing on how discourses are represented within the discursive event under study. This category is particularly relevant to the present analysis, in that most of the sampled papers include some form of represented discourse, either by incorporating original extracts and full Italian papers or by reporting and adapting their contents. As far as the representation of Italian natural philosophical discourse is concerned this was found to be neutral in most cases.

17th century   159

Indeed, 69 (79 %) out of the 87 papers primarily written in English were marked as being neutral, that is, lacking any forms of judgement or bias towards what was being reported or translated. Neutral papers included both translated papers and reporting papers.

In the case of neutral translated papers, the main tendency was that of inserting the translation with no form of framing other than the title. This can be seen in the following extract, where the title is immediately followed by a translation of the Italian letter:

> An Extract of a Letter Written from Rome, dated the 16*th*. Of November last, to Signior Sarotti, concerning a Discovery made upon the Inundation of the Tevere. Translated out of the Italian.
> I Believe you have already heard how the Inundation of our River has done several considerable Mischiefs all about this City, spoiling several fine Houses, and very large Aqueducts, by breaking down their conducts, &c. It has in several Places, (especially without this City) by breaking the Ground, discovered Vaults unknown before, great part of them full of earthen Urnes, and Sepulchers, but of no great consideration, by the Inscriptions they had upon them [...]. (*Phil. Trans.* 1686: 227, translated)

In fewer cases instead, the translation was preceded by a short introductory paragraph. In these papers, the publishers either maintained a neutral stance towards what they were presenting or displayed a positive attitude towards the discursive event:

> Observations concerning the Comet that was seen in Brasil, An. 1668. in March, by P. Valentin Estancel *a Jesuit, and by him sent to* Rome; *where they were printed in the 9*th Italian Giornale de Letterati, Septemb. 31. 1673.
> This being the same Comet with that, taken Notice of in Numb. 35. of these Tracts, and our Account, then given of the same, being likely to receive from Advantage and Light from these Observations, made in *Brasil*, and but lately come to our Hands, we thought it would not be un-acceptable to the Curious in England, to see them English'd out of the *Italian*, and here inserted. They are as followeth;
> There hath not been a Phenomenon this great while, of which, as of this, I am now going to describe, we have had Observations from all Parts of the World. Those from Europe and Asia may be seen in our 3d and 4*th* *Giornale*. (*Phil. Trans.* 1674: 91–93, translated with introduction)

In the above extract the first paragraph represents the editor's introduction, who humbly expresses his desire to publish the paper in question given its intertextual relation to another *PTRS* paper on the same subject. Overall, about 18 % of the papers (16 out of 87) positively represented Italian discourse:

> Remarks concerning factitious Salts; drawn from a Discourse written by Sen. Francisco Redi.

160   Discourse features

> The Happy Genius of the Cardinal *de Medici* favouring and promoting Mathematical and Philosophical Studies, as well as others, makes him among his most weighty Affairs, not pass by such things as may serve the Virtuoso's as well for Private as Publick Advantage, hence it is that Seignior *Francis Redi* has been induced to collect divers Writings and Observations made some Years past in *Florence*, about Vegetable Salts; which being not ready to be published, you will here receive an Extract of them, for the Satisfaction of the Curious, and the Improvement of Natural Knowledge, being hereby conducted into the Manner of extracting the Salts, their Quantity and Different Figures, as likewise their Virtue and Purging Quality. (*Phil. Trans.* 1698: 281–289, translated with introduction)

In translated papers the positive attitude on the Society's part was shown in the introduction as can be seen from the above extract which represents the introduction to a paper; while in reporting papers positive stance is generally found throughout the paper.

Returning to neutral discourse representation, reported papers marked as neutral, although lacking forms of judgement or bias, tended to make extensive use of attribution. Attribution itself can be used to influence readers' opinion towards the subject of the discursive event. This strategy generally contributed to maintaining the author's neutral stance towards the Italian piece of research described making use of public verbs such as *maintain, says, pretends, affirms* and avoiding personal comments and evaluative language. An example can be seen in the following extract:

> But, more particularly, in the *former* part **he explains**, How many ways *Light* is propagated or diffused, *viz*. Not only *directly*, and by *refraction*, and *reflexion*, but also by *diffraction*; which last, **according to him**, is done, when the parts of Light, separated by a manifold dissection, do in *the same medium* proceed in different ways. *Next*, **he considers** the Nature of Light, as also Diaphaneity, and Opacity; and **taketh notice**, that most Bodies, whether Solid or Fluid, are *porous*; on which occasion **he ventures to explain** almost the whole Philosophy of *Magneticks*. (*Phil. Trans.* 1671: 3064–3074, reported)

This extract comes from a book account, which was classified as neutral-positive, in that the presentation of the book – *Physico-Mathesis de Lumine, Coloribus et Iride*, by Francesco Maria Grimaldi (1665) – started off with a positive introduction[199] but then continued into a neutral description of Grimaldi's work. In a few cases however, the use of attribution was rather frequent and while, when moderately used, attribution can be considered a neutral discourse

---

199 "This learned Treatise was not to be altogether omitted in these *Philosophical Occurences*, though an Account of it hath been deferr'd (too long,) it being but lately fallen into the *Publisher's* hands." (*Phil. Trans.* 1671: 3064–3074)

representation device, its repetitiveness in certain papers appears to transform it into a negative device, since it seems to show a desire on the authors' part to distance themselves from what they are reporting and thus showing weak commitment, or possibly even scepticism, towards the reported discourse. An example is seen in the following extract of a paper, which was marked as showing a positive attitude towards Italian research, in that there is praise towards the work being described; however, at a certain point the author increases the number of attributive expressions (highlighted in bold) as if trying to distance himself from the information reported about the existence of a second ear-drum. The author's doubtful stance is confirmed by two (underlined) comments in brackets, which point out to a different version of facts in the literature:

> And whereas among the suppositions of Musick it hath been received for an undoubted Axiom, that *Consonance* is made by the frequent union of two Sounds in striking the *External* Drum of the Ear, (for **he pretends** there is another Drum) at one and the same time; **he affirms** to have discover'd this to be utterly false, and maketh it his business to prove it in the 4th and 17th speculation of this Book. In the making of which Discovery **he relates** to have been assisted by taking an exact view of the *Organ of Hearing* it self; **he and his Anatomical friends** having there taken particular notice. How the three little bones are fastned to one another and to the *two Drums*, the External and Internal. (Anatomists having hitherto spoken but of *one* only.) and to the little Cavern and the mouth thereof […]. (*Phil. Trans.* 1673: 6194–7002, reported)

Another example, where reiterative attribution seems to mark author distance and negatively represent discourse, is seen in the following section on discourse evaluation. Yet, out of 87 papers written in English, only one (1.1 %) appears to negatively represent Italian discourse.

Hence, these instances show that while the Fellows were appreciative of Italian research – given the display of positive stance in over a quarter of the papers – they were also cautious when presenting it to the *PTRS*'s readership as is shown by the tendency to neutrally present Italian research (in 79 % of cases), especially through the use of attribution by means of which the authors kept a certain distance from Italian research.

4.1.2.2 Meeting minutes and letter exchanges

It has already been stressed that *PTRS* papers were discussed at meetings and judged worthy of publication beforehand; this means that the published paper was more likely to show appreciation rather than a lack of it. Hence, as part of a critical approach to discourse study, a selection of intertextually related sources from the same contextual background were perused in order to view whether

162   Discourse features

differing perspectives emerged on what is found published in the journal. The sources considered were meeting minutes and letter exchanges between Italian scholars and English Fellows. What emerged from the former is that the meeting minutes tended to simply report bullet points of what was discussed, decided and presented at meetings. Language use is generally concise, hence free of any forms of involvement on the secretary's part. An example can be seen in the following extracts relating on a meeting in 1680 and another in 1686:

> Dr. Croune produced a letter of Mr. Pulleyn, dated at Rome, giving an account of a new physico-mathematical academy founded by Signor Champigni; and mentioning, that he would correspond with the Society: that he had sent some tracts as a present, and would continue to send such pieces, as should be published, that were curious, as BORELLI's book, when finished. This letter took notice of some things printing at Rome: of a balsam made by Signor Champigni: of a parcel of rare seeds sent by Mr. Upton for Oxford from Boccone: of the poverty of Boccone, and what would be an encouragement to him to make farther search for natural curiosities: and of a sort of lithan-thrax, sent among other things, and found in a crack of the Apennine mountains. This letter promised likewise to settle a correspondence with Signor Pittiati and Signor Tomaso Cornelio, two learned men at Naples. (Thomas Birch's *History of The Royal Society* vol. IV 1757: 55)
>
> After the reading of the minutes of the last meeting, Dr. Vossius's interpretation of the inscription found on the base of a great pillar at Rome, read at the preceding meeting, was produced and read; and the inscription and commentary were ordered to be inserted in the first *Philosophical Transactions.*
> A letter in Latin from Signor Francesco Spoleto to the society, dated at Venice Aug 26. 1686, was presented by Signor Sarotti; which letter being chiefly to request the Society's opinion of a book of his sent to them, the said book was recommended to Mr. Hooke to peruse, and give an account of it. (Thomas Birch's *History of The Royal Society* vol. IV 1757: 499)

Letter exchanges instead reveal that published papers were in most cases representative of the content of the original letter, only (nearly) reduced to the essence, free of unnecessary content such as social formalities. Elimination of unnecessary material however does not always occur; sometimes papers go straight to the core of the original letter, other times they maintain some of the introductory and concluding formalities contained in the original. The reasons behind this are unclear, although frequently what is maintained includes reverence towards the Society; hence a little pride and desire for recognition might be involved. More differences were instead seen between published paper and source letter when the topic concerned controversies or delicate matters. Examples of this are seen in the following section.

4.1.2.3 Evaluation in the discourse of the PTRS

Evaluation – i.e. positive or negative assessments which the author makes on his own behalf either explicitly or implicitly – should be considered in a critical approach to discourse analysis in that the author's opinions "might influence or position readers/listeners/viewers to take a negative or positive view of the people, events and states of affairs being depicted in the text" (White 2004). The high presence of an involved language production on the authors' part suggests that evaluation of Italian scholars and their work did indeed occur. It will be seen that evaluation was generally explicit and positive. As was mentioned earlier, also the original letter exchanges reveal mostly positive encomiastic relations between English and Italian natural philosophers, and between English natural philosophers speaking about Italians.

Explicit evaluation is mostly expressed through praise and positive evaluative adjectives or, in DHA terms, through the extensive use of nomination and predication strategies that positively construct objects, events, processes and especially social actors. Referring to Italian scholars, for instance, phrases such as the following are frequent: "the *ingenious* [Paolo] Boccone", "the famous astronomer of Bononia", "two excellent persons of the Florentine Virtuosi", "that *great* anatomist seignor Antonio Marchetti", and "that *learned* anatomist" referring to Lorenzo Bellini (my italics). The qualifying adjectives employed tend to recur; among the most common are *ingenious, learned, excellent, famous, noble, illustrious* and *inquisitive*. Translated papers – hence the writings of Italian scholars – also reveal examples of praise and positive evaluation towards their colleagues and their work. For instance, LANCISI refers to MALPIGHI as the "*incomparable* Malpighi, who naturally applied himself only to *serious* studies" and later "this *worthy* and *learned* man", "this *illustrious* person", "this *most learned* man". Italians also acknowledged the work of English and French scholars, as is seen in the following extract:

> [The study of the nature of oviparous insects] he [ANTONIO FELICE MARSILI] owes is with much glory pursued by the Royal Society of *London*, as well as that of *Paris* and *Florence*, and particularly makes very honourable mention of Mr. *Lister* for his history of Spiders and Snails; and of Mr. *Ray*, for his being the first Discoverer of their being all Hermaphrodites, or Androgine, as he terms it. (*Phil. Trans.* 1683: 356–359, 356, reported)

Other than the person, the Fellows also positively evaluated the studies of Italians with expressions like "a *curious* observation"; "many *notable* experiments"; "his book of vipers, which for several years passes in this country *almost for an undoubted truth*"; "these dissertations be all *ingenious* and *learned*, each in their kind". Here too, the same qualifiers tend to recur with the most common being *curious*, *ingenious*, and *learned*. Evaluation towards the author was however much more abundant than evaluation towards the study, once again reflecting how the agent played a central role in 17th-century natural philosophical discourses. These encomiastic strategies could be considered merely as part of early modern genteel manners; however, reverential forms of this kind were not employed in speaking of all scientists. In the case of unknown Italian men, for instance, English writers would generally speak of "an Italian" or provide whatever piece of information was available regarding their profession and/or reputation. For instance, one of the first papers mentioning REDI's work, which will later become very well known in England, refers to him simply as "a curious Italian" and "this Italian philosopher";[200] and in another paper the author introduces DOMENICO GUGLIELMINI by saying that "he is esteemed an excellent mathematician",[201] thus hedging his statement through the use of the passivized verb *esteem* and attributing the opinion to others. Great admiration was also displayed when speaking of Italian universities and academies; the author of *Phil. Trans.* 1672: 5125–5130 for instance, refers to the Bolognese Accademia dei Gelati as "the Illustrious Academy of Bologna". Finally, descriptions of Italy, and specifically of its artistic and natural beauties, are always positively constructed.

Explicit negative evaluation of Italy, Italians or their work was not found; only two minor cases of potentially face-threatening discourse were present among the whole 17th-century corpus, that is, only in two papers out of 102 (1.9 %). The first was a case of implicit negative evaluation of a synopsis that the Society had recently received of a book by FRANCESCO TRAVAGINO. The book was entitled *Nova Philosophia et Medicina* and, from the description made in the article, the book would seem to reflect a medieval summa, i.e. a compendium of all knowledge and sciences. In the extract below, the verbs and expressions that the author of the paper uses to attribute what is being said to TRAVAGINO are in bold:

> That **this Author hath compos'd** a System of Natural Philosophy by Observations and Experiments, accommodated to the benefit of Humane Life, and Subservient to Physick and other subalternate Arts; which Philosophy **he pretends** to have raised on Principles that are certain Bodies drawn out of Mixts, which, <u>though in themselves invisible and incoagulable, yet</u> become, **according to him**, visible by their Contrariety and mutual Operation […] And from their various <u>Complication (in which he places the whole business and moment of Philosophy)</u> **he holds**, that […] In particular, **he deduceth** from the said Principles the cause of Ferments and their variety, the Nature of Generations,

---

200 *Phil. Trans.* 1665: 160–162.
201 *Phil. Trans.* 1700: 613–614.

Concretions, Putrefactions, Precipitations, &c. and **sheweth** how those principles run through all Minerals, Vegetables, and Animals, by their manifold Combinations, and various ways of acting on one another [...] And having raised this Structure of his **as far as he judgeth** it sufficient for the Subordinate Arts, **he proceeds** to adapt it to the Art of Physick. And applying it to Animal Bodies, he thence **draws** the diversity of Humours and Tempers, the beginning and duration of Vital Heat, the motion of the Limbs, the faculties of Entrails, the origin, vitality, and properties of the Blood [...] concluding with an Indication of the proper Remedies (**as he conceives**) of many Diseases. Whether this Philosophy be new, is easy to judge. (*Phil. Trans.* 1665:555–556)

The extract is made up by frequent attributions, repetition of the *and* conjunction and by the presence of several sceptical parenthetical remarks such as "though in themselves invisible and incoaguable, yet..."; "(in which he places the whole business and moment of Philosophy)" or "(as he conceives)". By saying "in which he places the whole business and moment of philosophy" the author is also exaggerating and oversimplifying TRAVAGINO's book intentions (intensification strategy). The perlocutionary force produced by this utterance contributes to the overall understating process set in motion in the passage. Moreover, the text includes a series of lists of the numerous topics that the author discusses in the book. All of these strategies together seem to create a slight mockery of the book being described. The writer of the paper finally concludes with the rhetorical comment "whether this philosophy be new, is easy to judge", thus implying that the book was certainly not a novelty. Notice also the use of the indicative in the main clause (*is easy to judge*), which presents the writer's opinion as fact rather than suggestion.

An important step in a critical study of discourse is to interpret results taking into account the relevant context knowledge and any intertextually related sources. Hence, whenever some form of opinion, tension or evaluation appeared to transpire from a given discursive event, the original writing that lead to the publication and/or any related letter exchanges were consulted in order to view how original discourses were represented in the *Transactions* and to compare different perspectives of the same argument. From a closer look into TRAVAGINO's correspondence with OLDENBURG, it emerges that it was TRAVAGINO himself who had asked OLDENBURG to review his synopsis.[202] The Society satisfied TRAVAGINO's request by reviewing the synopsis and publishing the review in

---

202 Like many of his contemporaries, TRAVAGINO saw the Society as an authority in natural philosophical matters: "As it is the chief object of your Society to judge of the causes and effects in physics discovered by art and through art, and to promote discoveries sent to you from any quarter, no matter who makes them, I beg you again and again – or if you not yourself then whoever acts for you in this duty, but I assume you to be the most likely person – to take the laws into account and examine the

the *PTRS*. OLDENBURG moreover privately replied to TRAVAGINO's letter, but the tone of the letter is quite different from that of the published paper[203] going back to the traditional encomiastic formality of 17th-century letter exchanges:

> The Royal Society thinks highly of your remarkable deference to it, and instructed me to inform you of its great goodwill towards you and your endeavours. Indeed, nothing more pleasing to them could occur, than the news from my place on the globe that there are men who strive earnestly to promote science by reliance on observation and experiment and who, neither feigning nor formulating hypotheses on nature's actions, seek out the thing itself. And as they gather from the synopsis you submitted that you are a follower of the experimental method of philosophy, and more especially because the opportunities for exploring nature's hidden byways are so vast, they congratulate you upon your undertakings and labours, praying for your happy success in them. They desire you to supply what you so kindly offer in your letter (namely, the communication of the schedule of your experiments), when you conveniently can. When the work upon which you are engaged shall be published, it will assuredly furnish the Society with a further occasion for disclosing its judgement of yourself and your work. (Hall & Hall 1966: 415–416)

Hence, OLDENBURG's reply to TRAVAGINO positively congratulates him on his work and approach to nature and, although it mentions his synopsis and the Society's judgement of it, no actual judgement appears to be expressed. Moreover, it should be borne in mind that the *PTRS* paper was in English while the epistolary exchanges between the two men were carried out in Latin. OLDENBURG may thus have been trying to avoid expressing an actual judgement on the synopsis in order to maintain a good relationship with TRAVAGINO.

The second case of potentially face-threatening discourse is a claim of authorship of an experiment. The paper includes a letter, whose anonymous author subtly points out that an experiment made by the Italian physician, CARLO FRACASSATI, had been previously performed in a very similar manner by himself, and that possibly FRACASSATI

> may have had some imperfect Rumour of our Experiment without knowing whence it came, and so may, without any disingenuity, have thence taken a hint to make and publish what now is English'd in the *Transactions*. (*Phil. Trans.* 1666: 551–552)

> Synopsis or Idea of a new Physics, a Practice which I have discovered through my experiments. This is a new task of mine; many, as you know, have tried it before me, but all in vain. So I fear that the same may befall me [...]. However, it is certain that unless you too free me from my fear that the opposite is true I cannot ever consent to publish it, for all who know me, not to condemn me." (Hall & Hall 1966: 302)

203 The identity of the author of the paper is not stated.

The author of the letter appears to mitigate his claim of authorship through extensive hedging given by the modal verb *may* and the use of negation found in *imperfect* and *without* (mitigation strategy). Moreover, the author chooses words that have a connotation of smallness as in "an *imperfect rumor* of our experiment" – which suggests that FRACASSATI only heard a little of the experiment – and a *hint*. The author thus opts for negative politeness in order to put forward his claim in a more indirect, respectful and less imposing manner. The letter is framed by OLDENBURG's introduction and notes. In the paper, OLDENBURG acts as a referee providing evidence and thus confirming what is stated in the letter.

Writers sometimes temporarily abandoned their neutral stances making comments on what they were presenting to the English readership. This involved displays of opinion, which in most cases were positive, as was seen with above examples of evaluation, and showed agreement with the Italian author's opinion:

> The Fifth Observation is concerning Yellow Amber, or *Succinum*, and its Original. He endeavours by many Arguments to prove, that Amber is nothing else but Naphtha, or Petroleum coagulated or condensed. I was told by a Chymist at Montpellier, That Oleum Petroleum was the same with Oyl of Jet or Gagates, and not to be distinguished by Colour, Taste, Smell, Consistency, Vertues, or any other Accident, as he had by Experience found, which renders Signior Boccone's Opinion probable, there being great Affinity between Jet and Amber. (*Phil. Trans.* 1699: 53–67, 53–54, review)

The above extract comes from a review of a book by PAOLO BOCCONE, *Museo di fisica & di esperienze* (Venice, 1697). By means of an *argumentum ab auctoritate* (argument from authority) (Wodak & Reisigl 2016) with the scheme: if both *a* and *b* say *x*, then *x* must be true, where *a* is the Italian author and *b* is a chemist who was evidently considered reliable, the opinion of BOCCONE is given some credit. This kind of discursive strategy, although not fully in line with the Society's principles of studying nature by means of experimentation, was frequently employed by early modern scholars (see Section 4.1.4).

Other times however, writers of the *PTRS* were more dubious, showing some reservations towards the reported subject. The author of the same review also expresses some disbelief towards other subjects treated in BOCCONE's book:

> The Seventeenth Observation is of the *Tarantola* of *Apulia*, which is a beaten Subject, and of which more hath been said than is true. Notwithstanding what our Author hath written, I am not fully satisfied, that the Dancing of the *Tarantati* to certain Tunes and Instruments, and that these Fits continue to recurre Yearly, as long as the *Tarantola* that bit them lives, and then cease, are any other than acting Fictions and Tricks to get Money. The Symptoms that attend the biting of the *Tarantola* of *Apulia*, as also the manner of Cure and Remedits, are the same with those mentioned in the precedent

168 Discourse features

> Observation. The Stinging of a Scorpion produces the same Effects with the biting of a *Tarantola*. If a *Tarantola* be removed out of its natural Place, *v. g. to Naples, Rome &c.* and there admitted to bite, it doth no harm at all; which is very unlikely; but that the *Tarantola* bred at *Rome* are innocent, is probable. *The same being experienced in the stinging of Scorpions, which in* Africa *is deadly, but in* Italy, *if they are bred there, Innocent: and I doubt not but that we in* England *have the same Species of Spider with the* Tarantola. (*Phil. Trans.* 53–67, 57–58)

Here, without appealing to any persuasive argumentation strategies and without supporting his claims, the author overtly expresses that he does not fully believe in the truthfulness of BOCCONE's claims ("I am not fully satisfied" and "is very unlikely"). The author continues describing BOCCONE's book and commenting on his findings and then finally concludes his review with a relatively positive judgement and a little remark on the "excessive Civilty" of Italians:

> This Work contains great Variety of Matter, and a multitude of Medicines, simple and compound, for almost all Diseases and Infirmities. The Author shews himself to be a Man of great Candor and Ingenuity, speaking evil of no Man, nor detracting from any; without Emulation giving a fair Character of every one that deserves it; and that rather beyond than short of their Merit, according to the excessive Civility of his Nation. (*Phil. Trans.* 53–67, 63)

4.1.3 Reporting disputes and disagreements

As the Society was seen as an authority in matters of natural philosophy, the Fellows were often consulted when it came to disputes between scholars. Two examples of this have already been seen with the case of FRACASSATI's blood experiments and the insinuation of plagiarism, and TRAVAGINO's request for approval of his book (see Section 4.1.2.3). Disputes were reported about in the *PTRS*, and the present section will explore how the Fellows dealt with disagreements.

News about four different disputes and cases of disagreement – DEGLI ANGELI-RICCIOLI; AUZOUT-CAMPANI; CAMPANI-DIVINI; and REDI-CHARAS – is found in the sampled corpus for the 17th century. The first case regards:

> An Account of a Controversy Betwixt Stephano de Angelis, Professor of the Mathematicks in Padua, and Joh. Baptista Riccioli Jesuite as It Was Communicated Out of Their Lately Printed Books, by That Learned Mathematician Mr. Jacob Gregory, a Fellow of the R. Society (*Phil. Trans.* 1668: 693–698, 693, reported)

The dispute arose from RICCIOLI's *New Almagest* (Bologna 1651), in which he argues against Copernican heliocentrism. In the paper, STEFANO DEGLI ANGELI responds to Riccioli's arguments[204] in order to "let the world see that they [the arguments] are not more esteem'd in Italy, than they are in other places" and [Michele] Manfredi, a pupil of Riccioli's (Borgato 2016), intervenes in favour of Riccioli. Jacob Gregory, the author of the paper, reports some of Riccioli's, Degli Angeli's and Manfredi's arguments.[205] Apart from a brief compliment to Degli Angeli's "excellent" illustrations of examples of motion, Gregory reports arguments and counterarguments impartially, making use of neutrally connoted public verbs and without adding positively or negatively connoted words that could display bias. See for instance the following extract:

> *Riccioli* in his *Almagestum Novum* pretends to have found out several new demonstrative Arguments against the Motion of the Earth, *Steph. De Angelis*, conceiving his Arguments to be none of the strongest, taketh occasion to let the world see, that they are not more esteem'd in *Italy*, than in other places, *Manfredi*, in behalf of *Riccioli*, endeavours to answer the Objections of *Angeli*, and this latter replyes to *Manfredi*'s Answer. The substance of their discourse is this following.
> Although the Arguments of *Riccioli* be many, yet the strength of them consists chiefly in these *three*:
>
> The *first*.
>
> *Multa corpora gravia, dimissa per Aerem, in Plano Æquatoris existentem, descenderent ad Terram cum Velocitatis Incremento reali & notabili, non tantùm apparenti* […]. (*Phil. Trans.* 1668: 693–698, reported)

The controversy appears to have had some impact on the discourse community of 17th-century astronomers, since, according to Borgato (2016), it appears to have inspired a famous exchange between Newton and Hooke, which eventually influenced the writing of Newton's *Principia* and led to the confirmation of the Copernican system.[206]

---

204 Giovanni Alfonso Borelli also intervened against Riccioli's work, but this is not reported in the *PTRS* paper.
205 The *PTRS* paper simply refers to a "Manfredi", which should not be confused with the later astronomer Eustachio Manfredi, who was also in contact with the Royal Society. We learn the full name of Manfredi, in the following work: *Seconde considerationi sopra la forza dell'argomento fisicomattematico del m. reu. p. Gio. Battista Riccioli della Compagnia di Giesù, contro il moto diurno della terra. Spiegato dal sig. Michiel Manfredi nelle sue risposte, e riflessioni sopra le prime considerationi di F. Stefano de gl'Angeli…* (Padova 1668). For more on the dispute see Koyré (1955).
206 "Alla fama di questa controversia in campo europeo (James Gregory ne fece una relazione alla Royal Society pubblicata nelle *Philosophical Transactions* del 1668) si attribuisce l'origine di un celebre carteggio scambiato tra Isaac Newton e Robert Hooke (1679–80), le cui speculazioni condussero, cinque anni più tardi, alla composizione

The second case was more a debate rather than a dispute and it involved the Italian lens maker GIUSEPPE CAMPANI and the French astronomer ADRIEN AUZOUT. News had been given to the Society by their French correspondents about CAMPANI's advances in the making of telescopes (*Phil. Trans.* 1665: 1–2, 131–132 and *Phil. Trans.* 1668: 791–796). With his telescopes CAMPANI made a series of observations on Saturn and Jupiter which he published in his *Ragguaglio di due nuove osservazioni* (Rome 1664). He moreover claimed to make his lenses not by means of moulds but by means of a lathe. AUZOUT wrote some letters in which he commented on CAMPANI's work. The letters were then published in *Lettre a M. L'Abbé Charles, sur le Ragguaglio di due nuove osservazioni da Giuseppe Campani* (Paris, 1665). Some of AUZOUT's opinions on CAMPANI's work were reported in the *PTRS* (*Phil. Trans.* 1665: 70–75). Here AUZOUT raises "some scruple" regarding CAMPANI's claim to manufacture optic glasses without moulds and points out what he believes to be some inaccuracies in CAMPANI's observations of shadows in the ring of Saturn and in the proposed length and breadth of the ring. He then regrets having previously drawn a different conclusion regarding two spots on the belt of Jupiter observed by CASSINI since the latest news of CASSINI's and CAMPANI's observations showed that these spots were shadows of satellites, like CASSINI had suggested in the first place. Finally, the paper concludes by reporting that AUZOUT did not "doubt the excellency" of the telescopes used by CASSINI and CAMPANI. The *PTRS* also reports CAMPANI's answer to AUZOUT and AUZOUT's answer to CAMPANI (*Phil. Trans.* 1665: 75–77); all of which had been published in AUZOUT's tract. In short, CAMPANI justifies some of his observations and responds to some of AUZOUT's criticism by pointing out that the telescope used by AUZOUT did not allow him to view as much detail as CAMPANI's. The paper finishes with AUZOUT's answers to CAMPANI. What interests us here, however, is not so much the content of their exchange but rather the way in which the two scientists spoke to each other and how this exchange was reported in the *PTRS*. Although the two men had rather contrasting views on some points, they expressed their opinions moderately by hedging their statements and accompanying their criticism with praise. See for instance the following extract, where the writer includes AUZOUT's praise ("having commended Campani's sincerity...") between two objections:

> First therefore, after that M *Auzout* had raised some scruple against the Contrivance of Signor *Campani* for making *Great Optick-Glasses* without *Moulds*, by the means of

dei Principia e alla definitiva affermazione del sistema copernicano secondo l'ipotesi formulata da Keplero" (Borgato 2016).

a *Turn-lath*, he examines the *Observations*, made with such *Glasses:* Where, having commended *Campani's* sincerity in relating what he thought to have seen in *Saturn*, without accomodating it to M.*Hugens's Hypothesis*, he affirms, that supposing, there be a *Ring* about *Saturn*, Signor *Campani* could not see in all those different times, that he observed it, *the same Appearances*, which he notes to have *actually* seen. (*Phil. Trans.* 1665: 70-75, 71)

As to the reporting of the exchange, this is always neutral. The author's presence is only perceived in some attributing phrases in brackets, which remind the reader that the opinions reported are not his own but those of AUZOUT and CAMPANI:

Which difference yet does not appear to M. *Auzout* to be so great; but that M. *Hugens* perhaps will impute it to the Optical reason, which he (*Auzout*) hath alleged of the Advance of the light upon the obscure space; although he is of Opinion, he should not have concluded so great a Length, if he had not seen the Breadth spread out more, than he hath done: for (*saith he*) if the Length of the *Ring* be to the body of *Saturn*, as 2½ to 1, and the *Inclination* be 23 deg. 30' the *Ring* will be just as large, as the body, without spreading out; but if the *Ring* be bigger, it will a little spread out; and if it were treble, it must needs spread out the half of its breadth, which hath not so appeared to him. (*Phil. Trans.* 1665: 75-77, 76)

CAMPANI was also in competition with lens maker EUSTACHIO DIVINI and news about the trials made with their respective telescopes – to view whose telescopes were best – was given in the *Transactions* (*Phil. Trans.* 1665: 131-132 and 209-210). In one paper, based on French intelligence, it is reported that CAMPANI's lenses excelled those of DIVINI (1665: 131-132); while the other paper reports DIVINI's opinion. DIVINI claimed that his telescopes were better than CAMPANI's and that they had proven to be so in all the tests they had run. DIVINI's opinions are reported neutrally, there are however two comments made by the writer: one, at the end of the paper (1665: 209-210), is a positive comment on one of DIVINI's claims; namely that he was able to tell a good lens merely by looking at it and without needing to run tests. This comment is however unrelated to the competition between the two telescope makers. The second comment, instead, is a side note related to a claim of primacy made by DIVINI:

Eustachio de Divinis (saith the *Informer*,) has written a large Letter, wherein he pretends, that the Permanent Spot in *Jupiter* hath been first of all discovered with *his* Glasses; and that the P. *Gotignies* is the first that hath thence deduced the Motion of *Jupiter* about his *Axis*; and that Signior *Cassini* opposed it at first; to whom the said *Gotignies* wrote a letter of complaint there-upon.

See Numb. 1. *of these* Transactions; *by the date whereof it will appear, that that* Spot *was observed in* England, *a good while before any such thing was so much as heard of.* (*Phil. Trans.* 1665: 209-210, 209)

Here OLDENBURG – the writer of the paper, replying to DIVINI's claim of primacy over the discovery of a spot on Jupiter – momentarily breaks his neutral stance to include England in the debate and make his own claim of primacy (on England's behalf) over the finding; he moreover provides evidence by referencing an earlier *PTRS* paper. The original letter behind this publication (AUZOUT to OLDENBURG, 1666, in Hall & Hall 1966: 102–103) shows that the paper is quite representative of the original wording; though some parts of it, where the parties' stances were somewhat more judgemental, have been toned down by the editor. This can be seen by comparing the above extract from the *PTRS* with AUZOUT's original letter:

> Concerning Italy I can tell you that Divini has put together a thick letter in which he claims that the permanent spot in Jupiter was first discovered with his lenses, and that it was Father Gottignies who first worked out the rotation of Jupiter, at which Cassini initially laughed. Father Gottignies has written a letter on this point in which he complains of Cassini, but I haven't seen Cassini's reply. Divini also claims that his large telescopes excel those of Campani, and that they have done better in all the trials that have been carried out; and that Campani has always refused to do what was needed to compare the two properly, that is to give each identical eye lenses or exchange the eye lenses. There is one splendid innovation which I don't know: a way Divini says he has discovered of telling upon seeing an objective whether good or not, without trying it out. He explains nothing further. I confess I don't possess this secret; if one of you gentlemen can guess it, it will be particularly useful if one can apply it when the glass is still cemented down. (AUZOUT to OLDENBURG, 1666, in Hall & Hall 1966: 103)

Notice how the adjective *thick* has been substituted with the more neutrally connoted *large*, and how the comment on how CASSINI "laughed" at DIVINI's claim has been rephrased with "opposed it at first".

The last dispute involved the French apothecary MOYSE CHARAS and the physician FRANCESCO REDI. Both REDI's *Osservationi intorno alle vipere* (Florence, 1664) and CHARAS' *Nouvelle experiences sur la vipere* (Paris, 1669) were reviewed in the *Transactions* (*Phil. Trans.* 1665: 160–162 and *Phil. Trans.* 1669: 1086–1097). It is in CHARAS' review that his opposing views on the causes of viper venom are reported. While REDI believed that it was the "yellow liquor" to be venomous, CHARAS thought that the cause was to be found in the vipers' "vexed and enraged spirits". Moroever, another paper by THOMAS PLATT (*Phil. Trans.* 1672: 5060–5066) reports about multiple experiments that were carried out in order to corroborate REDI's findings (for more see Schickore 2010; Baldwin 1995; see also Section 3.1.4). While PLATT's paper and the review of REDI's book show a positive stance towards REDI's claims, the account of CHARAS' book and opinions are fully neutral.

In sum, the presentation of disputes and disagreements in the *Transactions* was generally found to be neutral with some occasional comments by the editor

(OLDENBURG) who acted as a referee and provided evidence for what was being stated in the articles whenever he had some further knowledge on the subject. OLDENBURG's comments moreover frequently served the purpose of encouraging further research on a given subject.[207]

4.1.4 Witnessing

Both in the case of translated papers and reported papers, great importance was given to the presence of notable gentlemen, or *virtuosi*, that witnessed what was now being reported in the article:

> Being opened, the Spectators were surprised to find his blood not curdled, but on the contrary more thin and florid than ordinary. (*Phil. Trans.* 1666: 490–491, translated)

> Signor *Steno*, who honour'd me with his visit, saw the administration of it [autopsy of a not-completely formed baby], which I had before made in the presence of many Noblemen and Physitians at my House. (*Phil. Trans.* 1670: 1188–1189, translated)

In both the above extracts, the writers specify that a number of witnesses were present at the running of the experiments. While in the first case the author speaks more generically of a number of spectators, in the second case an individual space is given to the Danish scientist NICOLAS STENO[208] – possibly regarded as more newsworthy – and then more vaguely, but still relevant, to "many Noblemen and Physicians".

There was moreover a tendency to specify who the witnesses were in terms of their profession, reputation and/or social standing. Notice in the following extract, how the writer lists the names of the persons present at the running of series of experiments and how, for each one of them, he provides brief biographic notes:

> Some few days after, a rendezvous [of experiments to see the effect of viper poison on pigeons] was made in Sign. *Magalotti's* Garden, where, besides the forenamed persons, met Mr. Thomas *Frederick*, Mr. *John Godscall* (two English Gentlemen), Abbot *Strozzi* (his Most Christian Majesties Publick Minister in this Court), Sign. *Paolo Falconieri* (the first Gentleman of the Bed chamber to the G. Duke), Sign. *Luigi del Riccio*, Mons. *Pelletier*, Mons. *Morelle* (the one Physitian, and the other Chirurgeon to the G. Dutchess), Dr. Gornia Physitian in Ordinary to His Highness, Dr. *Bellini* Professor of Anatomy at *Pisa*, Sign. *Lorenzo Lorenzini* a Mathematician, and Sign. *Pietro Salvetti* [...] who is one of the G. Dukes Musicians, and plays on all Bow instruments (*Phil. Trans.* 1672: 5060–5066).

---

207 For more on OLDENBURG's intrusions into the narratives see Gotti (2014: 165–166) and Atkinson (1999: 82 and 93).
208 NICOLAS STENO (1638–1686) settled in Italy in 1666 and converted to Catholicism in 1667.

Witnessing was common practice among early modern scientists and writers of science. Together with detailed recording and reporting of natural and experimental events, witnessing served the purpose of building a discourse of fact. In the absence of other forms of evidence, the presence of witnesses would ultimately give credibility to the truthfulness of the report (argument from authority). The existing literature on testimony in early modern science links the credibility attributed to witnesses to their social status; the higher the witness's status, the more credible was the report. However, Shapiro (2002) argues that the role of gentlemanly norms is overemphasised by historians.[209] She shows that gentle status was only one of the factors involved in assessing witness credibility and that witnesses were often not gentlemen. A more important aspect of the credibility of the testimony was the level of skill and experience. This appears to be confirmed by the author of the above-quoted paper when he writes:

> This is, Sir, what I can confidently affirm to have been an eye witness of; [....] but that, which urged me to make this repetition [to test the effects of viper poison], was the thought that it might be acceptable to you, to see his Assertions [FRANCESCO REDI's] confirmed by the Testimonies of so many Persons, that are the more able to be judges of them, because their understandings are such, that 'tis not possible to impose upon them. (*Phil. Trans.* 1672: 5060–5066)

Thus, according to the author, the referenced witnesses had a broader understanding of the subject and their opinion could not therefore be doubted.

4.1.5 Toponymy

The tendency to report in detail appears to lead writers to name the place where a piece of research was carried out or sent from and to specify the location where the event took place. This is especially true in the case of reporting papers and travel accounts. Various sampled articles report that the experiments were performed at the homes of specific amateur scientists. An example can be seen in the third extract quoted in the previous section (*Phil. Trans.* 1672: 5060–5066), where it is said that the experiments were carried out in LORENZO MAGALOTTI's garden. MAGALOTTI was not a physician but an intellectual and diplomat, who had visited the Royal Society and held regular correspondence with them.[210]

---

209 See also Fontes da Costa (2002b) and Gotti (2011).
210 MAGALOTTI, who had studied English, visited England and the Royal Society twice in 1667 and 1668 (Wis 1996: 343). Two of the main purposes of his visit were to bring back information about the Society to Italy and to encourage BOYLE to correspond with Italian scholars (Knowles-Middleton 1979:163).

Tab. 4.1: 17th-century cited place names and number of papers citing them

Toponym	Number of papers mentioning
Rome	22
Venice	11
Bononia (Bologna)	9
Florence	7
Padua	7
Sicily	7
Naples	4
Calabria	4
Messina	4
Genoa	3
Brescia	2
Leghorn (Livorno)	2
Palermo	2
Pisa	2
Milan	2
Other mentions:	Lombardy, Savoy, Bergamo, Val Sabbia (Brescia), Cozzo (Pavia), Piedmont, Valley of Lanzo, Le Langhe mountains, Piacenza, Udine, State of Modena, Reggio, Tuscany, Siena, Maiello, Fiorenzuola, Pontin Lakes (Forlì), Campagna Romana, Ronciglione, Viterbo, Pozzuoli, Otranto, Nicolosi, Catania, Channel of Messina, Melilli, Augusta, Mazzara, Siracusa, Emone, Noto, Troina, Randazzo, Nicosia, Castiglione, Francavilla, Linguaglossa, Mascali, Aidone, Acireale, Etna, Paterno, Adernò, Lentini, Carlentini, Licodia, Sortino, Cassaro, Agosta, Caltagirone, Militello, Trapani, Istria, Dalmatia, Pola, Zahara, Zebenico, St. Gioanni, la Fortezza Vecchia, Spalatro, Salona, Clissa, Lesina, Biondi, Trau, Ragusi, Antivari, Durazzo, Apollonia, Valona.

The most frequently cited toponyms reflected the locations of some of the main universities and academies of the late 17th century. For instance, a great deal of correspondence was kept between London and Tuscany, in that several Italian learned men lived in and moved about the Tuscan towns. Florence, moreover, was the home of the Accademia del Cimento, whose members – among whom also the DE MEDICI brothers, Prince LEOPOLD and the Grand Duke FERDINAND II, founders of the academy – had contacts with the Society.

Tab. 4.1 below lists all of the place names found in the papers with the number of papers mentioning them.

The papers mentioning Rome cover a number of topics, most importantly astronomy, optics, antiquaries and several book accounts. Rome was one of the favourite stops for travellers and the home of several scholars who held contacts with the Society, such as CIAMPINI, CAMPANI, BONOMO and FRANCESCO NAZZARI, the founder of the first *Giornale de' Letterati*, from which the Society occasionally retrieved papers. Although less active from a medical point of view (Cook 2004), Rome was also mentioned as the source of two LANCISI papers. Centres of natural philosophy at this time in Rome were the Jesuit Collegio Romano, where studies in astronomy were pursued and later, from 1678, the Accademia Fisico-Matematica founded by CIAMPINI under the patronage of Queen CHRISTINA of Sweden.

The papers coming from Bologna mainly concerned astronomy and the work of CASSINI, while those from Padua were mostly medical. The frequent mentions of Sicily, Naples and Messina instead are related to the volcanic eruptions and earthquakes that took place in the late 17th century.

Venice was home to another journal *de' Letterati* from which the Society often retrieved material for publication in the *PTRS*. It was also the place of residence of TRAVAGINO, GRANDI and of the English diplomat JOHN DODINGTON, who worked as an intermediary between Italians and the Society.

Finally, there are very few distinctions among the different Italian states. While place names are provided for the sake of exhaustiveness and factuality, scholars are generally referred to as being Italian, which would seem to suggest that the various states were seen as belonging to a unified socio-cultural entity, the Italian *Res publica litterarum*.

4.1.6 Interdiscursivity and intertextuality

Many of the astronomy papers stood in dialogic relation with one another and a few of them are particularly relevant to the purpose of this section. CASSINI, in one of his papers (*Phil. Trans.* 1676: 681–683), publicly invites astronomers to verify his hypotheses:

> The Configurations of the *Satellites of Jupiter*, which are observed this year 1676, and which may be observed the next year, are of so great importance to the verifying of their Hypotheses, that Signor *Cassini* thought fit to advertise Astronomers, not to let this occasion slip (which doth not present it self but twice in 12 years) of observing them with a singular care and attention. For, by comparing the Observations of this year with those of the next, they will find an apparent Inversion of the whole System of the Satellites, which will come to pass towards the end of *March* next, according to his

particular Hypotheses, which he proposes to verifie by comparing these Observations with those of *Galileus, Marius,* and *Hodierna,* who undertook to dress Tables of their Motions. (*Phil. Trans.* 1676: 681)

In the first volume of the *PTRS* we find papers on astronomical matters by CASSINI and an anonymous author (ADRIEN AUZOUT) that are written in reply to one another (*Phil. Trans.* 1665-1666: 17-8 and 1665-1666: 18-20). The dialogue concerns their observations (from Rome and Paris) on the motion of two comets. Both letters are translated into English. NICHOLAS MERCATOR references CASSINI in "Some Considerations of Mr. Nic. Mercator, Concerning the Geometrick and Direct Method of Signior Cassini" (*Phil. Trans.* 1670: 1168-1175). Further, one paper reports on the different observations of a number of astronomers – "Mr. Flamsteads, Mr. Townlyes, Mr. Haltons, Signor Cassini's and Monsieur Hevelius's, Observations of the Late Eclipse of the Sun" (*Phil. Trans.* 1684: 662–667, in Latin). And, again, another dialogue is entertained by CASSINI and JOHN FLAMSTEED over a moon eclipse (*Phil. Trans.* 1675: 390 and 1676: 561-565). These instances serve to show the clear dialogic and cooperative relationship that stood among 17th-century scientists. Indeed, CASSINI and FLAMSTEED had a whole conversation through the *Transactions*; and OLDENBURG played his role as intermediary too.

As has been seen, papers abounded with references to other scholars' research and to other *PTRS* papers that had intertextual (explicit) or interdiscursive (implicit) relations with them; and OLDENBURG encouraged scientists to cooperate, communicate and carry out further research on a same subject. The following section will provide a more detailed description of a group of papers that were intertextually related to one another in order to show how references, comments, literature reviews and wording choices (*asbestos, amianthus, salamandra, salamander's wool,* and *linum*) created diachronic and diatopic intertextual connections between *PTRS* papers.

4.1.6.1 *Dialogicity in the discourse on* amianthus

MARCO ANTONIO CASTAGNA, who was briefly introduced earlier for his work with mines, attracted the Fellows' attention for his experiments made with *amianthus* (today commonly known as *asbestos*). In a paper that was originally published in the *Giornale de' Letterati* of Venice in 1671 he says that he is able to render *amianthus* so malleable as to make it as thin as a white lamb's skin or a white sheet of paper. With this skin (and then the paper) he made several experiments; he covered it with burning coals and, although it did catch fire, it went back to its previous state as soon as the fire went out – "without the least change of its first whiteness, fineness, or softness" (*Phil. Trans.* 1671: 2168). The article finishes with CASTAGNA's desire to produce what he would have called the "Book of Eternity", made with *amianthus* sheets, bound with *amianthus* thread and skin and written in golden letters. This would have meant creating a book resistant to fire and all other elements – therefore everlasting.

With the same mineral CASTAGNA also made candlewick, which he would have liked to combine with an incombustible oil he had heard of in order to create an "everlasting light, so much celebrated by the ancients" (1671: 2168). Something reflecting this description is discussed in a 1685 *PTRS* article.[211] The paper talks about an incombustible cloth, brought to the Society by a merchant, which apparently was made from an Indian tree, from which also a "liquor, which not consuming, is used with a wick made of the same material with the cloth, to burn in their Temples to Posterity" (*Phil. Trans.* 1685: 1049–1050). The article was a letter and in the same *PTRS* volume the reply to it is present.[212] In this letter, we learn that there were doubts as to the truthfulness of the transforming *amianthus* into incombustible linen. The writer therefore draws on the existing literature to provide further testimonies on this subject:

> That this *Linnen* was very well known to the *Ancients*, beside that of *Pliny*, we have the further testimony of *Calius Rhodiginus*, who agrees with the *Letter*, placing both the *materials* and *manufacture* of it in *India*; and *Paulus Venetus* more particularly in *Tartary*, the *Emperour* whereof, he says, sent a piece of it to Pope *Alexander*. It is also mention'd by *Varro*; and *Turnebus* in his *Commentary* upon him, *de Lingua Lat.* and by all of them as a thing inconsumable by fire. In these latter ages: *Geo. Agricola* tells us, that there was a *Mantle* of this *Linnen* at *Vereburg* in *Saxony*; and *Simon Majolus* says, he saw another of it at *Lovain* exposed to the fire. *Salmuth* also acquaints us that one *Podocattarus* a *Cyprian* Knight shewed it publickly at *Venice*, throwing it into the *fire* without scruple or hurt; and Mr. *Lassells* saw a piece of it in the curious *Cabinet of Manfred Septalla*, Canon of *Milan*. Mr. *Ray* was shewed a purse of it by the Prince *Palatin* at *Heidleberg*, which he saw put into a pan of burning *Charcoal* till it was thoroughly ignite, which when taken out and cool, he could not perceive had receiv'd any harm; and we are told in the *Burgundian Philosopy*, of a long *Rope* of it, sent from *Signior Bocconi* to the *French King* & kept by Monsieur *Marchand* in the Kings gardens in *Paris*, which though steeped *in* oyle & put in the *fire*, is not consumed. To which add, that we have now seen a piece of

---

211 *Phil. Trans.* 1685: 1049–1051. A letter from Mr. Nich. Waite merchant of London, to Dr. Rob. Plot; concerning some incombustible cloth, lately exposed to the fire before the Royal Society.
212 *Phil. Trans.* 1685: 1051–1062.

this *Linnen*, pass the fierry triall both at *London*, & *Oxford*. So that it seems to have been known in all ages, all describing it after the same manner, as a thing so insuperable by *fire*, that it only *cleanses* and makes it better. (*Phil. Trans.* 1685: 1053–1054)

The writer first draws on the knowledge left to us by the ancients and then moves on to his own contemporaries. The relatively numerous quantity of testimonies he provides reveals that there must have been considerable interest in the topic all over Europe including England and Italy. Evidence for the existence of *amianthus* provided, the author continues to consider this mineral in terms of its names; places of formation; natural principles; manufacture of thread cloth with it; the uses that have been made of it; and the reasons behind its resistance to fire. As far as the names are concerned, it was known then under different names, among which are *amianthus* (<gr. *amiantos*, not stained), *asbestos* (<gr. inextinguishable), *salamandra*, and in English *salamander's wool*. This last name, according to the author, was possibly related to the use of asbestos fibre as candlewick. It was also often called *linum* accompanied by either the name of the place where it was found or some epithet referring to its properties. The mineral was first known as coming from the east, namely from China, India, Tartary, and Cyprus; but further deposits of it were now being found in several parts of Europe, including England and, as was communicated to them by CASTAGNA, Italy (1685:1056). An earlier description of its transformation into thread and cloth was given by MARCO POLO and is reported by the author of the paper. Its uses included shrouds, clothing, candlewick, napkins, mantles, ropes and

> we are told that Septalla, canon of Millan, had thread, roaps, net-works and paper of it.[j] Marco Antonio Castagna, who lately found this mineral somewhere in Italy, knows how to prepare, and render it so tractable and soft, that it resembles well enough a very fine lambskin, which he can thicken and make thin to what degree he pleaseth, and make it thereby, like either to a very white skin, or a very white paper.[k] We have also made paper of our Welsh amianthus but lately here at Oxford, which will bear fire and ink well enough, the ink only turning red by the violence of the fire.[l]
> j. Musæum Regalis Societ. Part 3 Chapt. 5. k. Philosoph Transact. Num 72. l. Philosoph Transact. Numb 166. (*Phil. Trans.* 1685: 1060)

Hence, CASTAGNA's words are found nearly 15 years later, together with those of many other researchers of the past and of the writer's time. The result was a gradual build-up of information concerning the nature and use of a mineral, which was still very much unknown. Furthermore, as can be seen by the above-quoted extract, every piece of information that the writer inserts is carefully referenced. Thanks to the reference to the Society's catalogue,[213] MANFREDO SETTALA is revealed as another source of information on the subject.[214]

Further papers on the subject appear in 1684, 1701, 1710, and 1759,[215] reporting about the possible discovery of *amianthus* deposits in Wales, Scotland and France. Interestingly, in the 1701 article, the publisher includes multiple names for the mineral in the title of the paper, i.e. "Lapis Amianthus, Asbestos, or Linum Incombustibile", which could be a way of giving the reader more possibilities to recognise the mineral, since it was known with a variety of names. In the 1759 paper instead, the name *asbestos* seems to have taken over in the title, but in the paper the author still speaks of "asbestos, or amianthus".

A contribution to this subject was also made by the Italian archaeologist GIOVANNI GIUSTINO CIAMPINI, who wrote to a Fellow in 1691 reporting about four different types of asbestos and about how it was spun into cloth (EL/C2/26 and *Phil. Trans.* 1700: 911-913). Interestingly, CIAMPINI too seems to make some observations on the name of this mineral;[216] however, these observations are not reported in the abstract of his letter.[217] The four types of asbestos in his possession were sent to the author from different places: a "reddish" specimen from Corsica;[218] a "silverish lead" coloured one from Sestri Ponente in Liguria; the third came from Cyprus and, according to the writer of the abstract (RICHARD WALLER), it was "the worst of all" and is described as having a "blackish earth" colour and looking like scales or laminae one upon another, since CIAMPINI appears to have represented it "like an onion" in his original letter. The fourth piece came from the Pyreneans and was given to CIAMPINI by PAOLO BOCCONE; it had longer, thicker and rougher filaments than the others. After having run some experiments with *amianthus*, CIAMPINI concluded that the candlewick

---

213 GREW, NEHEMIAH. (1681). *Musaeum Regalis Societatis, or, A catalogue & description of the natural and artificial rarities belonging to the Royal Society and preserved at Gresham Colledge*. London.
214 Ibid. :313.
215 *Phil. Trans.* 1684: 187-189, *Phil Trans* 1701: 1004-1007, *Phil. Trans.* 1710: 434-436, *Phil. Trans.* 1759: 837-838.
216 In this paper, too, both the names asbestos and amianthus are used to refer to the substace.
217 The original letter was not found in the archives, but only the abstract (EL/C2/26) written by RICHARD WALLER.
218 In actual fact, CIAMPINI writes "from Corsica or Corfu"; possibly having some doubts as to where this first specimen came from; however, he then refers to it as the Corsican sample of amianthus.

made of *asbestos* was not worthwhile, in that he found that it would "go out, and not attract or continue up the oyl [*sic*] for the flame". Like others, he also saw that asbestos does not alter in fire, but it does wear out a little by handling; and that it would not preserve a stick wrapped in it from the fire. He then proceeds to describe how to make cloth out of it[219] and concludes that, out of the samples he had, the most suitable one for the purpose was the one coming from Corsica, while the less suitable was the Cypriot one.

To conclude, part of the fascination with this mineral appears to be related with the descriptions made of it by ancient writers. For instance, upon the discovery of an ancient urn in Rome containing a particular *materia oleosa*, the discoverers assume it to be "one of those perpetual lamps that the ancients mention" (*Phil. Trans.* 1686: 227). In *Phil. Trans.* 1685: 1051–1062 the writer refers to what PLINY, CAELIUS RHODIGINUS and MARCO POLO wrote about *amianthus*. In the 1684 paper, the author wants to determine whether the asbestos found in the isle of Anglesey "be the same kind with the asbestos of the Ancients". And WILSON, in *Phil. Trans.* 1701 is concerned that the features of the *asbestos* found in Scotland seem somewhat different from the description given of it by PLINY. Hence, different papers published over the course of three centuries and in different countries all subscribe to a same discourse that had its roots long before their time. The authors of these papers explicitly subscribe to the discourse of *amianthus* by means of simple references, literature reviews and jargon.

4.2 18th century

This section presents and discusses the results of the discourse analysis carried out on the sampled *PTRS* papers for the 18th century. Results are reported in the same order of the previous century (*textual dimension* and *discursive practice* features) with a different focus for Section 4.2.2.3, which compares an example published paper with the original dissertation found in the archives in order to view features of the Society's translation and editing practices.

---

219 "Lastly, he proceeds to shew the manner of spinning it, which he tried thus; first he laid the Stone in Water (if warm better) for some time to soak, then it is opened and divided with their hands, that the Earthy parts may fall out of it, which are whitish like Chalk, and hold the thread parts together; this makes the water thick and milky; this is repeated six or seven times with fresh water, where it is again opened and squeezed, till all the heterogeneous parts are washed out, and then the Flax-like parts are collected, and laid in a Sieve to dry. As to the making of Paper, he says in the washing the Stone, there will remain several short pieces in the bottom of the Water". (*Phil. Trans.* 1700: 9112–913).

The linguistic analysis was carried out on both translated and reported papers but not on papers fully written in Latin or French. For the papers written in Italian with an English translation in appendix, the English translation was analysed and comparisons were made between the Italian and English texts to see whether there were any peculiarities regarding translation practices.

4.2.1 Textual dimension

4.2.1.1 Macrostructural features

The presentation of papers in their full or abridged letter form continues to be a distinctive feature of the *Transactions* in the 18th century. Indeed, 73 papers (39.4 %) are presented as either full letters or extracts of letters. Moreover, six further papers are accompanied by a cover letter. Letters were however in most cases intended for publication and the personal features that characterize this genre are, in this century, very often limited to salutations and a humble request to present one's letter to the Society.

Irrespective of whether the papers were letters or not, recurring text types were identified. The main text types were: observations (63 papers); experimental reports (22); reports of natural phenomena (17); book accounts and lists of published books (15); reports of objects and findings (13); travel accounts (6); medical case reports (4); reports of autopsies (3); descriptions of new instruments (3, plus further amidst experimental reports); news reports (3); and other miscellaneous papers. Some papers were made up of different parts, e.g. a book account followed by an experimental report based on the book. Some papers were moreover replies to other papers.

Titles continue to be quite detailed and self-explanatory. These would generally indicate the form of the paper, such as an "account", "observations", a "letter" or "an extract of a letter". In the case of letters, they could indicate the names of both the sender and the addressee or either of them, often giving some details regarding their occupation and/or affiliation. Titles also generally reported who "communicated" the paper to the Society and, in fewer cases, the translated nature of the text with the name of the translator and the source language. Below is an example comprising most of the above mentioned features:

> An Account of What Happened at Bergemoletto, by the Tumbling down of Vast Heaps of Snow from the Mountains There, on March 19, 1755: As Taken by the Intendant of the Town and Province of Cuneo. Received from Dr. Joseph Bruni, Professor of Philosophy at Turin, and F. R. S. Communicated by Mr. Henry Baker, F. R. S. Translated from the Italian. (*Phil. Trans.* 1755:796)

Lengthwise papers tend to be longer than in the previous century, where the average paper was about a couple of pages long. The majority of 18th-century papers were classified as "medium length", i.e. between 5 and 14 pages. These amounted to 78 papers (42.1 %). A total of 56 papers (30.2 %) were classified as "short" in that they did not exceed 4 pages in length. A large group also included 51 (27.5 %) "long" papers, i.e. longer than 14 pages, with some of them going over 40.

For the 17th century it was observed that papers frequently had an introduction by the editor, together with occasional intrusions in the body of the text. In the 18th century, this practice seems to have been abandoned, as very few papers contained an introduction other than the one written by the writer of the paper. Translated papers especially are now only framed by the title with no further intrusions by the editor. In very few cases, footnotes were present with the translator's doubts as to the understanding of words and sentences contained in the source text.

From the structural point of view, papers from this century differ considerably from the previous, becoming more and more organised as the century proceeds, at least as far as the more purely scientific papers are concerned. Literature reviews become regular practice towards the end of the century, in some cases followed by the indication of a gap in the existing literature and therefore finding a space for the author's piece of research. Figure 4.1 below is taken from one of FONTANA's papers, in which the various steps of the article are highlighted by curly brackets:

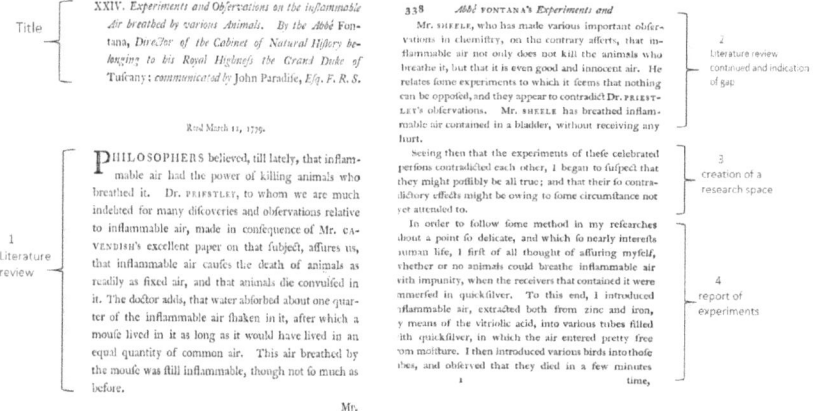

Fig. 4.1: Example of structural organisation of late 18th-century experimental reports. Reproduced with modification from *Phil. Trans.* (1779) XXIV. Experiments and observations on the inflammable air breathed by various animals. Vol. 69 337–361.

These first two pages of FONTANA's paper show the various moves that the author makes in order to introduce his experimental report: literature review; indication of an unsolved problem in the literature; consequent creation of a niche for FONTANA's experiments; report of the experiments. These moves very much resemble modern day practice in scientific writing. However, it ought to be reminded that this kind of structural organisation was only found in the more strictly scientific papers and especially experimental reports. Other papers, such as the general observations, travel accounts and the like, did not follow specific patterns, especially at the beginning of the century where they still tended to be quite miscellaneous in order, sometimes going off point. However, in all of the papers an increasing tendency to review the literature and describe the state of the art in a given area was observed. Very frequent in this century is also the use of footnotes to reference the works of peer scholars or other *PTRS* articles that had an intertextual or dialogic relation with one's own. Finally, very frequently also drawings were attached to papers in order to give the Fellows as clear an image as possible of what was described in the paper.

4.2.1.2 Language use

As far author positioning is concerned, the sampled 18th-century papers still display strong authorial presence throughout the century. Out of 185 papers, 92 (49.7 %) were marked as being characterised by an involved author-centred style, making use of first-person pronouns, active-tense verbs and private verbs. Of this group, 27 were translated, which means that the authorial presence was that of the Italian writer, while the remaining 65 were reported. Examples will not be provided here as they would be quite similar to the ones provided for the previous century (Section 4.1.1.2). However, one point that distinguishes the 17th century from the 18th as far as author-involvement is concerned, is that, although the new century is still characterised by author-centredness, many of the features that are part of this writing style were not as abundant as in the previous century. That is, while first-person pronouns, active-tenses and verbs of emotion were still present, the more involved features such as boosters, emphatics, insinuations, and open-hearted displays of opinion were hardly found.

Humble presentations of the papers to the Society were rather frequent; these tended to be worded in a similar manner both in translated and reported papers. Since the articles were generally presented in letters, the writer of the letter would ask the addressee to present the paper to the Society if the addressee believed it to be worthy of their attention. Encomia towards the Society was also very present in the letter exchanges but preserved in very few published papers. See for instance the following extracts:

the dissection of a young woman, I do myself the honour of communicating to this Society; nor do I doubt its being received with indulgence by them, when I consider that the Royal Society was instituted for the most laudable end of the improvement of arts and sciences, and that of rendering them serviceable to mankind. As the Philosophical Transactions are sufficient proofs of this, I cannot but believe that an uncommon anatomical observation must be acceptable to them. (*Phil. Trans.* 1765: 79, translated)

As you did me the honour to mention, in your last paper to the Royal Society, some remarks I had made on Mr. Volta's machine, I hope you will father oblige me by presenting this account to that learned Body, if you think it contains any thing deserving the attention of the curious. (*Phil. Trans.* 1777: 388, CAVALLO's original English)

Papers marked as being informational were 23 (12.4 %) and surprisingly only one paper showed very little use of abstract production – i.e. object-centredness given mainly by the use of passive constructions. The absence of abstract papers was unexpected in that linguistic research[220] shows that at least medical papers showed a discrete level of abstractedness throughout the existence of the journal, which was also confirmed in the analysis of the 17th century. This result led to a review of the papers and may be partly accounted for by the fact that many medical papers in this century were published in Latin.

A total of 27 papers (14.5 %) were marked as being midway between an involved production of language and an informational one. The majority of these papers had sections that tended to be involved – such as introductions and conclusions displaying reverence towards the Society and authors' thoughts on what is being put forward – and other sections that were instead purely informational (see Section 4.1.1.2 for examples).

Narrativity continues to be a distinctive feature in the *Transactions* as most papers report what was done by the author himself or were third-person reports. Indeed, 77 papers (41.6 %) were marked as being narrative or containing extensive narrative parts. Very frequent were also descriptive papers, especially given the high presence of reports of objects and instruments in this century. A total of 30 (16.2 %) papers were purely descriptive, but many papers marked as narrative contained descriptive sections too.[221]

Finally, in this century the presence of multiple articles by a same author made it possible to view recurring features of personal writing styles. For instance, WILLIAM HAMILTON tended to make extensive use of first-person pronouns – also

---

220 Atkinson (1992, 1999) and an unpublished piece of research I had carried out myself on medical papers in the *PTRS*.
221 Of the narrative papers 48 were reported while 29 were translated. Of the purely descriptive papers twenty-one were reported and nine translated.

because most of his papers reported personal experiences of nature (geology, volcanology and travel accounts) –; moderate displays of opinion; and frequent references to the works of others in the same field. He was humble in presenting his own opinions, as he was not a volcanologist himself. However, after years of sending papers on the volcanoes and geology of the two Sicilies, he eventually grew some confidence and started commenting on the excessive ease with which expert volcanologists made judgements on Vesuvius.

Two authors that were very methodical in structuring their papers were CAVALLO and FONTANA. Focusing on CAVALLO for instance, his papers, written between the 1770s and the early 1790s, all display some features of involvement (first-person pronouns and verbs of perception and emotion). However, a gradual reduction in the level of involvement was noticed and, while all of the earlier papers were marked as being involved, the later ones tended to limit author-centredness to certain parts of the paper, with the core – generally reports and results of experiments – becoming more informational. His later papers were moreover highly structured providing commentary on the state of the art in the field (electricity, magnetism and new instruments) complete with reviews of others' studies; creation of a research space; and then the explication of his own experiments, observations and instruments. His observations were generally based on experimental results and not on perceptions.

One paper that provides a good example to view different writing styles is "A Collection of the Observations of the Remarkable Red Lights Seen in the Air on Dec. 5. 1737. Sent from Different Places to the Royal Society" (*Phil. Trans.* 1739: 583–606). This paper, which has been partly reported in Tab. 4.2 below, is made up of six letters, of which one was written by D'ARAGONA from Naples, another by POLENI in Turin, another by ZANOTTI in Bologna, one by DE REVILLAS in Rome (left in Latin), one by JAMES SHORT in Edinburgh, and one by JOHN FULLER in Sussex. D'ARAGONA's letter starts with an involved introduction but the following narrative is then very informational displaying some features of abstractedness and the presence of the author is no longer felt until the end. POLENI's letter has a longer rather involved introduction and methods section, but the actual reporting of the phenomenon is then informational and slightly abstract. Differently from D'ARAGONA who limits himself to providing the observations, POLENI expresses his own personal opinions as to the nature of the phenomenon. ZANOTTI's letter is again involved in the introduction of the topic and more informational in the description. Both POLENI and ZANOTTI make more frequent use of qualitative adjectives in their descriptions. SHORT's account is instead more involved throughout; while FULLER's account is very informational – although it starts directly with the narrative and is much shorter than the others; there is moreover no introduction. The style of the five accounts can be compared in the following table in which the extracts contain the end of the introduction and the beginning of the narrative (except FULLER's, which did not contain an introduction):

Tab. 4.2: Comparison of different writing styles in a same paper (*Phil. Trans.* 1739: 583–606)

D'Aragona	Poleni	Zanotti
A Phænomenon of a fiery Meteor is my Motive for troubling you, Sir, with this other short Narrative; being persuaded that it will be as agreeable to you to persuaded, as it was to me to draw it up with all the Circumstances of Truth, to which I was an Eye-witness. *Dec. 16. 1737 (N. S.)* in the Evening, the Sun being about 25 Degrees below the Horizon, a Light was observed in the North, as if the Air was on Fire, and flashing; the intenseness of which gradually increasing at the Third Hour of the Night it spread Westward in such a Manner, that if a Perpendicular was let fall from the Polar Star, and afterwards [...]	And moreover, on the ensuing Evening, the same bright red Colour, appearing near the Horizon, deceived the common People into a Belief, that a new Phaenomenon, like the foregoing, was breaking out of the Horizon. Wherefore I am of Opinion, that in this Cafe there is a considerable Difference between the Aurora Borealis, and the Redness occasioned by the Sun's setting. About three quarters of an Hour after, the Length of the Zones was contracted, their Extremities having received about ten degrees from the East and West. The white lucid Part was not now so distinguishable from the red, as before: And this last Colour grew fainter almost everywhere else but at the Western Space from which the *Aurora* was withdrawn, there remained a brighter Space of Three or Four Degrees [...]	While the whole City was intent upon viewing this new Appearance, I and some Gentlemen were employed in calculating the *Ephemerides;* and, being apprised thereof, we jointly began to take Observations of it. This uncommon Light drew to the Observatory several others, that were used to come at other times. [...] 7h. 9. *p.m.* When we first perceived the *Aurora Borealis*, its Centre was near the North Pole. The Brightness extended along the Horizon about 70 Degrees, and its Height was judged 20 Degrees. The Sky was almost totally overcast with Clouds, but the Light was visible in several Parts, where the Sky was clear. The Two Stars [...] of the *Great Bear*, shone bright in the midst of the reddish Light of the *Aurora*.

(*Continued*)

Tab. 4.2: Continued

D'Aragona	Poleni	Zanotti
Short	Fuller	
Yesternight we were surprised upon looking out at the Windows, about Six o'Clock, to find the Sky, as it were, all in a Flame; but upon further Inquiry, it was nothing but the *Aurora Borealis*, composed of red Light. There was an Arch of this red Light reached from the West, over the *Zenith*, to the East; the Northern Border of this Light was tinged with somewhat of a blue Colour. This *Aurora*, as far as I saw, did not first form in the North, and after forming an Arch there, rise towards the *Zenith*, as they commonly use to do; neither did the Light shiver, and by sudden Jirks spread itself over the Hemisphere, as is common, but gradually and gently stole along the Face of the Sky, till it had covered the whole Hemisphere; which alarmed the Vulgar, and was indeed a strange Sight	It was a strong and very steady Light, as near as can be of the Colour of red *Okre*; it did not seem to dart or flash at all, but continued going on in a steady Course against the Wind, which blew fresh from the South-west. It began about North North-west, in Form of a Pillar of Light, at about 6h.15^i in the Evening; in about 10 Minutes, a Fourth Part of it divided from the rest, and never joined again; in 10 Minutes more it described an Arch, but did not join at Top; exactly at Seven, it formed a Bow, and soon after quite disappeared; it was all the while lightest and reddest at the Horizon: It gave as much Light as a Full Moon.	

4.2.2 Discursive practice

4.2.2.1 Discourse representation

Most of the sampled papers include some form of represented discourse, either by incorporating original extracts and full translated Italian papers or by reporting and adapting their contents. Papers were marked as "neutral" when there was no kind of judgement or bias towards what was being reported or, in the case of translated papers, when there was no form of framing and commentary, that is, the paper was only a direct translation of the Italian original. However, translated papers could themselves represent other discourses, both Italian and foreign. Papers marked as "neutral-positive" were mainly neutral but contained some minor forms of praise or positive evaluative language. Papers marked as "neutral-negative" contained minor negative or disagreeing comments, which were however irrelevant to the representation of discourse as a whole. And papers marked as "negative" contained criticism and/or disagreement. For clarity the results of the analysis have been summarised in the following table.

Tab. 4.3: 18th-century discourse representation results

	Tot.	Original English	Translated
Neutral	65 (35.1 %)	21	44
Neutral-positive	7 (3.7 %)	6	1
Positive	6 (3.2 %)	5	1
Neutral-negative	3 (1.6 %)	2	1
Negative	2 (1 %)	2	-
No discourse representation	63 (34 %)	63	-
Excluded from analysis	39 (21 %)	-	-

As can be seen from Tab. 4.3 the majority of papers tend to represent Italian discourse neutrally, especially Italian-translated papers, which are mostly reported as such, with no forms of framing (other than the title) and evaluative language. The frequent use of attribution that was observed in the previous century was moreover no longer present in the 18th and neither were the continuous intrusions by the editor reminding the reader that what was being written were the Italian author's words (distancing). Fewer papers compared to the previous century display a positive stance, which is related to the higher presence of neutrality.

What does not appear from the table instead is that Italian research is in this century much more debated, questioned and put to the test by further trials, which were then presented and discussed in consecutive publications. An example will be seen with the case of animal electricity (Section 4.2.3), where several papers on the subject were published and many experiments repeating the Italian ones had been carried out throughout Europe unsuccessfully. Another relevant example is given by two papers both reporting about some very small microscope lenses that DELLA TORRE had crafted and sent to the Royal Society as present. One of the papers written by FRANCIS HASKINS EYLES-STILES (*Phil. Trans.* 1765: 246–270), who together with DELLA TORRE in Italy had made several microscopical observations using those same lenses, displays a very positive opinion towards both DELLA TORRE and his lenses. However, once the lenses reached the Society, HENRY BAKER was asked to examine them and his observations and opinion were then reported in a paper (*Phil. Trans* 1766: 67–71). BAKER's opinion differed considerably from that of EYLES-STILES, he claimed that "Father di Torre's globules [i.e.lenses] are too small and make it difficult to view anything with them". His disbelief in the utility of the lenses led him to suggest that the observations made by EYLES-STILES may not have been made with the same lenses. He moreover stated that this kind of lens was outdated and had been replaced by convex lenses exactly because of the difficulties in managing them, i.e. correctly placing them on the microscope and putting objects into focus. However, BAKER did not use Wilson's microscope, for which the lenses were meant. Still, he was convinced that he did everything possible to be able to use them and was sure that their deficiency was not due to the microscope he placed them on. He further stated that, had the lenses been used by others, they might have possibly been blinded by looking into the microscope with them. His final judgement of the lenses was that they were "matters of curiosity rather than of real use".

Finally, while the representation of Italian discourse into the English discourse of the *Transactions* is mostly neutral, the representation of English discourse in the Italian papers is generally positive. This was signalled by positive evaluative language towards English Fellows and their work and will be exemplified in the following section.

4.2.2.2 Evaluation in the discourse of the PTRS

As for the previous century (Section 4.1.2.3), this section analyses the forms of evaluation made both by the Fellows towards Italian scholars and their research and by the Italians towards the Society and its work.

Starting from instances of positive evaluation, 33 papers (17.8 %) contain positive nomination and predication strategies. These were generally addressed to the Italian, English or other nationality researchers and – more frequently than in the previous century – to their work. The discursive construction of the Italian peninsula was mostly positive (see Section 3.2.5 for examples).

The Fellows' positive evaluation towards their Italian colleagues was generally constructed through nomination strategies which resemble those employed in the 17th century. Hence, we read of the "ingenious" and "indefatigable Signor Marsigli, who deserves all the encouragement that the world can possibly give him"; "the industrious and most inquisitive anatomist [ANTONIO PACCHIONI]"; the "nice", "celebrated" and "accurate" MALPIGHI; and the "genious, patience and resolution, which distinguish Dr. Donati". The adjectives employed tend to recur too, with *learned, inquisitive, curious, industrious, celebrated* and *ingenious* being the most frequent. The frequency of this kind of positive nomination strategies seems however to be more present in the first half of the century. One novelty, as far as the positive evaluation of persons and research is concerned, is that the Fellows now often spoke of the *accuracy* of their colleagues in carrying out their studies and experiments, which seems to anticipate the greater concern with applying correct methods that developed in the 19th century (Atkinson 1999).

Positive evaluation was found from the Italian side as well. This was shown in varied forms of appreciation towards the English Fellows – e.g. to SWINTON "which shew his acute judgement, and excellent learning in the oriental languages" and to PRIESTLEY "to whom we are much indebted for many discoveries and observations relative to inflammable air" – and through forms of encomia towards the Society and its presidents, which were more frequent in the letters, but were sometimes preserved in the published papers too, an example can be seen in the following extract:

> Sir, as your extraordinary Talents, and excellent Taste, in a true Examination of Natural Effects, and in Discoveries relating to Experimental Philosophy, are so well known, that you have, with Justice, been elected into the most celebrates Academics of *Europe*, and to the Presidentship of the Royal Society of *London*, in particular; I resolved, with good Reason, to offer you a short Account of the last great, dreadful and pernicious Eruption of our *Vesuvius* [...]. (*Phil. Trans.* 1739: 237–252, 247, translated)

Much more frequent in this century was positive evaluation towards others' work, especially towards the end of the century where praise of one's research seems to take over and substitute praise towards the author. Hence, both from Italian and English scholars remarks are found such as "this ingenious discourse"; "a considerable discovery"; a "learned Dissertation"; or CAVALLO's instruments "which are certainly of great use in many cases".

Negative evaluation towards the authors, apart from one case which will be described below (Section 4.2.3), was not present.[222] However, criticism

---

222 As has already been seen however, more explicit evaluation can be found in the epistolary exchanges between scholars, who had more personal relationships, speaking about third parties. Here, occasional instances of negative evaluation can be found; JOHN WOODWARD, for instance, in a letter to JOHANN JAKOB SCHEUCHZER, commented that GIORGIO BAGLIVI was a man too full of himself, which made him rush into subjects that he was not well informed about (letter quoted in De Beer 1948: 58).

towards another's work, methods and instruments were more frequent than in the previous century. A total of eleven papers (5.9 %) display some form of negative evaluation towards Italian researches, methods and instruments: for instance, PIVATI's experiments in the medical uses of electricity were described as having been "greatly exaggerated"; F. CORSINI's opinion on Parthian coins was judged as being inconsistent by SWINTON; and VENUTI's and MAFFEI's opinion in the same area was "deemed little better than vague conjectures, scarce meriting the attention of the learned"; some renovations on the Roman Pantheon were "unfortunately approved" and the Fellow reporting about them was disappointed of the "deplorable" condition of the finished work; CAVALLO's electrical instruments would have made "Volta's and Bennet's of little or no use"; and VOLTA, although he was the person who introduced GALVANI's researches to the Royal Society, also criticised them as being erroneous.

The Society became moreover very defensive whenever NEWTON's work was questioned and criticised; this occurred in two cases: one case regarded GIOVANNI RIZZETTI's criticism towards NEWTON in the field of optics (see Section 4.2.3), which led to a repeating of NEWTON's experiments to corroborate the latter's findings; the other was a minor case of criticism by FRISI, contained in a book which was overall positively judged by the Society, but as to the criticism FRISI addressed to NEWTON they could "not agree with this ingenious author" (*Phil. Trans.* 1753:10).

From all of the above mentioned instances of positive and negative evaluation one important aspect arises; and that is, that all forms of evaluation are related to the scientific achievements of the individual persons and not to their country of origin. The broad picture emerging from the discourse of the *Transactions* is that of a lively discourse community in which peer natural philosophers exchange their research and findings irrespective of their background.[223] There appears to be a slow gradual transition from the previous century, where *who* the person responsible for a given piece research was always relevant. Now greater importance is being given to the work, rather than the person. Moreover, hardly any instances were found of Italian scholars being defined by their Italianness. For example, Italian Fellows resident in England were in the previous century generally defined as "the Italian resident here"; while in this century this kind of

---

223 Only one minor case was found with some slight criticism towards Italians, this concerned JOHN STRANGE who criticised "the indolence of the Italians in not regarding" the "variety of remarkable zoophites", which their country offered; "the Italians have made no new account and only copy the ancients" (*Phil. Trans.* 1770: 182).

identification was not found. TIBERIO CAVALLO is never mentioned as being an Italian and indeed his writings and work are treated as any others', to the extent that one forgets he was not English.

4.2.2.3 Original, translation and publication: A brief comparison

A few 18th-century translated *PTRS* papers were compared with the original letters/papers sent to the Society in order to view translation and editorial practices. This section will provide and discuss an example by focusing on the paper "An extract of a letter of Signior Michele Pinelli, concerning the causes of Gout. Translated from the Italian by Joh. Gasp. Scheuchzer" (*Phil. Trans.* 1727: 491–494). The material available related to this paper were PINELLI's full dissertation written in Italian (RBO/13/85) and the English translation by SCHEUCHZER (RBO/13/86) both dated 1728. These documents were not the originals but exact copies that the Society's amanuenses transcribed in the Register Books (RBOs).[224] PINELLI's dissertation must have been accompanied by a cover letter which was not however transcribed with the dissertation. The style of the Italian text is rather elaborate in the introductory part and moderately author-centred throughout the dissertation. Occasionally he directly refers to his (unnamed) addressee, which confirms the epistolary nature of the paper; however there are no salutations at the beginning, which is what signals the existence of a possible cover letter. The results of the analysis are summarised in Tab. 4.4:

---

224 Also available was a further letter by DEREHAM, dated 1734, stating that he was shipping a book by PINELLI on the same topic (LBO/21/47).

Tab. 4.4: Comparison between Pinelli's 1728 original dissertation and the translated published paper

Pinelli's Italian dissertation	Scheuchzer's translation	PTRS paper 1728
4 manuscript pages	2 and a half manuscript pages	4 printed pages
	Translation of "an extract" of Pinelli's dissertation	Same text as the manuscript translation
Structure, wording and language use		
1-page long elaborate introduction eulogising great men of the distant and near past whose achievements continue to inspire men of learning to investigate nature	Removed	""
Two transition paragraphs to introduce topic: the example of the heroes of the past is what inspired Pinelli to search into the causes of Gout by reason and experiment	Removed apart for one sentence which is inserted in a new introductory paragraph in Scheuchzer's words	""
One paragraph explanation of methods (chemical analysis of the fluids of the body)	Translated the essence of this paragraph adapting its contents (second paragraph of Scheuchzer's translation)	""
Three paragraphs of observations based on former experiments and formulation of hypothesis. Further observations on the erroneous assumptions of previous scholars	Maintained main contents reducing to the essence	""
Reporting of experiments that were carried out to test his hypothesis	Maintained with minor adaptations	""
One paragraph results/conclusions (the cause of the Gout is attributed to an alkaline corrosive salt, which by corroding the sensible membranes about the joints, occasions the acute pains)	Maintained	""
One paragraph closing the letter suggesting corroboration of his opinion from the Society, as a body of wise and expert men. Further formal encomia, display of humility and salutation	Removed	""

As can be seen from the table, the translator reduced PINNELLI's dissertation to its main contents eliminating all unnecessary digressions and formalities. This was done prior to publication and most likely for economy; i.e. reducing reading time at the meeting where the letter was read and saving space in the *PTRS*. The part which was modified the most was the introduction, which was reduced to a very short and slightly more enticing paragraph, coloured by adjectives showing the importance and necessity of the piece of research.

The same kind of editing appears to have been carried out on personal letters as well. For instance, a letter sent in Italian by PAOLO MATTIA DORIA to PAOLO ROLLI was translated into English by the latter for the Society (EL/D2/ 53 and 54). The initial and final paragraphs of the original Italian letter have been crossed out. This is reflected in ROLLI's translation, in which those paragraphs are not present. ROLLI's translation is moreover characterised by deletions and a rearrangement of some sentences simplifying the original (an extract of Rolli's edited translation can be seen on the cover of this book).

Hence, this appeared to be the standard practice in translating and summarising letters and papers for publication; i.e. unnecessary parts to the understanding of the paper were eliminated, while the main parts were (almost) faithfully translated.

4.2.3 Reporting disputes and disagreements

While the Royal Society was the learned body to whom natural philosophers appealed for judgement and approval, the *Philosophical Transactions* occasionally became the place where news about disputes was provided and where defences of one's work could be made. Hence, some of the 18th-century disputes and disagreements between learned men found some space in the journal. It is the purpose of this section to briefly introduce them and investigate how they were dealt with in the *PTRS* and letter exchanges.

One case concerned the physicians JAMES JURIN and PIETRO ANTONIO MICHELOTTI, who both carried out studies on the movement and behaviour of fluids in the second decade of the century. In his *De separatione fluidorum in corpore animali* (Venice, 1721) MICHELOTTI criticised some of the observations JURIN had made in his study "De motu aquarum fluentium" (*Phil. Trans.* 1719: 748–766). Consequently, JURIN chose to defend his points in the *Transactions* with the paper "Defensio Dissertationis de Motu Aquarum Fluentium [...] Contra Animadversiones Viri Cl. Petri Antonii Michelotti" (*Phil. Trans.* 1722: 179–190). Interestingly, JURIN had received MICHELOTTI's book at the request of the latter, and although JURIN replied to his criticism in the *Trasactions*, he also sent a letter to MICHELOTTI, which displays a very cooperative approach, rather than an offended and defensive one:

> A few months ago I received, through the services of the distinguished Sherard, the splendid gift which you wished to be presented to me, through you hardly know me even by name: I mean your fine Meditations on Muscular Movement and on the Separation of Fluids in an Animal's Body. I studied these again and again, eagerly and with much pleasure, but among the numerous manifestations of your outstanding and acute intellect some points occurred which, to tell you the truth, I found less convincing. Do not think, most Learned Sir, in your integrity that the reason why some points displeased me is that I found myself mentioned and criticised at times in your work. If I know myself, my attitude is that if I ever happen to stay from the truth I would not resent but on the contrary consider it a kindness to be advised and led back on to the road by a friendly hand.
> Nevertheless, since it is a question of the truth, the love of which guides you, I believe, no less than myself, then you in your turn will be too civilised to resent it if, when I think my position can be defended, I put up a friendly, frank and vigorous defence, There are some of your comments on my work which, I think, you would have omitted if you had not failed to understand my argument. This is perhaps my own fault for not expressing it clearly enough. There are other points which, I think, can be demonstrated so that you yourself, if I am not mistaken, would concede them. I have explained my views of all these matters in the short Apologia which is presented to the public in the Philosophical Transactions now appearing. [...] I have given these Transactions to the Learned Sherard, who will have them sent on to you.
> I have been instructed, both by Royal Society and in particular by our illustrious President Newton, to inform you that the gift of copies of your works, was most welcome and acceptable to both recipients, and to send you most grateful thanks in their name. Goodbye, renowned Sir, and believe that I am your most devoted friend. (JURIN to MICHELOTTI 1722, in Rusnock 1996: 131–132)

While in the previous case it was an Italian criticising an English piece of research, the next case concerns an English Fellow questioning VALSALVA's findings. In 1725 an extract from the *Giornale de' Letterati* of Venice (1719) was published in the *PTRS* informing the public that

> The celebrated Anatomist Signor *Anton Maria Valsalva*, already known by this noble Treatise *De Aure humana*, has lately made a considerable Discovery. He has found the excretory Ducts of the *Glandulæ Renales*, or *Renes Succenturiati*, which discharge themselves into the Parts of Generation; that is to say, into the *Epididymides* in Men, and into the *Ovaria* in Women. (*Phil. Trans.* 1724–1725: 190)

VALSALVA's main claim was that the *glandulae renales* would play a role in the reproductive system by means of the said ducts, which connected them to the reproductive organs (Drummond & Monro 1737: 373). The discovery aroused the interest of JOHN RANBY (FRS 1724) who decided to dissect a human body in order to find the excretory ducts. He was however "not so happy as to discover any Duct of this Kind" (*Phil. Trans.* 1724: 270) but found instead, by opening the aorta, the arteries that go into the *glandulae renales* and from the *glandulae* proceed *"*down on both sides towards the Testicles *without supplying any of the neighbouring parts"* (my emphasis). Hence, he wonders whether VALSALVA could have confused these arteries with the supposed excretory ducts. However, he underlines that this is only a suggestion and not a statement "for want of further experiments". The whole paper was written with neutral tones, a display of reverence towards VALSALVA and by the use of hedges in putting forward the author's doubts. However, no further replies were made to RANBY in the *Transactions* and by the time his paper was published VALSALVA was deceased. GIOVANNI BATTISTA MORGAGNI however did attempt to explain VALSALVA's findings more fully, and news of it was published in *Medical Essays and Observations* (Drummond & Monro 1737) by the Philosophical Society of Edinburgh.

A paper written by DESAGULIERS reports of a dispute between GIOVANNI RIZZETTI and the same DESAGULIERS over the experiments in NEWTON's *Opticks* (*Phil. Trans.* 1728: 596–629). The paper traces all of the steps of the dispute with references to former *PTRS* issues. RIZZETTI had tried to reproduce NEWTON's optical experiments unsuccessfully and therefore concluded that NEWTON's theory of colours must have been false and invited the Fellows – both via an intermediary and through a letter directly addressed to the Royal Society – to repeat the experiments. Under DESAGULIERS' direction the experiments were repeated and corroborated in 1722. RIZZETTI was still not satisfied and repeated his claims in 1727 posing further questions, but DESAGULIERS, in the *PTRS* paper, claims that RIZZETTI should have first admitted to being wrong. In the first part of the article, where the dispute is reported, DESAGULIER maintains a neutral stance; however, the second part is an account of a book published by RIZZETTI, *De Luminis Affectionibus Specimen Physico-mathematicum* (Venice, 1727), which DESAGULIERS was asked to review. He thus proceeds with the review hoping that "no Body will blame me for making a faithful Report" (598). The tone of the paper here changes considerably:

> The Author in his Preface, and throughout the whole Book, in a most arrogant manner, has insulted the greatest Philosopher that this or any other Age ever bred, triumphing in what he thinks the Mistakes of Sir *Isaac Newton* and his own Discoveries. Had he modestly related the Facts as they appeared to him, and his Reasons for drawing Consequences different from those of Sir *Isaac Newton,* the World might have thought him urged on by the Love of Truth in his *ten Years Labours* [...]. (*Phil. Trans.* 1728: 598)

To show that he was not just speaking out of anger, DESAGULIERS quotes (in Latin) some of "Signior Rizzetti's Expressions against Sir Isaac Newton" (599) and responds to RIZZETTI's criticism making use of the witnessing strategy to reinforce his arguments. In the end, DESAGULIERS informs us that nothing more was heard from RIZZETTI, only that he was angry with Signor GIZLANZONI, who had acted as intermediary between them, and accused him of having "got into Sir Isaac Newton's Party" (597).[225] Hence, in this case the neutral stance which the Fellows generally maintained was abandoned. As is often the case with disputes, other Italians became interested in the RIZZETTI-DESAGULIERS discussion, and EUSTACHIO MANFREDI wrote a letter to DEREHAM in 1728, stating that he and some others had repeated the experiments contained in the *Opticks* and in a booklet by DESAGULIERS with the outcome that "if the Prisms were perfect and good, as some were [that] we had from England, they have always been found to correspond with the Doctrine and Principles laid down by Sir Isaac Newton" (RB0/13/90), hence siding with the English Fellows.[226]

Finally, another important case, not of a dispute but of an apparent refutation of another's research, was VOLTA's rejection of GALVANI's findings on the presence of an intrinsic electrical force in living organisms which was responsible for muscle contractions and nerve conduction. This case was represented in the *Transactions*, one-sidedly, with a paper published in French in 1793 based on two letters by VOLTA (pp. 10–44), "Account of Some Discoveries Made by Mr. Galvani, of Bologna; With Experiments and Observations on Them. In Two Letters from Mr. Alexander Volta". It was with this same paper

---

225 Hall (1991: 145–146) briefly sketches the Italian response to Newtonianism, explaining that "many Italian philosophers were, or would like to have been Newtonians, had it not been for the religious restrictions which weighed heavily on universities and academies. [...] Only after Newton's death did it become easier for Italians to praise Newton openly". Among the Italian Newtonians were GUIDO GRANDI, NICOLA CIRILLO, GIOVANNI POLENI, JACOPO RICCATI, and CELESTINO GALIANI. On the reception of Newtonianism in Italy see also Mazzotti (2013).

226 Another famous dispute worth mentioning and in which an Italian was involved was the LEIBNIZ-NEWTON controversy over the invention of calculus. Here, ANTONIO SCHINELLA CONTI acted as an intermediary between the two scientists (mainly on NEWTON's side), writing extensively to France and Italy. CONTI, however, eventually lost NEWTON's trust (Hall 1991: 138). One of CONTI's letters sent to LEIBNIZ and LEIBNIZ's reply were published in the *PTRS* in French (*Phil. Trans.* 1717: 923–928). Unusually, at least as far as the 18th century is concerned, the letters were in this case framed by editorial comments in English adding further information and details which appear to show a very slight negative stance towards LEIBNIZ.

that "animal electricity" was introduced into England.[227] According to VOLTA, GALVANI's *De viribus electricitatis in motu muscularius commentarius* (Bologna, 1791) contained "one of the most beautiful and surprising discoveries, and the germe of several others",[228] also referring to the further researches that the same VOLTA carried out in this field. After a brief introduction, the paper continues with VOLTA's description of one of GALVANI's experiments on the frog, which consisted in connecting the nerves of a dissected frog to its spinal cord using two different pieces of metal. The nerves of the frog were also insulated with metal coatings. Upon touching the metals to complete the circuit, the frog's leg twitched. Based on this and several other experiments, GALVANI concluded that the electricity that made the legs twitch was intrinsic to the frog with the metals acting as a conduit. From the beginning of the narrative, however, VOLTA prepares the ground for his subsequent objections. He starts by emphasising how GALVANI possibly became excessively enthusiastic over the results of his experiments and how the requisite circumstances were that the animal be in contact or very close to metal or other good conductors, "or what was still better, between two similar conductors"(Johnson 1794: 164).

> This, however, was the first step which led him to the grand and beautiful discovery of an animal electricity, properly so called, and which belongs not only to frogs and other animals of cold blood, but likewise to every animal of warm blood, quadrupeds, birds, &c.; a discovery which forms the subject of the third part of his book, a subject altogether new, and very interesting. It is thus he has opened to us an immense field, into which I propose to enter, and pursue my researches [...]. (*Phil. Trans.* 1793: 10–44, translated in Johnson 1794: 167)

VOLTA then reports of having successfully repeated GALVANI's experiments directing the electrical discharges from different kinds of conductors and measuring the power of the electricity by means of different electrometers (HENLEY's, CAVALLO's and BENNET's). He carries on discussing how GALVANI's many experiments on all kinds of animals *seemed* to prove the existence of a "real animal electricity" consisting of an electric fluid (galvanic fluid) passing through the nerves and leading to the muscles; meanwhile, however, he also anticipates how his own experiments "will be found considerably to extend the phenomena

---

227 The Society had already received news of GALVANI's book through a letter by the physician PIETRO MOSCATI sent to CAVALLO in 1792. GALVANI's experiments were immediately repeated by JAMES LIND successfully (Cavazza 2002: 19). However, VOLTA's letters were the first to be published in England on the subject.
228 An English translation of the paper was published in *Medical facts and observations* (Johnson 1794: 162–210).

attributed to this animal electricity" (Johnson 1794: 175). Indeed, he states that if GALVANI

> had a little varied his experiments, as I have done, he would have seen that this double contact of nerve and of muscle, this circuit which he imagines, is not always necessary. He would have found, as I have, that the same convulsions, the same movements may be excited in the legs and other limbs of frogs, and of every other animal, by placing metallic substances in contact with two parts of a nerve only, or with two muscles, or even with different parts of a single and simple muscle. (*Phil. Trans.* 1793: 10–44, translated in Johnson 1794: 176)

Finally, VOLTA comes to his point, by stating that, what GALVANI attributed to a spontaneous animal electricity, VOLTA attributes to the laws of common electricity: "they are really the effects of a very feeble artificial electricity, which is excited in a way never before suspected, by the simple application of two simple coatings of different metals" (Johnson 1794: 178). He then goes into the narrative of his own experiments to prove his point (concluding the first letter). The paper then continues with VOLTA's second letter, in which he reports of his own experiments on larger animals (rabbits, dogs, lambs and oxen) and with variations in the procedures and muscles tested. The reporting of the second letter is no longer focused on GALVANI but only on VOLTA's experiments and results. The writing is in this case more cautious making extensive use of hedging devices in reporting his observations.

VOLTA's paper succeeded in arousing interest on the subject and in 1795 another intertextually related paper by WILLIAM CHARLES WELLS was published in the *PTRS* – the first of a long series. Here, WELLS makes observations both on GALVANI's and VOLTA's experiments, stating that GALVANI's theory was erroneous (*Phil. Trans.* 1795: 248) and that VOLTA had "fallen into some mistakes himself". Since two years had passed from VOLTA's letters and – notwithstanding his promise to send more material – nothing more was heard from him, WELLS tried to account for the lacunae of VOLTA's researches. Without going into further detail, GALVANI's and VOLTA's experiments provoked a debate that lasted for about a decade, with both "physicists" and "physiologists" (in broad sense) entering the public dispute over the explanation of their experiments (Kipnis 1987: 116). The first immediate reaction to GALVANI's book from the international scientific community was disbelief, which then – after new experiments – turned into praise, although there was no unanimity about his theory. While some defended GALVANI, others sided with VOLTA. For some, but not for all, VOLTA's invention of the *pile* in 1800 was the proof that he was right. The pile was the final result of VOLTA's research to prove that electricity was not intrinsic to living beings but excited by the two different metals. In fact, the pile consisted of a number of plates of silver and zinc separated by cardboard plates soaked in salt water and arranged one on top of the other in sequence. This sequence or *pile* (<French, stack) of metal plates and water in contact produced electric shocks and sparks. But, returning to the outset of the debate, it was VOLTA who was eventually awarded the Copley medal by the Society in 1794 for his two letters. According to BANKS, VOLTA deserved the award, because, while GALVANI's findings were more likely "accidentally discovered", VOLTA was the one who subjected the experiments "to sound reasoning and accurate investigation" and explained them "to the whole of Europe with infinite acuteness of judgement and solidity of argument" (BANKS quoted in Cavazza 2002: 20). Although he continued to carry out his experiments, GALVANI, who had a reserved character and confined life, kept a low profile and was reluctant to intervene in the controversy associated with his name. Instead, his nephew and assistant GIOVANNI ALDINI continued GALVANI's research in animal electricity; he travelled extensively throughout Europe defending his uncle against the attacks and becoming the most ardent propagandist of animal electricity, or galvanism, as he called it (Parent 2004). Animal electricity also became popular among the wider public of non-scientists, providing both shock and amusement; *The British critic* commented on the Italian experiments with the following:

> Before we quit this article, we must enter our protest against the *horrid cruelties*, accompanying many of these experiments of the Italian anatomists, and express our apprehension lest the dissection of living human subjects, a practice of the ancient Egyptians, should by one step further in philosophical apathy, be renewed; nor can we conceive how Mr. Volta could consider the noise of a grasshopper, excited by tortures, as an amusement. (Rivington 1793: 91)

To conclude, news of disputes frequently reached the Fellows, and the judgement of the Royal Society was often requested by natural philosophers. When the Society or its Fellows were directly involved – as was seen with the JURIN-MICHELOTTI and RIZZETTI-DESAGULIERS cases – the Society always cooperatively replied trying to clarify any misunderstandings (e.g. by reproducing experiments). Instead, when the Society was not directly involved, its behaviour could differ, especially depending on how much was known upon a given subject (or person, earlier in the century). Indeed, the Fellows sometimes even refused to give their judgement. This was, for instance, the case with a book sent by PAOLO MATTIA DORIA in the early 1730s. The book in question was *Duplicationis cubi demonstratio* (Venice, 1730) and DORIA had repeatedly asked the Fellows to express their judgement on it. While a committee was appointed to review the book, we learn from one of DORIA's rather disappointed letters that the Society was not going "to pronounce any Judgement upon new Conventions, either by approving or by condemning them. [...] The learned Gentlemen, to whom the Examination of my new Invention has been referred, should give their Opinion, not in the name of the Society, but only as their own private Sentiment" (DORIA to ROLLI, translated by the latter, 1732, EL/D2/58).[229]

4.2.4 Witnessing

In the previous century the strategy of witnessing to give credibility to a report or an experiment was found to be frequent. It was also observed that there was a tendency to indicate who the witnesses were in terms of their occupation and/or social standing. These strategies continued to be employed in the 18th century as well with some variations as the century proceeded. In fact, in this century, identification of the witnesses or the person who carried out an experiment or made observations were more focused on their field of interest and affiliations to Italian universities and academies, rather than their social standing. For instance, JOHN SWINTON is always introduced in the titles of his papers as "the Rev. John Swinton, B. D. F. R. S. Custos Archivorum of the University of Oxford, Member of the Academy Degli Apatisti at Florence, and of the Etruscan Academy of Cortona in Tuscany".

Frequently, moreover, new scholars were introduced in discourse by emphasising their skills – which had been observed in their writings or experimental procedures – and scientific achievements. LAZZARO MORO, for instance, is introduced in a review of his book *De Crostacei e altri marini Corpi* (Venice, 1740) as

> a Clergyman; [who] never entered into any ecclesiastical Community, nor into any University as Professor; to be out of the Way of Envy: However he keeps a Boarding-School for young Men. He has published the Book in Question at his own Expence; which has brought him into some Trouble, and render'd the Book at first very scarce. He shews a great Conformity to the Principles of Sir *Isaac Newton*, and other modern

---

229 See also the case of the Bologna-Ferrara dispute on the flow of the Rhine river into the Po, reported by Cavazza (2002: 7–8). Also in this case, one of the Fellows (EDMOND HALLEY) was appointed to review the book where the dispute is reported, which he did, concluding however that no definitive judgement could be made without a visit to the place and an exact calculation of the velocity of the two rivers. The Fellows of the Society as an institution did not want to give an opinion and NEWTON replied that the Society "never gives its own opinion in doubtful matters" (quoted in Cavazza 2002: 8).

Philosophers, not very common in *Italy*, grounding himself upon Experience, and mathematical Proofs. (*Phil. Trans.* 1746: 163)

Hence, given the lack of any affiliation to some respectable institution, the author is subjected to the Society's consideration for his Newtonian ideas and principles of experience and proof, which were in line with the Society's principles. Father DELLA TORRE, instead, was regarded as trustworthy by EYLES-STILES, not only for his personal qualities, but also for his skill in using the microscope:

> this most worthy Father is so esteemed here [in Naples] for his excellent and amiable private character that there can be no room for the least suspicion of the veracity of the remarks in this book; and in respect of his care and exactness in making his observations, I have had such an experience of it in several meetings he has favoured me with for the purpose, that I should in any case as readily rely on his eyes as my own and indeed with better dependence; as, by the habit of observing, he is enabled to pronounce immediately with certainty upon appearances that cost me much time and inspection to examine and comprehend. (*Phil. Trans.* 1765: 247)

EYLES-STILES is here moreover placing himself as a testimony of DELLA TORRE's abilities and credibility.

The practice of witnessing is very interestingly used in ABBE NOLLET's paper "An Examination of certain Phenomena in Electricity" (*Phil. Trans.* 1749: 368). This paper reports on a series of experiments that had been run by Italian scholars of electricity and NOLLET himself. In the first pages NOLLET explains that he had travelled to Italy in 1749 specifically because he had attempted several times to run the electricity experiments of GIOVANNI FRANCESCO PIVATI – which had been heard of throughout Europe as being very successful for the treatment of distempers – without obtaining any results.[230] He therefore doubted the truthfulness of the claimed results and was eager to discover whether his doubts would be confirmed or disproved. In Turin with BIANCHINI and in Venice with the famous PIVATI, the experiments were carried out. These consisted in making different natural aromas (scammony, gamboge, balsam of Peru and several others) pass from the inside of an electrified glass cylinder to its external surface and consequently enter in contact with the person holding the cylinder. The result from all of the trials was that no particular effects were felt by any of the persons who underwent the experiments, including NOLLET himself. In all of these experiments NOLLET lists the names of the persons who witnessed the events as well as the names of those who underwent the experiments. Moreover, since the first experiment was carried out on people who were not considered

---

230 Throughout the paper NOLLET references in footnotes each of the works he refers to.

worthy of much credit (such as two servants and a teenage girl), the following trials were made on persons considered to be more reliable subjects (notable physicians and noblemen of the time). To add further evidence to his argument NOLLET quotes letter extracts from other experimenters in electricity, including GIUSEPPE VERATTI, showing that they all found that the experiments carried out by them had none of the effects that had been claimed in the writings of PIVATI. Further,

> I learned nothing in the other Cities of *Italy*, which did not strengthen my Doubts in relation to those electrical Phænomena, which I had a Desire to verify in the course of my Travels. Pere *La Torre*, Professor of Philosophy at *Naples*; M. *De la Garde*, Director of the Coinage at *Florence*, one who has been much engaged in these Inquiries; M. *Guadagni*, Professor of experimental Philosophy at *Pisa*; the Marquis *Maffei*, at *Verona*; Dr. *Cornelio*, at *Placentia*; Pere *Garo*, at *Turin*; all these, I say, with very excellent and well-contrived Machines, and with a great Desire of succeeding, have attempted many times to transmit the Odours, as well as the Powers of Drugs closed (carefully) in Tubes or Spheres of Glass, by electrizing them: all these have attempted to purge a Number of Persons; and, according to the Accounts they gave me, have never gained their Point; or the little Success they had, appeared too equivocal to draw therefrom Consequences conformable to those M. Pivati had believed to have seen in his Experiments. (*Phil. Trans.* 1749: 391)

He moreover quotes the work of PIVATI[231] highlighting that PIVATI himself had stated that his experiments had been found to be beneficial in only a few subjects, who, according to NOLLET, were sick and therefore "prejudiced perhaps by too great Hope, and possessed by a kind of Enthusiasm, have said themselves, and made others believe, more than really was the case". Hence, according to NOLLET, based on his and others' experiments and observations "the Facts have been greatly exaggerated" and "a great Part of the Cures of *Turin* have been no other than temporary Shadows, which have been taken with a little too much Precipitation, or Complaisance for Realities" (*Phil. Trans.* 1749: 383).

Hence, NOLLET uses witnessing to strengthen the credibility of his claims; however, at the beginning of the paper he uses the same strategy for a different purpose, which was to account for the great attention that was given to Italian research in electricity:

> Effects no less wonderful than these were published every Day, by Writings printed, and printed again, or by particular Letters and Memoirs in Manuscripts addressed to the Ingenious all over *Europe*. They were also confirmed by respectable Witnesses, and by such as were capable of imposing them upon Persons the most guarded against the Exaggerations, which never fail accompanying the Relations of interesting Novelties.

---

231 *Della Elettricita, Lettera del chiarissimo Signor Francisco Pivati, &c.* (Lucca 1747).

The importance of the facts themselves, and the Appearance of Authenticity which attended them, demanded that they should be considered; and indeed they roused every-where the Attention of those Philosophers, who had for any time returned their Thoughts to theses Enquiries. Every one of them was desirous of repeating what Mr. Pivati said had been done at Venice, Mr. Verati at Bologna, and Mr. Bianchi at Turin [...]. (Phil Trans. 1749: 371–372)

NOLLET thus shows that precisely because the work of the Italian "electricians" had been confirmed by respectable witnesses, it gained an *appearance* of authenticity. The credibility attributed to Italian research was then dismantled by some of the same witnesses throughout the course of NOLLET's paper. NOLLET's account did eventually convince the Fellows in England that "what the Italians printed upon the transmission of odours thro' the pores of glass, and upon the subject of medical electricity, [was] a too hasty publication" (*Phil. Trans.* 1751: 399–400); however, their interest on the subject did not wane and further accounts were published. The Fellows in England too had tried the experiments finding themselves disappointed; their final judgement on the inefficacy of administering medicine through the power of electricity was made in a paper published in *PTRS* in 1751, which was a book account on FORTUNATO BIANCHINI's *Recueil d'Experiences Faites a Venise sur le Medicine Electrique* (1750) (*Phil. Trans.* 1751: 399–406).[232] BIANCHINI's book reported on a number of experiments that had been carried out in Venice following all of the oppositions that had been raised against the first Italian experiments. The author of the paper, WILLIAM WATSON, explains that this time the experiments

> have been ingeniously imagined, sensibly conducted, ranged in proper order, robbed of all superfluous reasoning, and made just in the same manner as those of the academy *del Cimento*, the value of which every one present, I presume, is not now to be apprized of. The truth of this publication is not to be suspected; it comes from the very place, where medical electricity took its rise; and is not the production of one person, who might be suspected too slightly to have admitted what might tend to favour his own opinions. These are facts consider'd in themselves independently of all application, decisions of the unanimous voice of a number of every sensible men, and in the face of a great number of witnesses, many of them prejudiced to the contrary, and but here forced to be convinced by the evidence of facts. (*Phil. Trans.* 1751: 399–406, 400)

---

232 This was the French translation sent to the Society by NOLLET of the *Saggio d'esperienze intorno la medicina elettrica fatte in Venezia da alcuni amatori di fisica al signor abate Nollet ... Descritte dal dottor Gio. Fortunato Bianchini professore di medicina napoletano* (Venice 1749).

Hence, once again the presence of "several curious and learned men" served the purpose of giving credibility to the account, which in this case confuted the first Italian researches and corroborated NOLLET's opinion.

In sum, the strategy of witnessing was still widespread in the 18th century, as was also that of identifying the witnesses in terms of their occupation and affiliations. However, while there still were cases in which witnesses were introduced by their social status, there appears to be a growing tendency to classifying them according to their abilities and scientific achievements.

4.2.5 Toponymy

Part of the rhetorical strategies for the construction of a discourse of fact included the specification of the time and place where a given phenomenon was observed or where an experiment was carried out. Hence, writers in the 18th century continued to meticulously indicate the place names of where events, phenomena and experiments took place. The tracking of toponyms also becomes useful from a cultural and historical point of view, in that it allows us to view more precisely what were the main areas from which the Society received information. Tab. 4.5 below reports only the *main* place names mentioned in the journal and the number of papers citing them:

Tab. 4.5: 18th-century cited place names and number of papers citing them

Toponym	Number of papers mentioning	
Naples	48	
Rome	33	+ 3 Roman State(s)
Venice	20	+ 2 Venetian State/Republic
Bologna	19	
Padua	13	
Turin	10	
Florence	10	
Pisa	9	
Calabria	7	
Sicily	7	
Tuscany	6	
Leghorn	5	
Milan	3	
Apulia	3	
Viterbo	3	
Messina	3	
Lombardy	2	
Lucca	2	
Other mentions:	Verona and Veronese district, Vicenza, Treviso, Ancona, Piedmont, Cuneo, Vercelli, Cesena, Como, Abruzzo, Pesaro, Corsica, Appenines, Alps, Euganean Hills, Palermo, Messina (and many other towns of the Kingdoms of Naples and Sicily).	

As can be seen from the table, results differ considerably from those of the previous century. Naples dominates the *Transactions* in the 18th century with 48 papers referring to it, the majority of which concerned eruptions, earthquakes and the excavations at Herculaneum. Rome now moves to second place contributing papers mainly in astronomy and archaeology with few contributions in the medical field as well. Venice continued to be an important centre of diplomacy; the main topics of research from the Venetian Republic included electricity, medicine and geology. Bologna and the Istituto delle Scienze, which was founded in 1714 on the model of the Royal Society, were important centres for the study of electricity, astronomy and medicine. Padua continued to be a source of medical researches, but also contributed papers in astronomy and in the earth sciences (meteorology and geology). Turin and the region of Piedmont gained greater visibility in the 18th century with studies in electricity and earth sciences. The papers from Tuscany and, in particular, Florence and Livorno (Leghorn) were of a variety of subjects, including medicine, electricity, physics, antiquarianism and mathematics. The Kingdom of Sicily was a source of papers primarily in the earth sciences, especially in volcanology and seismology. The same subjects came from Calabria with the addition of a few papers in botany. From Lombardy and Milan came only a few papers in a variety of subjects (medicine, earth sciences, mathematics and a travel account).

While in the 17th century there appeared to be no specific distinctions made between the different Italian states, and very often scholars from the Peninsula were referred to simply as being Italian, in this century distinctions are present. In several papers mention is made of the various political entities, such as the Roman States, the Venetian Republic and the Kingdoms of Naples and Sicily. HAMILTON moreover communicated that to be able to leave Naples for a twenty-day visit "out of Italy", to Sicily and Calabria, he was furnished "by command of his Sicilian Majesty, with ample passports, and orders to the commanding officers of the different provinces to give me every assistance and protection in the pursuit of my object", which was to further his studies in volcanology and seismology (*Phil. Trans.* 1783: 174–175).

4.2.6 Interdiscursivity and intertextuality

Intertextual (explicit) and interdiscursive (implicit) relations between papers continue to be a feature of the *Transactions* as a journal. References to other *PTRS* papers, books and treatises were already frequent in the previous century and become even more frequent and detailed in the 18th. Indeed, in this century many papers make numerous and specific references to authors, titles and page numbers by means of footnotes. Some papers consisted entirely of literature reviews. Cooperative invitations to provide further contributions and perspectives on a given topic were also found, as was seen in the case of JURIN's meteorological project (Section 3.2.4), and various papers stood in dialogic relation with one another.

Moreover, by now the Italians seem to be fully aware of the manner in which the Society proceeded. And it was frequently the Italians themselves, who spontaneously sent material they thought the Fellows might be interested in. The purpose of publishing their reports was also clear to them; PASQUALE PEDINI for instance wrote an account of earthquakes and an *aurora borealis* that occurred in Livorno in 1742 and concluded his account by saying "I have related to you a true Exposition of all things as they really were; and there is now a way opened for Philosophical Observations and Inquiries" (*Phil. Trans.* 1742: 90). PEDINI therefore knew that accounts such as his own served a higher research purpose and would be considered together with others of the same kind to investigate into the subject. Indeed, PEDINI himself appended to his own letter two further accounts, by gentlemen worthy of credit, that reported further information on the same phenomena. The same kind of collaborative behaviour is seen in the letters of many Italians. Another example worth mentioning is that of NICOLÒ D'ARAGONA, who sent a paper on the eruption of Vesuvius in 1737 "to the End that, if you are pleased to investigate the Causes thereof, the Republic of Letters might reap some general Advantage, as it does daily, by means of its Members of the first Rank in Merit" (*Phil. Trans.* 1739: 237–252, 237). Indeed, the many individual accounts of earthquakes and eruptions were exploited by the Fellows according to their principles, i.e. investigating nature cooperatively in order to create an exact history of it. Consequently, STEPHEN HALES wrote a paper of considerations on the causes of earthquakes (*Phil. Trans.* 1749: 669–681). Here he takes into account many previous observations on these phenomena from England,[233] France and Italy highlighting the common aspects that the previous scholars had observed upon the subject.

Another example, where the existence of multiple observations on a same topic was exploited to achieve a more precise knowledge of it, was with "An Account of Some Human Bones Incrusted with Stone [...]" (*Phil. Trans.* 1744: 557–560). Here the author describes the body of a petrified man in the Villa Ludovisia in Rome, who was believed to have frozen in the Alps. He moreover includes in the paper a drawing of the man in order to provide an accurate representation for the Fellows in England. The author reports information from the previous accounts on the same petrified man – with intertextual references to the Society's Journal Book, to the published *Travels* of RICHARD LASSELS (Paris 1680), KIRCHER's *Mundus Subterraneus* (1665) and others – showing how they differed in what they reported about the man. For instance, one account reported that an arm had been broken off, while another claimed that it was a leg to have been broken.[234] The variety in the information reported by the various witnesses led the author of this paper to present his own account of it and to employ a painter for the drawing, so that the Fellows in England might have a realistic idea of the object.

---

233 Around 1750 there were earthquakes in London, Portsmouth and the Isle of White and several *PTRS* articles were dedicated to them.
234 The limb had apparently been broken off by the owners to show the public that the incrustations that covered the body were not manmade but spontaneous.

To conclude, many of the features that characterised the *Transactions* in the 17th century – witnessing strategies, factual discourse, a cooperative approach to the study of nature, author-centredness, interest in non-scientific subjects etc. – continued in the 18th. However, several developments were made including an increased accurateness and detail in reporting, fewer digressions, and an increased neutrality in the representation of discourse. The new rules set on Fellow elections did not considerably hinder admittance of the Italians to the Society; while the restrictions on publication did prevent the inclusion of several Italian papers in the journal; however, the number of excluded papers is relatively small compared to the great amount of both scientific and non-scientific Italian papers that were published.

4.3 19th century

This section presents and discusses the results of the discourse analysis carried out on the sampled *PTRS* papers for the 19th century, in the same order of the previous centuries. Unlike the previous centuries however, there will not be a sub-section on the reporting of disputes and disagreements because none of these were reported in the 19th-century papers – which is in itself a noticeable advancement compared to the previous centuries.

4.3.1 Textual dimension

4.3.1.1 Macrostructural features

The letter form, which was frequent in the previous centuries – 37 % in the 17th and 39.4 % in the 18th – represented a minority in the 19th with only six papers (11.5 %) published in this form.[235] This can be considered one of the first marks of the changes taking place in the Society's publications and development towards the modern scientific journal.

As to the text types, the majority of papers were experimental reports (sixteen papers) and observations (sixteen). Experimental reports were generally reports of multiple experiments of the same kind with minor variations between them. After each experiment the author would draw some preliminary conclusions. The last main text type was given by reports of autopsies (four papers plus two abstracts of the papers), which were either reports of single or multiple autopsies.

---

235 A further paper also contained a supplementary letter.

Titles in the first part of the century continue to resemble those of the 17th and 18th centuries, without however indicating the translated nature of the text and the name of the translator. Titles continue to state the topic, the name of the author, and the name of the person who communicated the paper to the Society. The persons named in the titles are also generally followed by their profession and affiliation to universities and academies.

Lengthwise, *Proceedings* papers ranged from mere titles to a few pages, generally no more than four, although towards the end of the century a couple of papers that were published only in this journal were around ten pages. *Transactions* papers instead reflect the general tendency that was observed in the 18th-century, namely that they were becoming longer, with fifteen papers having between 5 and 14 pages; ten papers exceeding 14 pages; and only two papers were shorter than four pages.

The general macrostructure of the papers was in twelve cases highly organised, featuring introductions, objectives, literature reviews with detailed references, reports of experiments and observations, and conclusions. Occasionally these papers were also divided into sections which were either numbered or preceded by subtitles. Most papers moreover tended to revise the existing literature, with some of them again being whole literature reviews on a given area of study. Many papers contained drawings and some also tables.

Translated *PTRS* papers never displayed any framing other than the title and there were no editorial interventions. On the contrary, *Proceedings* papers were frequently interspersed with additional paragraphs by the editor summarising long sections of the original paper and providing, in the case of the older papers, notices on subsequent developments on the piece of research. The editor often intruded into the author's narrative transforming it into third person and summarising the original authors' moves and intentions. In the case of the papers from Italy, brief biographical notes were also inserted in footnotes. However, as the century proceeds these interventions became fewer and the *Proceedings* became closer to the modern concept of scientific paper than the *PTRS*; this will be seen in the following section.

4.3.1.2 Language use

The linguistic features of the papers mark a very slow tendency towards becoming more informative and abstract; however, author-centredness and involvement continue to be present throughout the century. What decreases is the number of personal comments and, in some cases, the use of verbs of emotion. Often the papers display informative and slightly abstract sections alternated with more involved sections, or a mixture of all features throughout the paper. Papers in this century display a higher presence of connectors (e.g. *finally, in effect, therefore, in this view, thus, nevertheless* etc.) and passive voice, which mark a more informational and abstract production, and there was hardly any presence of forms of encomia towards the Society, and very little praise towards the Fellows and other researchers. Instead, humble presentations of papers are still present both in translated and original English papers (six instances).

Of the 26 *Transactions* papers, fifteen were marked as author centred. These still displayed strong authorial presence by means of first-person pronouns, active tenses, mild use of verbs of emotion, and questions. *Transactions* papers marked as midway between an involved and informational style were nine. Midway papers either displayed involved parts and informational and abstract parts, or were generally informative, sometimes displaying abstract features with occasional slips back into the author-centred style. Only one paper was marked as being purely informational.

The 26 *Proceedings* papers instead differ considerably from the *Transactions*. As anticipated above, these papers often displayed editorial interventions and summaries. In these cases, judgements were sometimes made on the piece of research reported, hence showing involvement on the editor's part. However, in other cases *Proceedings* papers were actual abstracts of the original papers, from which all of the involved parts were generally eliminated as unnecessary for the understanding of the paper. Hence, these papers became highly informational and abstract. See for instance the two extracts below taken from the same parts of SOMERVILLE's paper (on the magnetising of needles by exposure to the solar spectrum) published in both journals; the bold parts are instances of involved features while the italics show abstract and informational features:

> The sun was bright at the time, and in less than two hours **I had the gratification** to find that the end of the needle which *had been exposed* to the violet rays attracted the south pole of the magnetic needle, and repelled the north pole. It *had been previously ascertained* that there was no iron near to disturb the results. The experiment *was also repeated* on the same day, under **precisely** similar circumstances, with the view of detecting any source of error that might *have escaped observation* in a first attempt; but the result was the same as in the first.
> The season was **so** favourable that it afforded **me** daily opportunity of repeating the experiments, varying the size of the needles, **always taking especial care** to ascertain that they were free from magnetism. [...]
> **I was desirous** to ascertain whether this kind of glass suffered the chemical rays to pass and thereby occasion these changes in steel, *therefore* **I employed** a liquid holding silver muriate in suspension, as a test, in the following manner [...] (*Phil. Trans.* 1826: 133–134)

The sun *being* bright, in less than two hours the needle, which before the experiment showed no signs of polarity, *had become* magnetic; the exposed end *attracting* the south pole of a suspended magnetic needle, and *repelling* the north. No iron was near to disturb the experiment, which *was repeated* the same day, under similar circumstances, with a view to detect any source of *fallacy* in the first attempt, but with the same result. The season continuing favourable, afforded daily opportunities of repeating and varying the experiment. Needles of various size (all carefully ascertained to be free from polarity), and exposed in various position with regard to the magnetic dip and meridian, almost all became magnetic. […]
Experiments *were next instituted by transmission* of solar rays into *coloured media*. […]
The rays transmitted through the glass *employed* in this experiment, blackened muriate of silver as powerfully as those transmitted through uncoloured glass, proving it free *permeability* to the chemical rays. (*Proceedings* 1833: 263)

SOMERVILLE's *PTRS* paper exemplifies what was above reported as to the coexistence of authorial presence within the narrative together with features of abstractedness. Indeed, her paper is characterised by the use of first-person pronouns, expressions of emotion and boosters; while at the same time the experiments are reported by means of passives and the object of research in subject position. In the *Proceedings* paper, the features of involvement (in bold in the first extract) are eliminated and the verbal style is more nominalised and concise – e.g. "*to ascertain that they were* free from magnetism" becomes "*ascertained to be free* from polarity" and "detecting any *source of error that might have escaped observation*" becomes "detect any source of fallacy" – with the result that the paper now becomes highly informational and abstract.

Of the 26 *Proceedings* papers five were marked as being involved, six were marked as being both informational and abstract, four informational, and four midway between involved and informational. The joint results for the *Transactions* and *Proceedings* have been summarised in Tab. 4.6 below:

The results in the table show that up to the 19th century there was always a relatively high level of author-centredness, although the 17th-century papers displayed many more features of involvement than 18th- and 19th-century papers. As to the level of informativity and abstractness, these vary throughout the three centuries. One point that ought to be kept into consideration here is that in the last two centuries there are multiple articles by a same author, and their personal style could influence the overall numbers of the linguistic categories. For instance, in the 19th century there were 12 papers by MATTEUCCI (23 % of the corpus), whose style was always mildly involved, even though his papers were experimental reports, a genre which has been considered to increase in informativity and abstractedness in the course of the *Transactions*' existence.

Tab. 4.6: Classification of papers according to the presence of linguistic features of involvement, informativity and abstractedness

	19th century 52 papers	18th century 185 papers	17th century 102 papers
Involved	20 papers (38.4 %)	92 (49.7 %)	41 (40.1 %)
Midway between involved and informational	13 (25 %)	27 (14.5 %)	20 (19.6 %)
Informational	5 (9.6 %)	23 (12.4 %)	19 (18.6 %)
Informational and abstract	6 (11.5 %) (only *Proceedings*)	1 (0.5 %)	5 (4.9 %)
Papers excluded from this analysis	8 (15.38 %)	42 (22.7 %)	17 (16.6 %)

Focusing only on the 19th century, it was observed that the midway, informational and abstract papers were mostly those written directly in English,[236] while the twenty involved papers had a majority of translated papers (thirteen); i.e. papers that represented the Italian writing style rather than the English. This would seem to suggest that the English Fellows were moving faster towards an informational and abstract style than their Italian colleagues; however, the presence of MATTEUCCI's papers, which tended to be more involved, influences the overall results. It was also noticed, both in the 18th and 19th centuries, that the papers of the two Italian naturalised British Fellows, CAVALLO and GRANVILLE, tended to be more informational and structured than those of the other Foreign Members.

Finally, narrativity continues to be a distinctive feature with 35 papers (67.3 %) characterised by a narrative and/or descriptive style. This reflects the higher number of experimental and autopsy reports in the corpus.

4.3.2 Discursive practice

4.3.2.1 Discourse representation

Most of the sampled 19th-century papers include some form of represented discourse: (1) by incorporating original extracts of Italian writings – this only occurred in the case of the *Proceedings* –; or (2) by reproducing full translated

---

236 Informational and abstract papers were all of English pen; midway papers were all but one English; informational, all English; involved, thirteen Italian and seven English.

Italian papers; or (3) by reporting and adapting their contents (*Proceedings* only). Papers were marked as "neutral" when there was no kind of judgement or bias towards what was being reported or, in the case of translated papers, when there was no form of framing and commentary, that is, the paper was only a direct translation of the Italian original. However, translated papers could themselves represent other discourses, both Italian and foreign. Papers marked as "neutral-positive" were mainly neutral but contained some minor instances of positive evaluative language. Papers marked as "neutral-negative" contained minor negative or disagreeing comments, which were however irrelevant to the representation of discourse as a whole. And papers marked as "negative" contained criticism and/or disagreement. The results for discourse representation in the 19th century are summarised in table 4.7 below:

Tab. 4.7: 19th-century discourse representation results

	Tot.	Original English	Translated
Neutral	29 (55.7 %)	14	15
Neutral-positive	1 (1.9 %)	1	-
Positive	2 (3.8 %)	2	-
Neutral-negative	1 (1.9 %)	1	-
Negative	2 (3.8 %)	2	-
No discourse representation	10 (19.2 %)	10	-
Excluded from this analysis	7 (13.4 %)	-	-

As can be seen from the table, the tendency with translated papers is to report them neutrally, without any forms of framing and editorial intervention that could display the Fellows' stance towards them. Neutrality in reporting or referencing others' research is indeed the main tendency in this century and was already becoming so in the 18th. The two papers marked as negative – one from *Proceedings* and one from *PTRS* – contained negative judgement of an Italian reported paper and, the other, negative evaluation towards VOLTA's approach to research. The two positive papers instead were both published in the *Proceedings* and praise the works of NOBILI and SECCHI. The neutral-negative paper came from the *Proceedings* and pointed out the lack of plans and drawings in SECCHI's description of the Observatory of the Collegio Romano, which would have made his account more "intelligible". Apart from one minor statement of error in the research of EMIL DU BOIS REYMOND by MATTEUCCI, no negative representations of others' discourses were made by the Italians.

In the intertextually related papers, the research of Italians is taken into consideration mostly neutrally – and positively in the two above mentioned cases. The papers of this century did not present the Fellows' doubts and questions towards the Italian researches that were more frequent in the 18th century, but rather took them as starting point to expand knowledge in the field, as in the case of electricity, animal electricity and magnetism.

4.3.2.2 Evaluation in the discourse of the PTRS

The greater neutrality in reporting entails the presence of fewer forms of evaluation. Indeed, only five papers displayed positive evaluation towards the research of Italians – Galvanism, which was found to provide "a valuable means of relief" in patients suffering from asthma; NOBILI's plate of colours; the "fine experiments of VOLTA and MARIANNI"; and the "firmness of definition" and "priceless value" of SECCHI's drawing of the lunar spot Corpernicus. Negative evaluation was found in four cases, one concerned GRANDI's 18th-century book on Geometrical Flowers, which was judged to be "more curious and fanciful than anyway useful" (Hutton & C. 1809: 664). WILLIAM RITCHIE made several remarks on VOLTA – such as "claims to himself the honour of the discovery"; "taking for granted the truth of the experiment, the conclusion which Volta deduced by no means follows as a legitimate inference"; "a gratuitous supposition" etc. – which display an implicit dislike for VOLTA's boasting of his findings (*Phil. Trans.* 1829: 361–366). In this century, like in the 18th, the focus is more on the piece of research or the approach to the research, rather than on the researcher.

Another way to view the Fellows' evaluation of Italian research was through the referee reports. For instance, most of MATTEUCCI's papers were all welcomed for publication and were neutrally presented in the *Transactions*; the referee reports confirm that his papers were found "interesting" and "valuable", and only in one case the contents were commented as being somewhat similar to those published in previous papers (RR/3/201), nevertheless the paper was still recommended for publication. In another case, a paper was accepted for publication, but modifications were suggested (RR/3/199).

4.3.3 Witnessing and toponymy

The practice of mentioning the names and titles of witnesses present at experiments and autopsies was not found in the sampled papers for the 19th century, which suggests that this practice had either been abandoned or was in any case decreasing. On the contrary, the general impression arising from the papers is that of individualism in making science. While the authors all show to be part of an international discourse community, by describing and referencing the works of others, most of the papers report of experiments that were carried out by the author alone; and the frequent use of first-person pronouns further enhances this sense of individualism. The practice of stating who the author was in terms of their field of study and affiliations remained instead a feature of the *Transactions* in the 19th century as well.

As to the naming of place names, this practice too diminished considerably in the 19th century. Very few papers mention the location where the experiments and autopsies were carried out. However, it was still custom to state from what parts of Italy the Italian scholars came from. Tab. 4.8 below lists the most frequent place names and number of papers mentioning them:

Tab. 4.8: 19th-century cited place names and number of papers citing them

Toponym	Number of papers mentioning
Pisa	12
Rome	5
Milan	4
Pavia	2
Sicily, Etna	2
Naples	2
Puglia	2
Venice	2
Genoa	2 (same paper in the two journals)
Brescia	1

The table for the 19th century displays considerable differences in both the place names cited in the journal and the number of mentions. While a few reports on volcanology remain, Sicily and Campania are no longer the main sources of Italian research. Puglia this time is not mentioned for the tarantula but for seismology and cholera. Pisa in this century is at the top of the list because of the presence of MATTEUCCI's twelve papers in physiology. Rome and the Jesuit Roman College continue to be a source of research in astronomy, physics (MORICHINI), and one paper on zoology by GRASSI. Milan receives more attention in this century thanks to CARLINI's meteorological observations. Venice is mentioned for cholera and the disease that affected cattle in the 18th century. While Pavia enters the chart for the first time, thanks to the influence of VOLTA and LUIGI CONFIGLIACHI, professor of physics at the University of Pavia. Apart from the Collegio Romano and the Academy of Science at Naples, the

Italian academies and institutions that kept correspondence with the Society in the previous centuries are no longer mentioned in the 19th. Consequently, some of the main scientific centres of the previous centuries, Bologna, Florence and Padua, are no longer mentioned as sources of Italian science. Patterson (1983: 28), in reporting about the SOMERVILLES' trip to Italy in 1817–18, would seem to corroborate the general idea that emerges from the Society's relations with Italy in the 19th century, stating that "Unlike London, Edinburgh, or Paris, unlike the larger provincial towns of Britain and the Continent, the various states and cities of the Italian peninsula offered little science at the time, and that done by one or two isolated scientists rather than in specialist societies or institutions that characterised England and France".

4.3.4 Interdiscursivity and intertextuality

Although individualism is one of the features that seems to characterise 19th-century Italian research in the *PTRS*, all scholars, both Italian and English, subscribe to broader discourse communities. The practice of writing literature reviews appears to have become widespread in this century, with some papers being whole discussions of the literature in a given subject area. In many cases the reviews are done by carefully referencing the works, dates and places of publication; while in other cases there is only a mentioning of the authors, and only their surnames. However, the very fact that authors were mentioned only by their names is a further indication that researchers in a specific field knew the work of their peers and did not therefore need to receive detailed bibliographies. Another aspect that characterises some of the 19th-century papers – and still characterises present-day scientific writing – is the creation of a niche for the author's piece of research by finding lacunae in the research of others.

Chapter 5   General conclusions

5.1   Development of Anglo-Italian socio-cultural relations 219
5.2   Languages of international scientific communication and linguistic
       consequences 221
5.3   Development of Italian and English *PTRS* papers 223
5.4   Development of Italian discourse representation 224
5.5   Concluding remarks 224

5.1   Development of Anglo-Italian socio-cultural relations

From the analysis of the social and cultural aspects that emerged from the elections of Italians to the Royal Society, and the contributions made by them and by the non-elected contributors, it appears that the Society had a strong interest towards Italy in the 17th and 18th centuries and welcomed Italian research in the 19th century – although with the greater scientific orientation and the founding of specialist societies the publication of Italian researches was not a priority. Indeed, in the 17th century it was the Society, and OLDENBURG especially, who worked on the creation of a scientific correspondence network with Italy. Conversely, by the 18th century, it was more often the Italians who sought connections with the Society and frequently asked for the Society's judgement and approval of their works and its opinion in doubtful matters. In the 17th and 18th centuries the main Italian centres connected with the Society were set in Tuscany with the Accademia del Cimento and its universities; in Rome, with the Accademia Fisica Matematica and most importantly the Collegio Romano, whose members would be the only ones to maintain a regular correspondence well into the 19th century. The Accademia Pontificia dei Nuovi Lincei was also to revive contacts in the latter period. Exchanges with Rome were moreover eased by the presence of English residents and travellers throughout the whole period. New links were also established with the Senese Accademia dei Fisiocritici.[237] Venice was another source of contacts both for the presence of intellectual circles and as an important commercial and diplomatic centre. The connections with Padua and Bologna were related to their universities and later with the Bolognese Istituto delle Scienze e delle Arti, but then decreased in the 19th century. The Kingdoms

---

237   NLB/4/203, 231 and NLB/11/435.

of Naples and Sicily – with their learned academies on the one hand and natural and archaeological beauties on the other – continued to attract the Fellows for both scientific and cultural reasons throughout the whole period. Towards the end of the 18th century and throughout the 19th, there was a stronger interest in the research of the Italians from the northern parts of Italy, with Turin, Milan and Pavia gaining new prominence as sources of scientific advances.

However, while universities and academies provided a greater sense of eminence and reliability, what was most important at the end of the day were the efforts made by individual scholars. On both sides there were men of learning who did not just focus on discussing their own research but kept each side up to date by collecting and transmitting works and scientific intelligence from their respective countries. Key to the Society's relations with Italy were also English residents and travellers in Italy, who acted as intermediaries between the Fellows and the Italian scholars. It has moreover been seen with various examples that Italian Foreign Members were not necessarily of higher importance, in the Society's eyes, than Italian contributors who were not elected. From the Italian perspective, the Royal Society was for the Italians a model, with its focus on science "without regard for birth or religion".[238] On multiple occasions moreover the Italians eagerly collaborated on the projects of the Fellows. In turn, the Society rewarded them by sponsoring their research and making them known in the *Transactions*, by publishing their books and, in the 19th century, by occasionally providing financial support and/or awarding medals.

As to the topics of interest, Italian research covered – in the first two centuries – a wide range of subjects including non-scientific writings such as travel accounts and papers on antiquities. Botanical, geological and natural historical papers were always present, with a peak of papers on earthquakes and volcanos in the mid-18th century; however, medicine and astronomy were generally more frequent in the early *Transactions*. In the late 18th century and through the 19th, less space was given to non-scientific subjects, and physical research was given greater importance, although medical subject matter was also still very present.[239]

---

238 MARSILI quoted in McConnel (1993: 190).
239 The Society's lack of specialisation, the abundance of medical men among its Fellows and the publishing of medical material, were among the points of criticism in the 1820–1830s.

5.2 Languages of international scientific communication and linguistic consequences

As has been seen, Latin was the primary language used in the letter exchanges between England and Italy up until the 18th century. Gradually its use started diminishing and more material was sent to the Society in French or Italian and translated into English. Moving into the 19th century Latin was eventually completely abandoned and substituted by French both in the letter exchanges and in many of the papers sent for publication. However, while many Latin papers were kept in the source language – to make them more widely available to the *Transactions*' international audience – the papers in Italian and French were generally translated into English. As far as the epistolary exchanges are concerned, compared to Italian French was more commonly known both by Italians and Englishmen and therefore represented the suitable compromise for their correspondence. Finally, there was very little evidence of English scholars studying Italian, while there was increasing evidence of Italians studying English, which testifies to the greater cultural and political importance that Britain had gained over the centuries.

Contact between two cultures generally results in the exchange of linguistic material, and this was also a result of Anglo-Italian scientific relations. Indeed, while Italian papers were translated into English, a number of Italianisms were maintained in translation, or willingly reproduced by English reporters, with some of them eventually becoming appropriated by the English language. Indeed, loanwords are generally welcomed into a language in th absence of lexical items to express a particular meaning. The period under study here is characterised by the fashion of the *Grand Tour* and many of the Italian borrowings that entered the English language around this time are related to the Italian territories, leisure activities, foodstuffs, art, architecture and, finally, also science – although the *OED* appears to date most of the scientific borrowings from the 19th century. The scientific borrowings are related especially to the fields of biochemistry, chemistry, physics, geology and mineralogy.[240] The higher presence of scientific borrowings further shows how science played an important role in Anglo-Italian relations between the late 17th and 19th centuries.

---

240 Pinnavaia (2001) has carried out a study on Italian loanwords in the *OED*; here she shows that while between the 14th and the 20th centuries there was always an intake of Italian lexical matter in the English language, the adoption of Italian borrowings varied in quantity and quality throughout the centuries. See also Iamartino (2001 and 2002).

During the analysis any Italianisms were noted and Tab. 6.4 in the appendix provides a list of the borrowings from the Italian language in the *Philosophical Transactions* dividing them per century.[241] The variety of Italian borrowings in the 17th century (fourteen in total) reflects the more miscellaneous character of the first *Transactions*, with Italian lexical items ranging from geology, measures and instruments to musical and literary terms. In the 18th century, 28 borrowings again from a variety of semantic domains were found. The higher influence of the Neapolitan area and the papers on volcanology is also reflected in the various words that were taken from Italian and repeated in many papers – *volcano, lava, rapilli, tuffa, tartana, speronara, felucca*. The experience of Italy from the perspective of travellers is seen in the use of borrowings such as *Via Appia, banditti, sbirri, soldi*; and the interest in architecture and antiquities resulted in the use of several Italian technical words – *intaglios, stucco, cupolas, basso rilievo, granitello*. In the 19th century the borrowings suddenly drop in numbers (nine loanwords) and semantic variety. These reflect both the Society's specialisation to more strictly scientific subjects and the high influence that galvanism had on late 18th- and early 19th-century Europe with various borrowings related to GALVANI's name.[242]

Italian place names were also reported in a variety of forms, especially in the earlier centuries. These could be reported with either their Latin, English or Italian name, and occasionally anglicised by morphological adaptation, e.g. *Leghorne/Livorno/Livorne; Strombilo/Strombulo/Stromboli; Puzzolo/Pozzuoli; Padua/Patavii/Padova; Roman campagne* (Campagna Romana); *Vesuve/Vesuvio/Vesuvius* etc.

None of these borrowings were reported in the papers with a negative connotation, which confirms the positive relations that were kept in the Anglo-Italian scientific exchanges. Moreover, this brief analysis testifies to the importance of the *Philosophical Transactions* as a cultural and linguistic repository, and

---

241 The analysis of the borrowings was of course one-sided as the object of study was the Royal Society's journal, which thus records only the lexical matter that the English took from Italy and not what was taken by the Italians.

242 The words related to galvanism reported in the *OED* are traced back to the French language, rather than the Italian. This is of course true, in that the Italians who wrote to the Society in the late 18th and early 19th century wrote to the Society in French. However, the first news on GALVANI's researches that reached the Society came from Italy (see Section 4.2.3).

the influence that the Royal Society had on British culture in general.[243] At the same time the Italian borrowings – added to the influence of the various Italian scientists and Italian research that have been discussed throughout the three main chapters of this dissertation – show that Italy was by no means in decline in the eyes of the British, at least from a scientific perspective.

5.3 Development of Italian and English *PTRS* papers

The results of the discourse analysis carried out on the corpus showed that, from a structural point of view, there was a gradual increase in the level of organisation of the papers. While the first papers were frequently characterised by editorial framing and comments, the later papers are free of any forms of editorial intervention and often organise contents following a more modern pattern given by introductions, literature reviews, narration of experiments/autopsies/experiences/procedures, tables, drawings and conclusions. This was found both in the Italian translated papers and in the papers originally written in English. Also, while in the early period papers generally consisted of letters, gradually the letters were separated from the actual papers; i.e. the Italians sent papers in their own right accompanied by presentation letters and only the former were then published in the journals.

The analysis of linguistic features showed that there was authorial presence in most of the papers, however the display of personal comments and general prolixity appeared to decrease. Features of informativity and abstractedness instead were found to increase. Papers moreover maintained a generally narrative style throughout the whole period.

Early papers tended to provide extensive detail as to the places and specific locations of the running of experiments and observations of phenomena; also, a tendency to name witnesses was common to give credibility to the accounts. By the 19th century these practices no longer represent a distinctive feature of the sampled writings; and credibility, in the case of experiments, is based purely on their results and the possibility of successfully replicating them.

From the intertextual point of view, papers frequently displayed (mostly positive or neutral) intertextual relations with one another. A cooperative dialogic approach was given in the early *Transactions* by acknowledgements of others' research, calls for further research, and observations of phenomena and

---

243 Various loanwords reported in the *OED* are moreover dated later than the instances found in the *Transactions*. The antedatings of Italian borrowings in the *OED* will be the object of further study.

repeating of experiments from scholars of different regions and countries. In the course of the 18th century, intertextual ties continue by means of literature reviews although there were cases in which Italian research was questioned and doubted. In the 19th century, the only features of intertextual ties are extensive literature reviews and the furthering of Italian research. The naming of scholars by their surname only, moreover, shows the existence of discourse communities within specific subject areas and whose members were well acquainted with the works of their peers and did not therefore need to provide detailed references.

5.4 Development of Italian discourse representation

The representation of Italian research in the *Transactions*, and of English research in the Italian papers, showed that in the early *PTRS* the works of the two countries were either neutrally or positively presented. Scholars of both countries addressed each other by means of forms of encomia, which were also a consequence of the papers consisting of letters. These formalities then gradually decreased throughout the period of study, although humble presentations of writings continued in order to maintain polite and constructive relations. By the 19th century, the representation of Italian discourse became almost completely neutral. Forms of evaluation moved from being primarily focused on the "performer" of science, to the quality and results of the research, its worth, and methodological procedures.

5.5 Concluding remarks

It is hoped that by joining historical and critical linguistic analysis this study has produced a more objective account of scientific relations between Italy and the Royal Society in diachrony. Objectivity nowadays heavily relies on quantitative data, which were here provided only by means of manual counts of the presence of linguistic features and discursive strategies, while the overall analysis was mostly qualitative. However, social relations cannot be fully analysed by means of statistics and computational techniques alone; results of linguistic analyses always need to be re-contextualised and individually considered by the analyst.

One point that has come through in the course of this study is that the linguistic analysis of the *PTRS* papers was rendered less straightforward by a series of cultural and linguistic hindrances. In the present research, DA was not limited to a linguistic study of the English language and style *per se* but served as a means to an end; that is, understanding how relations between English and Italian scientists were negotiated. Therefore, a paper that has been translated into

English is actually transposing the linguistic strategies of the Italian language and the personal opinions of the Italians. Instead, papers written by Englishmen reporting and referencing what has been done by Italians display English writing strategies and opinions.

This study also presents some limitations; at the outset of this research, it had not been foreseen just how much Italian material was to be found in the *Transactions* and in the Society's archives. The intention was to systematically deal with all the exchanges between Italian and English scientists, but the very high number of primary sources and especially the variety of fields of specialisation made it impossible to go into great detail. It was either a matter of choosing specific subject areas or of keeping it general. Since historical accounts on specific cases of Anglo-Italian scientific relations have already been researched, and will surely be further researched, the choice, from the historical point of view, was to provide a general picture of Anglo-Italian relations in the context of natural philosophy in diachrony. Indeed, while more narrowly focused studies in synchrony exist, a broader study on the development of these scientific relations over the course of three centuries does not.

A second limitation is related to the mostly one-sided perspective of the study: by focusing on the *Transactions*, the view of these relations is mostly from the Society's, i.e. English, point of view, and less from the Italian. Still, it was attempted to provide as much detail as possible from the Italian perspective as well by focusing on the Italian views represented in the translated papers and in the Italian letters. To make the picture of Italy-Royal Society relations more complete, this study could be integrated for instance by further research on Italian journals of Italian academies together with their archival material. A final limitation was the exclusion of Latin papers from the critical linguistic analysis, which could have provided further insight, for instance as to the differences between the Latin and vernacular approach to writing science and further information on the Italian side of the relations, given that most of the Latin papers were written by Italians. Finally, a time limit was set on the analysis (1665–1900), which was related to the greater impact of the socio-historical events that took place in the 20th and 21st centuries.

Despite these limitations, the study hopefully has succeeded in providing a general picture of the development of scientific relations between Italy and the Royal Society in the early and late modern periods. These relations can be said to have been successful and consequently led to many notable discoveries and advances, the results of which are still tangible today.

Appendix

Tables of Fellows and contributors

Biographical data has been retrieved from the Royal Society's Fellow directory and the *Dizionario Biografico degli Italiani*. For each Fellow/contributor are provided: date of election, birth and death, main towns where the person lived and worked, and the number of related contributions in the *PTRS*.

Tab. 6.1: 17th-century Fellows and contributors. Fellows are listed in order of their date of election, non-Fellow contributors are listed immediately after.

ELECTION	BIRTH-DEATH	NAME	GEOGR. AREAS	FIELD	Related *PTRS* papers
1667	1614–1687	GASCOIGNE, SIR BERNARD [BERNARDO GUASCONI]	Florence, Milan, England	Military and diplomat. Fought on the royalists' side during the English Civil War.[244]	–
1667	–	UBALDINI, CARLO	Marche, England	Count of Montefeltri, converted to Protestantism.	–
1669	1628–1694	MALPIGHI, MARCELLO	Bologna, Pisa, Messina, Roma	Anatomist, embryologist, papal physician. **Correspondent**.	2 letter extracts Latin
1672	1625–1712	CASSINI, GIOVANNI DOMENICO	Perinaldo (Liguria), Bologna, Paris	Mathematician, astronomer. **Correspondent**.	2 art. in Eng. (translated) published in France first.
1674	1641–1695	PACICHELLI, GIOVANNI BATTISTA	Rome	Historiographer, Abbot, catholic.	–
1676	1623–1678	TRAVAGINO, FRANCESCO	Venice	Alchemist and physician. **Correspondent**.	1 account of one of his books.
1679	?–1714	SAROTTI, GIOVANNI AMBROSIO	Venice, London,	Venetian resident in London.[245]	2 English
1680	1647–1682	PIGHIUS, JACOBUS [PIGHI, GIACOMO]	Verona, Padua	Physician, Professor of anatomy at Padua.	–

244 The Fellowship was not attributed to GASCOIGNE for any particular scientific merits but more likely for his importance as a diplomatic figure. (Villani, S. 2003. Guasconi, Bernardo. In *Dizionario Biografico degli Italiani*. Vol. 60. Retrieved from http://www.treccani.it/enciclopedia/bernardo-guasconi_(Dizionario-Biografico)/). He was an active member from inside the Royal Society.

245 In a report of a meeting GIOVANNI AMBROSIO SAROTTI is introduced as the "son of the Venitian resident" (Birch 1968, vol 3: 510). He may therefore have been the son of PAOLO SAROTTI, a representative of the Republic of Venice in London.

Appendix 229

1681	1630–1701	LETI, GREGORIO	Milan, England (1680–1683)	Polemic and historical writer.[246]	–
1682	1660–1729	BORGHESE, MARCANTONIO	Rome	Prince of Sulmona and Rossano,	–
1690	1646–1691	GRANDI, JACOBUS	Modena, Venice	Professor of anatomy at Venice. **Correspondent**.	1 letter extract Englished out
1691	1658–1730	MARSILI, COUNT LUIGI FERDINANDO	Bologna	Natural historian, emissary, soldier. Founder Academy of Sciences of the Institute of Bologna (1712). **Correspondent**.	2
1695	1641–1721	BOTTONI, DOMENICO	Sicily	Physician at the University of Messina. **Correspondent**.	1
1695	1652–1739	DEL BENE, TOMMASO	Florence, Pisa	Courtier, envoy.	–
1696	1628–1697	FORNASARI, IPPOLITO	Bologna	Barrister and Clergyman.	–
1696	–1712	SPOLETI, FRANCESCO	Padua	Physician and professor of Medicine at Padua.	–
1696	1622–1703	VIVIANI, VINCENZO	Florence	Mathematician. **Correspondent**.[247]	–
1696	1637–1696	BONFIGLIOLI, SILVESTRO	Bologna	Clergyman and physician.	–
1698	1655–1710	GUGLIELMINI, DOMENICO	Bologna, Padua	Physician, mathematician.	1 Latin

(*Continued*)

246 LETI went with his family to England in 1680 and later decided to live in London. He dedicated two of his works to JAMES STUART Duke of York and the Royal Society. In 1681 he was commissioned to write a history of England. The book, *Del Teatro Britannico*, was published in London in 1682; however, his description of CHARLES' II relationship with his wife KATHERINE OF BRAGANZA and other observations on members of the aristocracy were not happily received. Every copy of the book was eventually destroyed and LETI was expelled from the country. See Bufacchi, E. (2005). Leti, Gregorio. In *Dizionario Biografico degli Italiani*. Vol. 64. Retrieved from http://www.treccani.it/enciclopedia/gregorio-leti_(Dizionario-Biografico)/
247 Signor [SILVESTRO] BONFIGLIOLO, he transmitted MALPIGHI's posthumous works (CMO/2/113).

Tab. 6.1: Continued

ELECTION	BIRTH-DEATH	NAME	GEOGR. AREAS	FIELD	Related *PTRS* papers
1698	1668–1707	BAGLIVI, GIORGIO	Dubrovnik, Naples, Bologna, Padua, Rome	Armenio-Italian physician.	–
CONTRIBUTORS					
–	1626–1697	REDI, FRANCESCO	Florence, Pisa	Physician, biologist.	2 Eng
–	1608–1679	BORELLI, GIOVANNI ALFONSO	Naples, Rome	Physiologist, physicist, and mathematician. **Correspondent.**	2 Eng + 1 book review
–	1630–1672	FRACASSATI, CARLO	Pisa, Bologna, Messina	Physician	3 Eng
–	1633–1687	MONTANARI, GEMINIANO	Florence, Bologna	Astronomer. Correspondent.	–
–	1635–1715	CAMPANI, GIUSEPPE	Rome	Lens maker, astronomer.	5 Eng 1 Latin
–	1631–1687	LANA, FRANCESCO	Brescia	Jesuit priest, mathematician, aeronautics.	2 Eng from *GdL*, 1 book account
–		CASTAGNA, MARCANTONIO	Venice	Possibly mineralogist.	2 Eng from the *GdL*
–	1600–1680	SEPTALIUS, MANFREDUS / SIGNOR SETTALLA [MANFREDO SETTALA]	Milan	Canon of St. Nazarus Church in Milan[248] **Correspondent.**[249]	2 Eng

Source: *Travels Through the Low-Countries, Germany, Italy and France, with Curious Observations, Natural, Topographical, Moral, Physiological [...] Also a Catalogue of Plants, Found Spontaneously Growing in Those Parts, and Their Virtues*, by FRANCIS WILLOUGHBY (1738).

248 Hall and Hall (1966: 440–1)

249

Appendix 231

–		RICCIOLI, JOH. BAPTISTA AND DEGLI ANGELI, STEPHANO	Bologna	Astronomy Mathematics.	1 Eng 1668
1594–1647		DONI, GIOVANNI BATTISTA	Florence	Humanist.	1 Eng accompt from the *GdL*, 1670
1633–1704		BOCCONE, PAOLO	Sicily	Botanist. **Correspondent.**	1 paper 1673 + 1 book account
–		GUATTINI, MICHAEL ANGELO and DIONIGI, CARLO	Piacenza, Congo	Missionaries.	1 art from JDS 1677
(1709)		MAGALOTTI, LORENZO	Rome, Florence	Humanist and diplomat. **Correspondent.**	1
1637–1712					
1617–1675		LEOPOLD DE MEDICI	Florence	Prince of Tuscany, Cardinal, promoter of sciences and founder of the Accademia del Cimento (1657).	–
1633–1698		CIAMPINI, [GIOVANNI GIUSTINO]	Rome	Historian, mathematician and clergyman, who was very involved in scientific circles and corresponded both with the Académie de France and the Royal Society.	2 Eng.
1633–1684		GORNIA, GIOVANNI BATTISTA	Pisa	Professor of Medicine and personal physician to Cosimo de Medici. **Correspondent**	–
–		SPOLETO, FRANCESCO	Padova	Professor of Medicine	–

Tab. 6.2: 18th-century Fellows and contributors. Fellows are listed in order of their date of election, non-Fellow contributors are listed immediately after.

ELECTION	BIRTH-DEATH	NAME	GEOGR. AREAS	FIELD	Related *PTRS* papers
1703	1661–1730	VALLISNERI, ANTONIO	Padua	Physician and naturalist. **Correspondent**	–[250]
1703	1665–1741	TIMONE, EMANUELE	Born in Greece of Italian parents. Travelled throughout the Ottoman Empire; practised in Constantinople. Travelled to England in 1703.	Physician. He reported to the Society about the inoculation against smallpox in the Ottoman Empire.	2 Lat., 1 Eng., several books
1706	1654–1720	LANCISI, GIOVANNI MARIA	Rome	Physician, epidemiologist, and anatomist	1 Lat., 1 Eng.
1706	–	GALLUCCI GATUCCI, GATUZZI or GALLUCI	–	–	–
1706	–1709	[?] CARRON DI TOMMASO, COUNT OF BRIANCON	Duchy of Savoy	Envoy of the Duke of Savoy	–

[250] One classified paper (CLP) entitled "Lezione Academica intorno l'origine delle Fontane" (Academic lesson on the origin of the fountains) is available in the Society's archives.

1708	1655–1740	TILLI, MICHAELANGELO	Pisa	Physician and botanist.[251] **Correspondent**	1 Eng. and Lat.
1708	1670–1734	CORNARO, FRANCESCO	Venice	Venetian Ambassador	1 Eng.
1709	1637–1712	MAGALOTTI, LORENZO	Rome, Florence	Humanist and diplomat. **Correspondent**	1 Lat.
1709	1671–1742	GRANDI, GUIDO	Pisa, Florence	Monk, mathematician, philosopher, and engineer	2 Latin
1710	1667–1738	BIANCHI, VENDRAMINO	Venice, Milan	Diplomat	–
1710	1683–1761	POLENI, GIOVANNI	Venice Padua	Marquis, physicist, mathematician and antiquarian. **Correspondent**	5 Lat.
1712	1662–1738	AVERANI, GIUSEPPE	Pisa, Florence	Jurist and naturalist	–
1712	? –1743	DULIOLO, RINALDO	Bologna	Professor of Medicine	–
1712	1677–1752	GRIMANI, PIETRO	Venice	Statesman, diplomat	–
1713	1662–1729	BIANCHINI, FRANCESCO	Verona, Roma	Philosopher and astronomer. Travelled to England 1713	Referenced in several papers and letters.

(Continued)

251 Tilli travelled to the Balearic Islands, Constantinople and the Aegean islands as a naval surgeon in Cosimo's III fleet. Throughout his travels he studied and collected plant species, which were transferred back to Pisa. In 1685 he became head of the Botanical Garden of Pisa and published in 1723 the *Catalogus Plantarum Horti Pisani*, where he listed over 4000 Tuscan plant species and a description of the Botanical Garden. Most importantly, one of Tilli's most notable contributions to science is the use of greenhouses to cultivate tropical plants in Europe.

Tab. 6.2: Continued

ELECTION	BIRTH-DEATH	NAME	GEOGR. AREAS	FIELD	Related *PTRS* papers
1713	1654–1725	BALDINI, CONTE GIOVANNI ANTONIO	Parma and several other places in Europe	Ambassador Extraordinary of the Duke of Parma to Spain, Vienna and to London[252]	–
1715	1656–1743	TOZZI, BRUNO	Florence	Monk, botanist and mycologist	–
1715	1677–1749	CONTI, ANTONIO SCHINELLA / ABBE CONTI	Padua, France (1713–1715 and 1718–1726), England (1715–1718)	Historian, mathematician, philosopher and physicist. **Correspondent.** Acted as intermediary in the Leibniz-Newton calculus controversy	1 Lat./Fr./Eng.
1715	1685–1772	TRON, NICOLÒ /TRONI NICOLO	Padova, Venezia	Venetian Ambassador to Britain; Patron of JAMES STIRLING (FRS 1726). Had business savvy and tried to import foreign technologies to Italy	–
1716	1652–1733	(D')ORSI, MARCHESE GIOVANNI GIUSEPPE / D'ORCI JOSEPH	Bologna, France, Modena Rome	Author, poet. Started a private academy, which met in his home twice a week to discuss literature	–
1716	1653–1729	SALVINI, ANTONIO MARIA	Florence	Clergyman and classicist. Spoke eight languages including English	–

252 Delegate at the Congress of Utrecht (1713); went to Amsterdam, where he added a collection of Indian and Chinese objects to his cabinet of curiosities, which greatly impressed Antonio Vallisneri (FRS 1703) when he visited it in Piacenza (1719); died of apoplexy.

1717	1658–1741	TORTI, FRANCESCO	Modena, Turin, Padua	Physician, Professor of medicine at Modena	–
1717	1672–1750	MURATORI, LUDOVICO ANTONIO	Modena	Clergyman, librarian, historian. Envoy of the Duke of Modena to England	–
1718	1680–1740	MICHELOTTI, PIETRO ANTONIO	Venice	Medicine, physiologist. **Correspondent**	2 Lat. 1 Eng.[253]
1718	–1746	RIPA, LUDOVICUS / RIVA LUDOVICO	Padua	Physician, botanist, Professor of Astronomy and Meteorology at Padua	–
1720	1687–1734	RECANATI, GIAMBATTISTA	Venice	Collectionist	–
1722	1682–1771	MORGAGNI, GIOVANNI BATTISTA	Padua	Anatomist, Professor of Anatomy at the University of Padua. **Correspondent.**	1 Eng.
1723	1682–1766	FAGNANO DEI TOSCHI, GIULIO CARLO	Senigallia (Marche)	Mathematician	–
1723	–1744	FERRARI, DOMENICO	Naples, England	Barrister, librarian. Converted to Anglicanism; Librarian to the Earl of Leicester; left a valuable library to the Earl of Leicester	–

(*Continued*)

---

[253] There are more unpublished papers in the archives: "Account of a Disease resembling a Dropsy of the Breast" (1735, LBO/21/100); a "Paper on an Ischuria [retention of urine] by a vice of the kidneys" (1734, LBO/20/95); a "Case on the Cure of Madness" (1734, LBO/21/54); discussed curing haemorrhages with ice and snow and cures for gout (1734, EL/M3/56).

Tab. 6.2: Continued

ELECTION	BIRTH-DEATH	NAME	GEOGR. AREAS	FIELD	Related *PTRS* papers
1723	1676–1727	D'ARAGONA, NICOLO ALERBO, PRINCE OF CASSANO / NICOLAUS MICHAEL, [D'ARAGONA, NICOLÒ MICHELE D'AYERBE E TRIVULZIO, III PRINCIPE DI CASSANO?]	Naples	Nobleman	2 Eng.
1727	1671–1735	CIRILLO, NICOLA	Naples	Professor of Natural Philosophy and Medicine at Naples. Regularly received and read *Phil. Trans.* in English.	3 Lat. 1 Eng.
1728	1674–1739	MANFREDI, EUSTACHIO/ MANSREDI	Bologna	Mathematician, astronomer. **Correspondent**	6 Lat.
1728	1682–1766	BECCARI, IACOPO BARTOLOMMEO	Bologna	Professor of Medicine, chemistry and experimental physics at Bologna. President of the Bolognese Accademia delle scienze e delle arti. Discovered gluten. **Correspondent**	2 Eng.
1729	1687–1767	ROLLI, PAOLO ANTONIO	Rome, England, Todi (Perugia)	Italian teacher and literate. Italian tutor to the prince of Wales and the Royal Princesses (1715–1744); translator of classics and translated letters for the Royal Society.	1 Eng.

1729	1694–1750	**CARBONE, JOAN BATTISTA**	Brindisi, Naples, Chieti, Lecce, Portugal	Neapolitan Jesuit, astronomer. Established an observatory in Portugal (Udías 2014)	14 Lat, (only 4 included in the corpus)
1731	–1744	**GIUNTINI, JERONIMO/ HIERONYMUS**	Florence	Physician[254]	–
1733	1679–1747	**TAGLINI, CARLO**	Pisa	Professor of philosophy at Pisa.	–
1734	1685–1746	**LEPROTTI, ANTONIO**	Rome	Physician. Papal physician. **Correspondent**	1 Lat.
1734	1690–1743	**CRIVELLI, GIOVANNI FRANCESCO**	Venice	Clergyman and mathematician	–
1735	1681–1753	**GALIANI, CELESTINO**	Naples	Archbishop of Thessalonica & Capellan Mayor of the kingdom of Naples	–
1735	1698–1780	**D'ESTE, FRANCESCO MARIA, PRINCE OF MODENA**	Modena	Prince of Modena. He was present at one of the Society's meetings (Fontes Da Costa 2002a:152)	–
1735	–	**JATTICA, JACOBUS**	Modena, England	Physician to the Prince of Modena, with whom he came to England.	–
1736	1663–1748	**CERVI, JOSEPH**	Parma, Spain	Physician. Practised in Parma; became Physician to Philip V of Spain; founded the Academy of Medicine, Seville	–

(Continued)

---

254 Source: Baker, J and Warner, T. (1731). *The Political State of Great Britain*, Vol. 41. p. 301.

Tab. 6.2: Continued

ELECTION	BIRTH-DEATH	NAME	GEOGR. AREAS	FIELD	Related PTRS papers
1736	1675–1755	SCIPIONE, FRANCESCO MARCHESE DI MAFFEI	Verona	Writer, playwright, art critic, archaeologist and antiquarian. Correspondent	1 Eng.
1736	1676–1755	MARINONI, GIOVANNI GIACOMO	Udine, Vienna	Astronomer. His Imperial Majesty's Mathematician and Professor of Astronomy at Vienna	–
1736	(1690–1769)	CERATI, GASPARE	Florence, Pisa, spent some time in England	Clergyman and Provisor General of the University of Pisa & Councellor of State to the Great Duke of Tuscany	1 Eng.
1736	1695–1758	COCCHI, ANTONIO	Campania, Tuscany. Spent three years in England	Physician, naturalist and writer	–[255]
1736	1712–1764	ALGAROTTI, COUNT FRANCESCO	Venice, Bologna, Paris, London and more	Polymath intellectual, anglophile, travelled extensively, wrote criticism on Newton	–
1738	1692–1777	GORI, ANTONIO FRANCESCO	Florence	Antiquarian, priest	–

---

255 One paper present in archives: "An account of a book ed. by Dr Cocchi published in Florence 1754, containing several old Greek and Latin surgical treatises, such as those of Soranus and Oribarius" by Robert Watson" (L&P/2/585).

1740	–	GIACOMELLI AGNOLA, MICHEL (ANGELO)	Pistoia	Clergyman	–
1740	1709–1782	ZANOTTI, EUSTACHIO	Bologna	Astronomer and engineer	3 Eng.
1740	–1757?	SACCHETTI, GIULIO	Rome	Clergyman. Camerier d'Onore to the Pope Clement the xii, Canon of St Peters in the Vatican, and a Commander of the Religious and Military Order of St John of Jerusalem (EC/1740/11)	–
1741	1692–1777	ZANNOTTI, FRANCESCO MARIA / [ZANOTTI FRANCESCO MARIA]	Bologna	Philosopher and literate. Secretary to the Institute of Science of Bologna. **Correspondent**. Informed the Royal Society about the goings-on of the Bolognese academy and the Royal Society sent him the *PTRS*	–
1744	1702–1763	CAPELLO, PIETRO ANDREA	Venice	Venetian ambassador (Hall 1982.77)	–
1744	1724–1775	BRUNI, GIUSEPPE LORENZO	Turin	"Doctor in Physick, and one of the Colledge of Physicians at Turin. Being a Gentleman well Accomplish'd in Mathematical and other Learning" (EC/1743/15). **Correspondent**	5 Eng.

(*Continued*)

Tab. 6.2: Continued

ELECTION	BIRTH-DEATH	NAME	GEOGR. AREAS	FIELD	Related PTRS papers[256]
1745	1708–1791	SALVEMINI DI CASTIGLIONE, GIOVANNI FRANCESCO MAURO MELCHIORRI	–	Mathematician	2 Lat.[256]
1747	1694–1780	PASSERI, GIOVANNI BATTISTA	Farnese, Orvieto, Todi, Pesaro	Archaelogist and literate. "One of the most eminent persons of his Country for Learning and knowledge in the Greek & Roman Antiquities, and one who no less diligently applies himself to Philosophical Studies." (EC/1747/10)	–
1747	1701–1769	NICOLINI, ANTONIO	Florence	Nobleman, abbot and jurist	1 Eng.
1747	1719–1760	RINUCCINI, FOLCO	Florence, visited England	Nobleman	–
1748	–	OSORIO, CAVALIERE II	Kingdom of Sardinia	Envoy of the King of Sardinia	–
1749	1702–1764	MOLINELLI, PIER PAOLO	Bologna	Physician and surgeon. **Correspondent**	–
1749	–	BAILLOU, CAVALIER DE	Florence	Surveyor general of the Duke of Tuscany (Hall 1982: 77)	–

256 Not included in the corpus as they were found after having carried out the analysis.

1750	GUASCO, OTTAVIO DE	Piedmont, London	Antiquary of Piedmont (Hall 1982:78)	–
1751	VENTURI, MARSILIO	Parma, Madrid	First Physician to the Queen Dowager of Spain (EC/1751/02)	–
1755	BAIARDI, OTTAVIO ANTONIO	Parma, Rome	Clergyman	–
1755	BECCARIA, GIOVANNI BATTISTA	Mondovì (Piedmont)	Physicist and literate. **Correspondent**	6 Lat. 1 Eng.
1755	PADERNI, CAMILLO	Naples	Painter and art restorer. **Correspondent**	8 Eng.
1756	MANETTI, SAVERIO/ XAVERIUS	Florence	Physician, botanist and ornithologist	–
1756	PANCRAZI, GIUSEPPE-MARIA	Cortona, Florence, Sicily, Naples	Nobleman, clergyman, antiquary (Carlino 2010)	referenced in one paper
1757	VENUTI, RIDOLFINO	Cortona, Roma	Clergyman, antiquary. **Correspondent**	4 Eng.
1757	DONATI, VITALIANO	Padua, Turin, Asia Minor	Professor of Botany in Turin. **Correspondent**	3 Eng.
1757	CELESIA, PIETRO PAOLO	Genoa, England	Diplomat. "Genoese Patrician, and Minister at the British Court from the Most Serene Republic of Genoa" (EC/1757/08)	–

(Continued)

Tab. 6.2: Continued

ELECTION	BIRTH-DEATH	NAME	GEOGR. AREAS	FIELD	Related *PTRS* papers
1757	1728–84	Frisi, Paolo	Milan, Pisa	Mathematician and astronomer. Professor of Mathematics at Pisa and Milan, travelled to England. Correspondent	1 Eng.
1758	1727–1795	Marsili, Giovanni	Padua, Bologna, Florence, England (spoke English)	Botanist	–
1758	1728–1804	Allioni, Carlo	Turin	Physician and professor of botany	1 Eng.
1759	1696–1763	Foscarini, Marco	Venetian	Diplomat and historiographer	–
1759	1709–1769	Venuti, Filippo	Cortona, Livorno	Archaeologist	1 Eng.
1759	1715–1768	Carafa, Giovanni	Naples	Duke of Noia, antiquary	–
1760	1717–1788	Albertini, Giambattista	Regno di Napoli	Prince of Cimitile, diplomat	–
1760	1734–1810	Saluzzo, Guiseppe Angelo	Turin	Artillery General; Member and President of the Academy of Sciences of Turin; Chemist	–
1763	1714–1793	Morosini, Francesco Lorenzo	Venice	Venetian ambassador	–
1763	1730–1779	Matani, Antonio	Pistoia, Pisa	Professor of Medicine at Pisa	–
1764	1733–1824	Stratico, Simone	Padua, Pavia, England	Italian-Greek mathematician and a nautical science expert	–
1764	1734–1790	Cigna, Giovanni Francesco	Turin	Physician and member of the Academy of Sciences of Turin	–
1765	1715–1789	Caraccioli, Domenico	Naples, England	Diplomat, marquis of Villamaina	–

Appendix

1765	CARBURI, GIOVANNI BATTISTA	Piedmont, Padua, France	Physician to the daughter of the King of Sardinia and Piedmont. Professor of Medicine at Padua	–
1768				–[257]
1768	SPALLANZANI, LAZZARO	Scandiano, Pavia	Catholic priest, biologist and physiologist	
1772	CALDANI, LEOPOLDO MARCO ANTONIO	Bologna, Padua	Anatomist and physiologist	–[258]
1774	PAOLI, FILIPPO ANTONIO PASQUALE DI	Corsica, England	Corsican patriot and leader	–
1776	REZZONICO, ABONDIO	Venice, Rome	Nobleman	–
1777	TOALDO, GIUSEPPE	Padua	Teacher of Grammar, Rhetoric, Philosophy and Mathematics' Professor of Astronomy and Meteorology, Padua (1762); constructed an observatory at Padua (1767–1774)	2 Eng.
1779	POLI, GIUSEPPE SAVERIO	Naples	Physicist, biologist, naturalist	–
1779	CAVALLO, TIBERIUS	Naples, England	Physicist and natural philosopher. **Bakerian medal** 1780–1792.[259]	16 Eng.

(Continued)

---

257 A paper related to SPALLANZANI's work is however present in the archives (L&P/5/91).

258 One paper present in archives: "Inequalities in human uterers; with abstract by William Hunter" (1775, L&P/6/203).

259 "The Bakerian Medal and Lecture is the premier lecture in physical sciences. The lectureship was established through a bequest by Henry Baker FRS of £100 for 'an oration or discourse on such part of natural history or experimental philosophy, at such time and in such manner as the President and Council of the Society for the time being shall please to order and appoint.' The lecture series began in 1775. The medal is of silver gilt, is awarded annually and is accompanied by a gift of £10,000" (https://royalsociety.org/grants-schemes-awards/awards/bakerian-lecture/).

Tab. 6.2: Continued

ELECTION	BIRTH-DEATH	NAME	GEOGR. AREAS	FIELD	Related *PTRS* papers
1781	1728–1801	BARBIANO DI BELGIOIOSO, LUIGI CARLO MARIA COUNT OF	Duchy of Milan	Diplomat of the Holy Roman Empire	–
1783	1736–1795	VENANZIO D'AQUINO, FRANCESCO MARIA, PRINCE OF CARAMANICO	Naples Palermo	Ambassador to London and Paris for the Kingdom of Naples and later viceroy of Sicily.	–
1784	1754–1784	MALASPINA DI SANNAZZARO, LUIGI DI, MARQUIS	Pavia, England	Marquis with interests in architecture	–[260]
1788	1735–1796	LORGNA, ANTONIO MARIA	Verona	Brigadier; Governor of the Military School, Verona; Mathematician, founder of the Accademia nazionale delle scienze (1782)	–
1791	1736–1813	LAGRANGE, JOSEPH-LOUIS / LAGRANGIA, GIUSEPPE LODOVICO	Turin, Berlin, France	Mathematician and astronomer	–
1791	1745–1827	VOLTA, ALESSANDRO	Como, Milan	Physicist, chemist, invented Voltaic pile in 1799. **Copley medal** (1794). **Correspondent**	3 Eng. 1 Ital. 2 Fr.

---

260 But one archived paper was found in the archives: "Geometrical and Analytical Deductions concerning the thickness of insulated arches composed of four parts or wedges" (AP/3/16).

1791	1752–1832	SCARPA, ANTONIO	Modena, Pavia, travelled to England in 1781	Anatomist. Ordinary Professor of Anatomy and theoretical surgery at Modena and later at Pavia	–
1795	1735–1803	FONTANA, GREGORIO	Pavia	Mathematician, chair of mathematics at Pavia	–
1795	1741–1803	FORTIS, ALBERTO	Padua, Bologna	Augustinian abbot; Librarian, Bologna; Professor of Natural History, Padua University; travelled in Dalmatia (1770–1774); travels to investigate river sources funded by Frederick Augustus Hervey (FRS 1782)	–
1795	1752–1832	ORIANI, BARNABA	Milan	Astronomer; Director of the Observatory, Milan (after 1802); made Senator of the Kingdom of Italy by Napoleon	–
CONTRIBUTORS					
	1666–1696	BONOMO, [GIOVANNI COSIMO]	Livorno, Pisa, Rome	Physician, discovered the scabies mite.	1 Eng.
–	1633–1714	MAGLIABECHI, ANTONIO	Florence	Erudite and bibliophile. **Correspondent**	1 Eng.
–	1719–1779	BIANCHINI, FORTUNATO	Fermo, Udine, Padova	Physician	1 Eng.
–	1739–1799	CIRILLO, DOMINICO	Naples, England	Physician and botanist	1 Eng.
–	1636–1714	VALLETTA, JOSEPH	Naples	Philospher and jurist. **Correspondent**	1 Lat.

(Continued)

Tab. 6.2: Continued

ELECTION	BIRTH-DEATH	NAME	GEOGR. AREAS	FIELD	Related PTRS papers
–	1659–1718	Pylarinum, Jacobum [Pilarini Giacomo]	Cefalonia,[261] Padova, Costantinople	Physician; the first who practiced inoculation of smallpox outside Asia; tutor of Emanuel Timonious	1 Lat.
–	1685–1756	Benevoli, Antonio	Norcia, Florence	Physician and surgeon	1 Eng.[262]
–	–	Crusio, Carlo	Naples	Medical topic	1 Eng.
–	1703–1788	Paitoni, Giovanni Battista	Venice	Physician	1 Eng.
–	–	Pasquale R. Pedini	Florence	Clergyman and Professor in the University of Pisa	1 Eng.
–	1730 – 1805	Fontana, Felice (brother of Gregorio)	Florence	Physicist; director of the Cabinet of Natural History Belonging to His Royal Highness the Grand Duke of Tuscany	2 Eng. 1 Ital. and Eng.[263]
–	1648–1735	Del Papa, Giuseppe	Empoli (Tuscany)	Physician	5 Eng.

261 The Greek island of Cefalonia was part of the Republic of Venice in the 18th century.
262 Plus two more papers in the archives: "'Treatment of cataracts by Antonio Benevoli, surgeon in the hospital of St Maria Nova in Florence, to Dr Valsalva" (Cl.P/12ii/8) and "Nuova proposizione intorno alla caruncola dell'uretra detta carnosita' by A Benevoli" (Cl.P/22ii/16).
263 Plus one archived paper, "Physical Researches on fixed Air, by Signor Felice Fontana - translated from the Italian, printed at Florence in the Year 1775", concerning "fixed Air, which the celebrated Dr Priestley has made so many curious experiments and observations, has of late years attracted the attention of all modern Philosophers. Many singular qualities have already been discovered in that fluid; but great obscurity and several doubts still remain concerning its real nature and inherent properties" (AP/4/8).

	1737–1798	GALVANI, LUIGI	Bologna	Physician and physicist	8 Eng.
	1687–1784	MORO, LAZZARO	Venice	Geologist and naturalist	1 Eng.
	–	IPPOLITO, FRANCESCO	Naples	Nobleman	1 Ital. and Eng.
	–1786	DA ROVATO, GIUSEPPE	Tibet	Missionary	1 Ital. and Eng.
	1736–1814	RIZZI ZANNONI, GIOVANNI ANTONIO	Padua, Naples, Sicily	Geographer	1 Eng.
(1804)	1746–1806	PIAZZI, JOSEPH [GIUSEPPE]	Ponte in Valtellina (Lombardy), Genoa, Malta, Ravenna, Palermo, Naples	Theatine monk; taught philosophy (Genoa), mathematics (Malta, Ravenna, Palermo), dogmatics (Rome); Director of the Observatory of Naples	1 Eng.
	–	PINELLI, FLAMINIO	–	Physician	1 Eng.
	1743–1831	BOVI, ROCCO	Calabria	Biologist, astronomer and physician	1 MS paper (MS/91)
	1710–1782	DELLA TORRE, FATHER GIOVANNI MARIA	Rome, Naples	Physicist and naturalist	2 Eng.
	–	V PUCCI	Tuscany, Bologna	Secretary to the Grand Duke of Tuscany	–

248 Appendix

Tab. 6.3: 19th-century Foreign Members and contributors. FMs are listed in order of their date of election, non-FM contributors are listed immediately after.

ELECTION	BIRTH-DEATH	NAME	GEOGR. AREAS	FIELD	Related PTRS papers
1804	1746–1826	PIAZZI, GIUSEPPE	Ponte in Valtellina, Palermo, Naples	Catholic priest mathematician, and astronomer. Established an observatory at Palermo[264]	1 Eng.
1817	1783–1872	GRANVILLE, AUGUSTUS BOZZI	Milan, London, Dover	Surgeon, interested in midwifery. Travelled extensively and worked as translator, interpreter and courier to the Foreign Office	3 Eng.
1827	1773–1836	PINI MORICHINI, DOMENICO	Aquila, Rome	Physician, chemist	2 Eng.
1827	1781–1864	PLANA, GIOVANNI ANTONIO AMEDEO	Voghera, Turin	Mathematician and astronomer. **Copley medal** 1834[265]	–
1832	1783–1862	CARLINI, FRANCESCO	Milan	Astronomer. Director of the Observatory in Milan	3 Fr.
1838	1797–1870	LEOPOLD II OF TUSCANY	Tuscany	Grand Duke of Tuscany	–

264 GIUSEPPE PIAZZI was also the Godfather of CHARLES PIAZZI SMYTH, also a Fellow of the Royal Society.
265 MORICHINI did not directly contribute papers, his work was however known and referenced in the Society's journals. The Society's interest was particularly in his discoveries of the magnetic properties of the violet rays of the solar spectrum.

1839	1798–1854	MELLONI, MACEDONIO	Naples	Physicist. "Celebrated for his discovery of the different scales of the diathermaneity of transparent & coloured media" (EC/1839/29). **Rumford medal** 1834[266]	–[267]
1856	1818–1878	SECCHI, PADRE ANGELO	Reggio Emilia, Roma	Astronomer. Director of the Roman observatory. **Correspondent**	4 Eng.
1879	1830–1903	CREMONA, LUIGI	Pavia, Milan, Bologna, Rome	Mathematician	1[268]
1889	1826–1910	CANNIZZARO, STANISLAO	Palermo, Marseilles, Alexandria, Geneva, Rome	Artillery Officer (during the Sicilian revolution); chemist); researched atomic and molecular weights; Italian Senator. **Copley medal** (1891)[269]	–

(*Continued*)

266 The Copley Medal is the Society's oldest and most prestigious award (1731). The medal is awarded annually for outstanding achievements in research in any branch of science. The award alternates between the physical and biological sciences.
267 For his discoveries relevant to radiant heat. The Rumford Medal is awarded for outstanding research in physics. The award was established following a donation by BENJAMIN THOMPSON FRS (PDF), Count RUMFORD of the Holy Roman Empire, an American-born former soldier, spy, statesman and scientist who would go on to found the Royal Institution. The first award was made in 1800. The medal is of silver gilt, is awarded annually and is accompanied by a gift of £2,000 (Royal Society).
268 He is however referenced by JOHN TYNDALL in various papers.
269 Intertextually related paper published in the 20th century.

ELECTION	BIRTH-DEATH	NAME	GEOGR. AREAS	FIELD	Related PTRS papers
1891	1838–1905	TACCHINI, PIETRO	Modena, Padova, Palermo, Rome	Astronomer. Osservatorio del Collegio Romano. Collaborated with HERBERT HALL TURNER. **Rumford medal** 1886 (NLB/2/767)[270]	–[271]
1896	1835–1910	SCHIAPARELLI, GIOVANNI	Cuneo, Milan	Astronomer and science historian	–
Contributors					
–	1811–1868	MATTEUCCI, CARLO	Forlì, Bologna, Florence, London, York, Livorno	Physicist and physiologist. **Copley medal**, 1844[272]	17 Eng.[273]
–	1864–1829	RUFFINI, ANGELO	Pretare, Siena, Blologna	Histologist and embryologist. **Correspondent**	–[274]

---

270 For his contributions to Chemical Philosophy especially for his application of Avogadro's theory (NLB/5/855).
271 "For important and long-continued investigations, which have largely advanced our knowledge of the physics of the Sun" (Royal Society).
272 But several referencing.
273 For his research on animal electricity (MS/581/107).
274 Only 12 included in the corpus.

	1846–1910	**Mosso, Angelo**	Turin, Leipzig	Professor of pharmacology (1876) and physiology (1879) at Turin. **Croonian lecturer** in physiology, 1892.[275] **Correspondent**	1 Fr. and 1 Eng.[276]
–	1797–1879	**Panizzi, Antonio**	Modena, London	Principal Librarian of the British Museum from 1856 to 1866 and Librarian to the Royal Society	–
–	1824–1897	**Brioschi, Francesco**	Milan, Pavia	Mathematician, politician and Dean of the University of Pavia	1 Fr.
–	1762–1834	**Aldini, Giovanni**	Bologna	Physicist	–[277]
–	–	**Piccolomini[-Aragona], Count Vincent**	Florence	–	1 Eng.

(Continued)

---

275 His book, *Sulla presenza di nuove forme di terminazioni nervose nello strato papillare e subpapillare della cute dell'uomo, con un contributo allo studio della struttura dei corpuscoli del Meissner* (Siena, 1898), was funded by the Royal Society; together with a paper published in the *Journal of Physiology* ("On the Minute Anatomy of the Neuromuscular Spindles of the Cat, and on their Physiological Significance" vol. 23, 1898) (Eccles 1975 and NLB/17/38).

276 "The Croonian Medal and Lecture is the premier lecture in the biological sciences. The lectureship was conceived by William Croone FRS (PDF), one of the original Fellows of the Society. Among the papers left on his death in 1684 were plans to endow two lectureships, one at the Royal Society and the other at the Royal College of Physicians. His widow later bequeathed the means to carry out the scheme. The lecture series began in 1738. The medal is of silver gilt, is awarded annually and is accompanied by a gift of £10,000." (Royal Society: https://royalsociety.org/grants-schemes-awards/awards/croonian-lecture/). Mosso's lecture was entitled: *Les phenomenes psychiques et la temperature du cerveau*.

277 The papers correspond to Mosso's Croonian Lecture published in its original French in the *PTRS*, and an abstract published in the *Proceedings* in English (*Phil. Trans. B* 1892: 299–309 and *Proceedings* 1892:83–85). Both papers have not been included in the corpus as they were found after the analysis was carried out.

Tab. 6.3: Continued

ELECTION	BIRTH-DEATH	NAME	GEOGR. AREAS	FIELD	Related *PTRS* papers
–	1784–1835	Nobili, Leopoldo	Florence, Reggio	Physicist, pioneer in electrochemistry	1 Eng.[278]
–	1817–1888	Mancini, Pasquale Stanislao	Naples	8th Marquess of Fusignano. Jurist and Stateman	1 AP[279]
–	1854–1925	Grassi, Giovanni Battista	Rovellasca, Catania, Rome	Zoologist and physician. **Darwin Medal** (1896).[280] **Correspondent.**	1 Eng.[281]
–	–	Argentati, Raffaele	Senigallia	Priest and inventor	1 AP[282]

278 Two papers are preserved in the archives: "Observations and experiments on galvanism" (Read 25 November 1802, L&P/12/39) and "Experimental Enquiries upon Gas Light on the Continent, with some observations upon the present state of the illumination of London" (c.1817–1821, AP/9/1).

279 "Nuove Idee sull' Elettricita applicare all'invenzione d'un Parairemuoto" (1837, AP/22/13) translated the following year, "New ideas on electricity, applied to the invention of an earthquake guard" (1838, RR/1/160).

280 For his research on the life history and societies of the *Termitidae*, and on the developmental relationship between *Leptocephalus* and the common eel and other *muraenidae*. "The Darwin Medal is awarded for work of acknowledged distinction in evolution, population biology, organismal biology and biological diversity. The Darwin Medal was created in memory of Charles Darwin FRS and was first awarded in 1890" (https://royalsociety.org/grants-schemes-awards/awards/darwin-medal/).

281 One other paper was sent by Grassi, "Relations between Malaria and Particular Insects" (NLB/18/745).

282 "Descrizione d'un Locomotore Aeres" (1840, illustrated, AP/24/1).

Tab. 6.4: Italian loanwords recorded in the *Transactions*, 17th–19th century

	Loanword	No of papers	*OED* etymology
17th century	Telescope	5	< Italian *telescopio* (1611 or earlier)
	Signor/Signior/Senior/Seignior	24	< Italian *signor*, variant of *signore* (16th century)
	Virtuoso	2	< Italian *virtuoso*, †*vertuoso* (noun) person who demonstrates special skill, knowledge, or accomplishment in the arts, sciences, or fine arts. †**a.** A learned person; a scholar; *esp.* a scientist, a natural philosopher. Also: *spec.* a member of the Royal Society. *Obsolete.* (early 17th cent.)
	Letterati	1	< Italian *letterato*, †*leterato* (also †*literato*, †*litterato*) person engaged in literary pursuits, writer (early 14th cent.), learned person, use as noun of *letterato* educated, cultured
	Stiletto	1	< Italian *stiletto*, diminutive of *stilo* dagger (early 17th cent.)
	Braccia	1	< Italian *braccio*, lit. "an arm", hence a measure of length (18th cent.)[283]
	Tarantula/Tarantola	2	< medieval Latin *tarantula* (*Onomast. Lat. Graec.*), Italian *tarantola*, < *Taranto* a town in modern Apulia (16th cent.) **a.** A large wolf-spider of Southern Europe, *Lycosa tarantula* (formerly *Tarantula Apuliæ*), named from the town in the region where it is commonly found, whose bite is slightly poisonous, and was fabled to cause "tarantism" n.

(*Continued*)

---

283 The *OED* provides as the earliest attestation an example from the *Philosophical Transactions* of 1761. The instance found in the corpus however was a much earlier instance, dated 1671.

Tab. 6.4: Continued

Loanword	No of papers	OED etymology
Volcan	1	< Italian *vulcano*, *volcano* (1555, earliest in translations of Spanish accounts of exploration of the Americas), ultimately < classical Latin *Vulcānus*, *Vulcānus* and place names of volcanic locations derived from this (see Vulcan *n.*), but in use as common noun after Spanish *volcán*
Alte	1	< Italian *alto* a high voice in polyphonic music (1580)
Grotto/Crotta	4	< Italian *grotta* (for which Dante has also *grotto*) = Old French *crote, croute*, Provençal *crota* … (early 17th cent.)
Mastro	1	< Italian *maestro* **a.** With capital initial. A title or form of address designating someone (originally esp. an Italian) who is a master of or who has achieved eminence in a skill or profession, esp. a musician. (13th cent.)
Motto	1	< Italian *motto*, *mutto* **1.** Originally: a word, sentence, or phrase attached to an impresa or emblematical design to explain or emphasize its significance. Later also: a short sentence or phrase inscribed on an object, expressing a reflection or sentiment considered appropriate to its purpose or destination; a maxim or saying adopted by a person, family, institution, etc., expressing a rule of conduct or philosophy of life (14th cent.)
Malmignatto	1	Probably < French *malmignatte* (1860 in the passage translated in quot. 1861) < Corsican *malmignatta* or Italian *malmignatta* (earliest attested in the 19th cent.) < *malo bad* (< classical Latin *malus* : see mal- *prefix*) + *mignatta* leech[284] A venomous spider of Mediterranean countries which belongs to a race of *Latrodectus mactans*, closely related to the American black widow

[284] The *OED* states that the earliest example of *malmignatta* as an Italian loanword was found in the 19th century; however, the example here provided is dated 1699.

Appendix

	Archipelago	1	< Italian *arcipelago*, < *arci-* chief, principal (arch- *prefix* 4) + *pélago* deep, abyss, gulf, pool… (16th cent.)
Other Italianisms	Sciarri (stones made of solidified lava); Dolci di Sale ("young wanton girls" who fall into "melancholly madness"); Coccio maligno; Volo; Lira (music).		
18th century	Senior/Signior/Signor/ Seignior	10	See above
	Virtuosi/Virtuoso	3	""
	Soldi of Venice	1	Italian < Latin *solidum* (16th cent.) An Italian coin and money of account, formerly the twentieth part of a lira
	Via Appia	1	The Appian way is not recorded in the OED but it may be considered a borrowing of the *Grand Tour* period, as it was commonly used by English travellers through Italy. This represented an ancient Roman road which connected Rome to Brindisi
	Tarantula	2	See above
	Porticos	2	< Italian *portico* (17th cent.) **2. a.** *Architecture.* A formal entrance to a classical temple, church, or other building, consisting of columns at regular intervals supporting a roof often in the form of a pediment; a covered colonnade in this style
	Volcano	9 and more	See above
	Canna	1	< Italian *canna* (13th cent.) In Italy, other parts of southern Europe, and the eastern Mediterranean: a unit of length used especially for measuring cloth, varying locally and with time, but typically between 1 and 3 metres (approx. 3.3 to 9.8 feet)
	Braccio	1	See above

(*Continued*)

Tab. 6.4: Continued

Loanword	No of papers	OED etymology
Lava/Lavas	11 and more	< Italian *lava* (< *lavare* to wash…), originally "a streame or gutter suddainly caused by raine" (Florio 1611), applied in the Neapolitan dialect to a lava-stream from Vesuvius; hence adopted in literary Italian, where it developed the senses represented by 2 and 3 below 2. The fluid or semi-fluid matter flowing from a volcano. (18th cent.) a. The substance that results from the cooling of the molten rock. (18th cent.)
Piazza	1	< Italian *piazza* square, marketplace …. With the form *piazza* perhaps compare Italian †*piaza* (15th cent. or earlier), variant of *piazza*. In plural form *piazza* after the Italian plural form.
Rilievo/Rilievos/Rilievi	4	< Italian *rilievo*, †*relevo*, †*relievo*, †*rilevo* (of a protuberance or elevated area or point) fact of standing out from a surface (*a*1342), moulding, carving, or stamping in which the design stands out from the surface, a work of art produced in this way, effect or appearance of three dimensions given on the plane surface of a painting or similar work of art (all second half of the 14th or early 15th cent.)
Rotolo	1	< Italian *rotolo* (*a*1292; plural *rotoli*) < Arabic *ruṭl* rottol *n.* Compare post-classical Latin *rotulus*, denoting a dry measure (12th cent. or earlier, especially in Italian sources); also *rotulum* (12th cent. in an Italian source), *rotula* (13th cent. in British sources; also in continental sources), both denoting a liquid measure. In plural form *rotoli* after the Italian plural form. In forms *rottolis*, *rottollies* with an English plural suffix added to the Italian plural form.
Sbirri	1	< Italian *sbirro*, whence French *sbirre*; compare Spanish *esbirro*. An Italian police officer. (17th cent.)
Cameo/Cameos	2	< Italian *caméo*, *cameo*… **1.** A precious stone having two layers of different colours, in the upper of which a figure is carved in relief, while the lower serves as a ground (16th cent.)

Intaglios	2	Italian, = engraving, engraved work, a carving (plural *intaglio*)… (17th cent.)
Stucco	2	< Italian *stucco* a relief moulding or ornament in plaster … In plural *stucchi* after the Italian plural form. (late 16th cent.)
Cupolas	1	< Italian *cupola* (also *cuppola, cuppula* in Florio)… **1.a.** *Architecture.* A rounded vault or dome forming the roof of any building or part of a building, or supported upon columns over a tomb, etc.
Rapilli	1	< Italian regional (Naples) *rapilli*, plural of *rapillo* (15th cent.), variant (by dissimilation) of Italian *lapillo* lapillo *n*. Compare earlier lapilli *n*. With *plural* agreement. Small fragments of rock, esp. pumice, ejected from a volcano. (18th cent.)
Bufolo/Buffalo	1	< Italian *buffalo* (Florio), *bufalo, bufolo* (Baretti)… (late 16th cent.)
Tartana	1	< Italian *tartana:* see tartan *n.*² A small one-masted vessel with a large lateen sail and a foresail, used in the Mediterranean (late 16th cent.)
Granitello	1	OED Granitell < French *granitell*, < Italian *granitello*, diminutive of *granito* granite *n*. *Geology*. A binary granite, or granular aggregate of two ingredients (late 18th cent.)
Maltese Speronara	1	Italian. A large rowing and sailing boat used in southern Italy and Malta. (late 18th cent.)
Neapolitan Felucca	1	< Italian *felu(c)ca* A small vessel propelled by oars or lateen sails, or both, used, chiefly in the Mediterranean, for coasting voyages. (early 17th cent.)

(*Continued*)

Tab. 6.4: Continued

Loanword	No of papers	OED etymology
Banditti	1	*OED* Bandit < Italian *bandito* "proclaimed, proscribed", in plural *banditi*, noun, "outlaws," past participle of *bandire* … Early spellings, as well as the current plural *banditti*, were apparently corrupted by form-assoc. with ditto *n.*, Italian *detto*, plural *detti*. The Italian singular *bandito* is not now used in English: *bandit* is also modern French. But the plural *banditti* (for Italian *banditi*) is more used than *bandits*, especially in reference to an organized band of robbers; in which sense it has also been used as a collective singular; in 17th cent. this was taken as an individual singular, with plural *-is, -ies* (late 16th cent.)
Basso rilievo	1	< Italian *basso-rilievo* (late 17th cent.)
Tuffa	1	< Italian *tufa, tufo* … *Geology*: **a.** A generic name for porous stones, formed of pulverulent matter consolidated and often stratified (late 18th cent.)
the Italian **Capuchins**	1	< 16th cent. French *capuchin* (now *capucin*), < Italian *cappuccino* **1.** A friar of the order of St. Francis, of the new rule of 1528. So called from the sharp-pointed capuche, adopted first in 1525, and confirmed to them by Pope Clement VII. in 1528.
Other Italianisms		Prataiuoli (type of mushroom); Opuntia, Tuna (both common names for the Ficus Indica Spinosa); Tre Taberne; Bocca in Capo or Prete (fish name); Mola or Sun Fish; Pesce Balestra; Pesce Pettine; Trombetta; the Centrina or Pesce Porco; Aquila (fish); Pinna; Tamburo; Francolinos (birds); Beccafigos (birds); Marmotta; lucciole, i.e. "small lights", farfalls "which are butterflies"; Gattopardo; Staiolo (measure); Ravenelle (< It. ravanello, radish); muffa (mould); Manna grassa; "Un fulmine! Un fulmine!"; "vorticoso, orizzontale, oscillatorio"; tomoli of corn; Ferilli (lightning)

19th century	Galvanism	many	< French *galvanisme*, < the name of Luigi *Galvani* who first described the phenomena in 1792 Electricity developed by chemical action. Also, the application of this for therapeutic purposes.
	Galvanic	many	< *galvan-* (in galvanism *n.*) + -ic *suffix*. Compare French *galvanique*. **a.** Of, pertaining to, or produced by galvanism. ***galvanic battery***, an apparatus constructed for the production of galvanic electricity. ***galvanic belt***, a belt containing a galvanic apparatus to be worn round the body for therapeutic purposes. ***galvanic electricity*** = galvanism *n*. ***galvanic pile***, a "pile" (see quot. 1802¹) for the production of galvanic electricity. ***galvanic skin response (or reflex)***, the rapid variation in the electrical conductivity of the skin as a measure of the effect of an emotional stimulus on autonomic activity (late 18th cent.)
	Galvanoscopic frog	4	*OED* Galvanoscope, n. < galvano- *comb. form* + Greek -σκόπος looker. An instrument for ascertaining the presence of galvanic electricity. **galvano'scopic** *adj*; pertaining to, or of the nature of, a galvanoscope; **galvanoscopic frog**, a frog used as a galvanoscope (early 19th cent.)
	Galvanometer	2	< galvano- *comb. form* + -meter *comb. form²*. Compare French *galvanomètre* (1802) An apparatus for detecting the existence and determining the direction and intensity of a galvanic current (early 19th cent.)
	Malpighian	1	< the name of Marcello *Malpighi* (1628–94), Italian physician and anatomist + -an *suffix* **1.** *Anatomy* and *Zoology*. Designating structures described by Malpighi, and others (especially of the kidney) connected with these. **a. Malpighian body** *n*. (*a*) a glomerulus of the kidney together with the Bowman's capsule enclosing it (now *rare*); (*b*) a lymphoid follicle of the spleen. (mid 19th cent.)
	Lava/lavas	2	See above

(*Continued*)

Tab. 6.4: Continued

Loanword		No of papers	OED etymology
	Volcano	1	See above
	Lapilli	1	Italian; < Latin *lapillus* Matter ejected from volcanoes in the form of lapilli (early 19th cent.)
	Signor/Signore	6	See above
Other Italianisms	Galvanoscopic nerve; Sal Inglese (Sulphate of magnesia used for medical purposes).		

Bibliography

Primary sources

17th century

Phil. Trans. (1665). An Account of the Tryalls, Made in Italy of Campani's New Optick Glasses. Vol. 1, 131–132.

Phil. Trans. (1665). A Further Account, Touching Signor Campani's Book and Performances about Optick-Glasses. Vol. 1, 70–75.

Phil. Trans. (1665). A Relation of the Raining of Ashes, in the Archipelago, upon the Eruption of Mount Vesuvius, Some Years Ago. Vol. 1, 377.

Phil. Trans. (1665). A Way of Preserving Ice and Snow by Chaffe. Vol. 1, 139–140.

Phil. Trans. (1665). An Accompt of the Improvement of Optick Glasses. Vol. 1, 1–2.

Phil. Trans. (1665). An Extract of a Letter not Long Since Written from Rome, Rectifying the Relation of Salamanders Living in Fire. Vol. 1, 377–378.

Phil. Trans. (1665). An Observation of Optick Glasses Made of Rock-Crystal. Vol. 1, 362.

Phil. Trans. (1665). Considerations of Monsieur Auzout upon Mr. Hook's New Instrument for Grinding of Optick-Glasses. Vol. 1, 57–63.

Phil. Trans. (1665). Extract of a Letter, Lately Written from Rome, Touching the Late Comet, and a New One. Vol. 1, 17–18.

Phil. Trans. (1665). Extract of a Letter, Lately Written from Venice, by the Learned Doctor Walter Pope, to the Reverend Dean of Rippon, Doctor John Wilkins, Concerning the Mines of Mercury in Friuli; And a Way of Producing Wind by the Fall of Water. Vol. 1, 21–26.

Phil. Trans. (1665). Extract of a Letter, Written from Paris, Containing Some Reflections on Part of the Precedent Roman Letter. Vol. 1, 18–20.

Phil. Trans. (1665). Observations Made in Italy, Confirming the Former, and Withall Fixing the Period of the Revolution of Mars. Vol. 1, 242–245.

Phil. Trans. (1665). Of a Means to Illuminate an Object in What Proportion One Pleaseth; And of the Distances Requisite to Burn Bodies by the Sun. Vol. 1, 69–70.

Phil. Trans. (1665). Of Some Philosophical and Curious Books, That are Shortly to Come Abroad. Vol. 1, 145–146.

Phil. Trans. (1665). Signor Campani's Answer: And Monsieur Auzout's Animadversions Thereon. Vol. 1, 75–77.

Phil. Trans. (1665). Some Observations of Vipers. Vol. 1, 160–162.

Phil. Trans. (1665). Some Particulars, Communicated from Forraign Parts, Concerning the Permanent Spott in Jupiter; and a Contest between Two Artists about Optick Glasses, &c. Vol. 1, 209–210.

Phil. Trans. (1665). Mr. Hook's Answer to Monsieur Auzout's Considerations, in a Letter to the Publisher of These Transactions. Vol. 1, 64–69.

Phil. Trans. (1666). An Account of Some Experiments of Injecting Liquors into the Veins of Animals, Lately Made in Italy by Signior Fracassati Professor of Anatomy at Pisa. Vol. 2, 490–491.

Phil. Trans. (1666). An Account of the Synopsis Novae Philosophiae & Medicinae Francisci Travagini Medici Veneti. Vol. 2, 555–556.

Phil. Trans. (1666). An Experiment of Signior Fracassati upon Bloud Grown Cold. Vol. 2, 492.

Phil. Trans. (1666). Some Observations Communicated by Signior Manfredus Septalius from Milan, Concerning Quicksilver Found at the Roots of Plants, and Shels Found upon In-Land Mountains. Vol. 2, 493.

Phil. Trans. (1666). A Confirmation of the Experiments Mentioned in Numb. 27. to Have Been Made by Signor Fracassati in Italy, by Injecting Acid Liquors into Blood. Vol. 2, 551–552.

Phil. Trans. (1666). An Extract of a Letter Written by Signor Cassini Professor of Astronomy in Bononia, to Monsieur Petit at Paris, and Englished out of the Journal des Scavans, Concerning Several Spots Lately Discover'd There in the Planet Venus. Vol. 2, 615–617.

Phil. Trans. (1666). An Account of Some Discoveries Concerning the Brain, and the Tongue, Made by Signior Malpighi, Professor of Physick in Sicily. Vol. 2, 491–492.

Phil. Trans. (1668). An Account of a Controversy Betwixt Stephano de Angelis, Professor of the Mathematicks in Padua, and Joh. Baptista Riccioli Jesuite as It Was Communicated Out of Their Lately Printed Books, by That Learned Mathematician Mr. Jacob Gregory, a Fellow of the R. Society. Vol. 3, 693–698.

Phil. Trans. (1668). An Account of Some Books. - […]. II. De infinitis spiralbus inversis, infintis que hyperbolis, aliisq; Geometricis, Auth. F. Stephano de Angelis, Veneto. Patavij, in 4°. - III. Michaelis Angeli ricci exercitatio geometrica; in 4° printed in Rome […]. Vol. 3(37), 736–740.

Phil. Trans. (1668). An Account of Two Books. I Saggi di Naturali Esperienze fatte nell' Academia del Cimento, in Firenze, A. 1667. In fol. II. Vera Circuli et Hyperbolæ Quadratura, in propria sua proportionis specie inventa & demonstrata, à Fac Gregorio Scoto, Patavii, in 4°. Vol. 3(33), 640–644.

Phil. Trans. (1668). Extracts of Three Letters; [...] The Third, Written from Paris, about the Polishing of Telescopical Glasses by a Turn-Lathe; As Also the Making of an Extraordinary Burning-Glass at Milan. Vol. 3, 791–796.

Phil. Trans. (1668). Some Observations Concerning the Comet, That Hath Lately Appear'd in Forain Parts, Communicated from Italy and Portugal. Vol. 3, 683–684.

Phil. Trans. (1668). Two Extracts out of the Italian Giornale de Letterati; The One, about Two Experiments of the Transfusion of Blood, made in Italy, the Other, Concerning a Microscope of a New Fashion, Discovering Animals Lesser Than Any Seen Hitherto. Vol. 3, 840–842.

Phil. Trans. (1669). An Answer to Some Inquiries Concerning the Eruptions of Mount Ætna, an. 1669. Communicated by Some Inquisitive English Merchants, Now Residing in Sicily. Vol. 4, 1028–1034.

Phil. Trans. (1670). An Extract of an Italian Letter Written from Venice by Signor Jacomo Grandi, to an Acquaintance of His in London, Concerning Some Anatomical Observations, and Two Odd Births: English'd by the Publisher, as Follows. Vol. 5, 1188–1189.

Phil. Trans. (1670). An Observation of M. Adrian Azout, a French Philosopher, Made in Rome (Where He Now is) about the Beginning of This Year 1670. Concerning the Declination of the Magnet: Out of an Italian Printed Paper, English'd by the Publisher. Vol. 5, 1184–1187.

Phil. Trans. (1670). An Accompt Given by a Florentin Patrician, Call'd Jo. Battista Donius, Concerning a Way of Restoring the Salubrity of the Country about Rome: Extracted Out of the Ninth Italian Giornale de Letterati; And English't as Follows. Vol. 5, 2017–2019.

Phil. Trans. (1670). Some Considerations of Mr. Nic. Mercator, Concerning the Geometrick and Direct Method of Signior Cassini for Finding the Apogees, Excentricities, and Anomalies of the Planets; As That was Printed in the Journal des Scavans of Septemb. 2. 1669: Which Considerations are Here Delivered in the Latine Tongue, Wherein they Were Written by the Author, as Chiefly Regarding the Learn'd in Astronomy, viz. Clarissimi Cassini Methodus Investigandi Apogea, Excentricitates & Anomalias Planetarum, Breviter Exposita & Demonstrata. Vol. 5, 1168–1175.

Phil. Trans. (1670). An Extract Out of a Lately Printed Epistolary Address, Made to the G. Duke of Toscany Touching Some Anatomical Engagements, of Laur. Bellini, Ord. Anat. Prof. at Pisa. Vol. 5, 2093–2095.

Phil. Trans. (1671). A Curious Relation, Taken out the Third Venetian Journal de Letterati, of March 15. 1671; Of a Substance Found in Great Quantities in Some Mines of Italy; Out of Which is Made a Kind of Incombustible Both Skin, Paper, and Candle Week, together with the Experiments Made Therewith. Vol. 6, 2167–2169.

Phil. Trans. (1671). An Account of Some Books. [...]III. Physico mathesis de lumine, coloribus & iride, &c. Auth. Franc. Maria Grimaldo s. J. Bononiæ, 1665.[...]. Vol. 6(79), 3064–3074.

Phil. Trans. (1671). An Extract of a Latin Letter, Written by the Learned Signior Malpighi to the Publisher, Concerning Some Anatomical Observations, about the Structure of the Lungs of Froggs, Tortoises, & c. and Perfecter Animals; As Also the Texture of the Spleen, & c. Vol. 6, 69–80.

Phil. Trans. (1671). An Extract Out of the 3d and 7th Concerning the Formation of Faetus's. Vol. 6, 2224–2227.

Phil. Trans. (1671). An Intimation of Divers Philosophical Particulars, Now Undertaken and Consider'd by Several Ingenious and Learned Men; Here Inserted to Excite Others to Joyn with Them in the Same or the Like Attempts and Observations. Vol. 6, 2216–2219.

Phil. Trans. (1671). An Observation and Experiment Concerning a Mineral Balsom, Found in a Mine of Italy by Signior Marc-Antonio Castagna; Inserted in the 7th. Giornale Veneto de Letterati of June 22. 1671, and Thence English'd as Follows. Vol. 6, 3059.

Phil. Trans. (1671). Some Experiments of Signor Carolo Rinaldini, Philosopher and Mathematician in the University of Padoua; Shewing the Difference of Ice Made without Air, from That Which Is Produced with Air: In the Same Venetian Journal. Vol. 6(72), 2169–2170.

Phil. Trans. (1671). Two Observations Made by P. Francesco Lana, the Author of the Prodromus Premised to Arte Maestra, Concerning Some of the Effects of the Burning Concave of Lions; And Also an Odd Salt Extracted out of a Metallic Substance: Taken Out of the Same Venetian Journal. Vol. 6, 3060.

Phil. Trans. (1672). An Accompt of Some Books. - I. Prose de signori academici di Bologna; in Bologna, 1672 in 4° [...]. Vol. 7(89), 5125–5130.

Phil. Trans. (1672). An Extract of a Letter, Written March 5. 1672 by Dr. Thomas Cornelio, a Neapolitan Philosopher and Physician, to John Dodington Esquire, His Majesties Resident at Venice; Concerning Some Observations Made of Persons Pretending to Be Stung by Tarantula's: Engilish'd Out of the Italian. Vol. 7(83), 4066–4067.

Phil. Trans. (1672). Two Observations about Stones Found, the One in the Bladder of a Dogg, the Other Fastned to the Back-Bone of a Horse: Both Mentioned in Two Roman Journals de letterati. Vol. 7(84), 4094–4095.

Phil. Trans. (1672). A Relation of the Return of a Great Permanent Spot in the Planet Jupiter, Observed by Signor Cassini, One of the Royal Parisian Academy of the Sciences. Vol. 7, 4039–4042.

Phil. Trans. (1672). An Account of the Aponensian Baths Near Padua; Communicated by the Foremention'd Inquisitive Gentleman, Mr. Dodington, in a Letter Written to the Publisher from Venice March 18. Vol. 7(83), 4067–4068.

Phil. Trans. (1672). An Extract of a Letter Written to the Publisher by Mr. Thomas Platt, from Florence, August 6. 1672. Concerning Some Experiments, There Made upon Vipers, Since Mons. Charas His Reply to the Letter Written by Signor Francesco Redi to Monsteur Bourdelet and Monsieur Morus. Vol. 7, 5060–5066.

Phil. Trans. (1672). Observations of a New Comet, Made at Paris in the Royal Observatory by Signor Cassini. Vol. 7, 4042–4050.

Phil. Trans. (1672). Reflections Made by P. Francisco Lana S. F. upon an Observation of Signor M. Antonio Castagna, Super Intendent of Some Mines in Italy, Concerning the Formation of Crystals: English'd Out of the XI. Venetian Giornale de Letterati. Vol. 7, 4068–4069.

Phil. Trans. (1673). A Discovery of Two New Planets about Saturn, Made in the Royal Parisian Observatory by Signor Cassini, Fellow of Both the Royal Societys, of England and France; English't Out of French. Vol. 8, 5178–5185.

Phil. Trans. (1673). An Account of Some of the Natural Things, with Which the Intelligent and Inquisitive Signor Paulo Boccone, of Sicily, Hath Lately Presented the Royal Society, and Enriched Their Repository. Vol. 8(99), 6158–6161.

Phil. Trans. (1674). Extracts of Two Letters, Written by Mr. Flamsteed to Mr. Collins; The One of Novemb. 25. 1624, Concerning an Instrument to Shew the Moon's True Place to a Minute or Two; as Also the Writer's Design of Correcting the Hitherto Assign'd Motions of the Sun: The Other, of Decemb. 14. 1674. Touching the Necessity of Making New Solar Numbers, Together with An Expedient for Making Trial, Whether the Refractions in Signor Cassini's Tables are Just. Vol. 9, 219–221.

Phil. Trans. (1674). Extracts of Two Letters, the One from Monsieur Christian Hugens de Zulichem, Touching His Thoughts of Mr. Hook's Observations for Proving the Motion of the Earth; The Other from Seignior Cassini, Concerning the Same Argument. Vol. 9, 90–91.

Phil. Trans. (1674). Observations Concerning the Comet That Was Seen in Brasil, An. 1668. in March, by P. Valentin Estancel a Jesuit, and by Him Sent to Rome; Where They Were Printed in the 9th Italian Giornale de Letterati, Septemb. 31. 1673. Vol. 9, 91–93.

Phil. Trans. (1675). A Letter Written to the Publisher by the Excellent Cassini, Relating to the Foregoing Observations. Vol. 10, 390.

Phil. Trans. (1675). Some Communications from Rome and Paris. Vol. 10, 309.

Phil. Trans. (1675). A Phytological Observation Concerning Orenges and Limons, Both Separately and in One Piece Produced on One and the Same Tree at Florence: Described by the Florentin Physitian Petrus Natus, and the Description Lately Communicated to the Publisher. Vol. 10, 313–314.

Phil. Trans. (1676). An Extract of a Letter Written by Signor Cassini to the Author of the Journal des Scavans, Containing Some Advertisements to Astronomers about the Configurations, by Him Given of the Satellites of Jupiter, for the Years 1676, and 1677, for the Verification of Their Hypotheses. Vol. 11, 681–683.

Phil. Trans. (1676). An Extract of Signor Cassini's Letter Concerning a Spot Lately Seen in the Sun; Together with a Remarkable Observation of Saturn, Made by the Same. Vol. 11, 689–690.

Phil. Trans. (1676). Extracts of Three Letters of Signor Cassini, Containing His Sentiment of Mr. Flamsted s Account of the Last Eclipse of the Moon; as Also His Own Observations of That Phaenomenon; and likewise An Observed Occultation of a Fixt Star by the Moon. Vol. 11, 561–565.

Phil. Trans. (1676). Mr. Francis Vernons Letter, Written to the Publisher Januar. 10th. 1675/6 giving a Short Account of Some of His Observations in His Travels from Venice Through Istria, Dalmatia, Greece, and the Archipelago, to Smyrna, Where This Letter Was Written. Vol. 11, 575–582.

Phil. Trans. (1676). Mr. Flamsteads, Mr. Townlyes, Mr. Haltons, Signor Cassini's and Monsieur Hevelius's, Observations of the Late Eclipse of the Sun. Vol. 11, 662–667.

Phil. Trans. (1677). An Extract of a Letter Written by Signior Borelli, about the Price of His Telescopes: Communicated to Sir Jonas Moore. Vol. 12, 1005.

Phil. Trans. (1677). Extract of a Letter Sent from Genoua to Sign. Sarotti, the Venetian Resident Here, and by Him Communicated to the Honourable Mr. Boyle. Vol. 12, 976.

Phil. Trans. (1677). Observations of Some Animals, and of a Strange Plant, Made in a Voyage into the Kingdom of Congo: by Michael Angelo De Guattini and Dionysius of Placenza, Missionaries Thither. Extracted out of the Journal des Scavans. Vol. 12, 977–978.

Phil. Trans. (1677). Some New Observations Made by Sig. Cassini and Deliver'd in the Journal des Scavans., Concerning the Two Planets about Saturn, Formerly Discover'd by the Same, as Appears in N. 92. of the se Tracts. Vol. 12, 831–833.

Phil. Trans. (1677). Signor Cassini's Letter, Giving Some Account of the Observations Made at Paris of the Late Comet. Vol. 12, 868.

Phil. Trans. (1683). An Account of a Book entituled Relatione de Ritrovamento dell' Vova di Chiocciole. Vol. 13, 356–358.

Phil. Trans. (1684). A Further Account of the Bridg at Pont St. Esprit, whereof before, Numb. 160. Together with a Parallel History of Some Other Bridges at Rome: In a Letter from the Ingenious Mr. Tankred Robinson to Dr. Martin Lister. Vol. 14, 712–713.

Phil. Trans. (1684). Observations of the Eclipse of the Sun on the 12 of July Last (New Style) Made at the Observatory at Paris 1684. in the Lower Apartment, by Messieurs Cassini and Sedileau; In the Upper, by Messieurs de la Hire and Pothenot. At the College of Lewis the Great, in the Presence of Monseigneur the Duke of Bourbon, by R. P. Fontenay; At Aix in Province; At Lyons; At the Bay of Roses; At Honfleur and at Pau; By Divers Other Learned Persons. Vol. 14, 715–720.

Phil. Trans. (1684). Praeclarissimo et Eruditissimo Viro D. Jacobo Sponio Medicinae Doctori, et Lugdunensi Anatomico Acuratissimo. Marcellus Malpighius S. P. Vol. 14, 601–608.

Phil. Trans. (1684). Praeclarissimo et Eruditissimo Viro D. Jacobo Sponio Medicinae Doctori, et Lugdunensi Anatomico Accuratissimo. Marcellus Malpighius S. P. Vol. 14, 630–646.

Phil. Trans. (1684). A Calculation of the Same Eclips, Juxta Tabulas Philolaicas with the Observations of the Learned Bullialdus and Cassini. Vol. 14, 693.

Phil. Trans. (1685). An Extract of a Letter from Senior Ciampini, to Dr. Croon, concerning a Late Comet Seen at Rome. Vol. 15, 920–921.

Phil. Trans. (1686). An Uncommon Inscription Lately Found on a Very Great Basis of A Pillar, Dug Up At Rome; With An Interpretation of the Same By the Learned Dr. Vossius. Vol. 16, 172–175.

Phil. Trans. (1686). A Letter of Monsieur Cassini to the Publisher, Giving His Corrections of the Theory of the Five Satellites of Saturn; With Tables of the Motions of Those Satellites, Adapted to the Meridian of London, and the Julian Account. Vol. 16, 299–306.

Phil. Trans. (1686). An Extract of a Letter Written from Rome, Dated the 16th. of November Last, to Signior Sarotti, Concerning a Discovery Made upon the Inundation of the Tevere. Translated out of the Italian. Vol. 16, 227.

Phil. Trans. (1686). An Extract of the Journal Des Scavans. of April 22 st. N. 1686. Giving an Account of Two New Satellites of Saturn, Discovered Lately by Mr. Cassini at the Royal Observatory at Paris. Vol. 16, 79–85.

Phil. Trans. (1693). A Letter from Mr. Martin Hartop at Naples, to the Publisher. Together with an Account of the Late Earthquake in Sicily. Vol. 17, 827–829.

Phil. Trans. (1693). An Extract of the Account Mentioned in the Foregoing Letter, Taken Out of an Italian Paper. Written by P. Alessandro Burgos. Printed First at Palermo, and Afterwards at Naples. Vol. 17, 830–838.

Phil. Trans. (1693). Observatio Solaris Eclipsis Habita Die 12. Julii, 1684. Bononiae, a Cl. V. Dono. Dominico Gulielmino. Vol. 17, 858–860.

Phil. Trans. (1694). An Account of Books. Osservazioni Naturali, ove si contengono materie Medico-Fisiche, &c. Natural Observations Containing Several Medico-Physiscal and Botanical Matters, with Divers Natural Productions, Several Sorts of Phosphori, Subterraneous Fires in Italy, and Other Curious Subjects, in Familiar Letters, by Signior Paul Boccone M.D. Printed at Bononia in 12o 1684. Vol. 18, 33–40.

Phil. Trans. (1694). An Account of the Earthquakes in Sicilia, on the Ninth and Eleuenth of January, 1692/3 Translated from an Italian Letter Wrote from Sicily by the Noble Vincentius Bonajutus, and Communicated to the Royal Society by the Learned Marcellus Malpighius, Physician to His Present Holiness. Vol. 18, 2–10.

Phil. Trans. (1694). Solutio Problematis Florentini de Testitudine Veliformi Quadrabili, a Davide Gregorio, M. D. ac R. S. S. Communicata. Vol. 18, 25–29.

Phil. Trans. (1694). Monsieur Cassini His New and Exact Tables for the Eclipses of the First Satellite of Jupiter, Reduced to the Julian Stile, and Meridian of London. Vol. 18, 237–256.

Phil. Trans. (1695). Extract of a Letter from Jean Marie Lancisi, Prof. Anat. Rom. To Mr. Bourdelot, Giving an Account of Mr. Malpighi, the Circumstances of His Death, and What Was Found Remarkable at the Opening of His Body. Being Art. I. of the 3d. Journal of Brunets Progres de la Medecine. Vol. 19, 467–471.

Phil. Trans. (1695). Part of a Letter from Mr. Octavian Pulleyn, Dated, Rome March 16 1696, Giving an Account of an Inscription There Found in the Language of the Palmereni; and Another in the Etruscan Language Found on an Old Vine. Vol. 19, 537–539.

Phil. Trans. (1698). An Extract of a Letter from Dr. Robert St. Clair, to Dr. Rob. Hook, F. R. S. Giving an Account of a Very Odd Eruption of Fire out of a Spot in the Earth Near Fierenzola in Italy, with an Easy Cantrivance of a Lamp to be Kept Always Full Whilst It Burns. Vol. 20, 378–381.

Phil. Trans. (1698). A Catalogue of Books Lately Printed in Italy. Vol. 20, 426–428.

Phil. Trans. (1698). An Account of Books now Printing beyond Sea. Vol. 20, 388.

Phil. Trans. (1698). Remarks Concerning Factitious Salts; Drawn from a Discourse Written by Sen Francisco Redi. Vol. 20, 281–289.

Phil. Trans. (1699). An Extract of a Letter from Leghorn to Dr. Martin Lister, November 24. 1698, concerning Seignior Redi's Manuscripts, and the Generation of Fleas. Vol. 21, 42–43.

Phil. Trans. (1699). IX. An Account of Books. I. Museo di Fisica & di Esperienze, &c. By Signior Boccone ; with additional Remarks by Mr. John Ray, F. R. S. 2. An Account of Paradisus Batavus, Contiens plus Centurn Plantas, &c. with additional Remarks by Mr. John Ray, F. R. S. Vol. 21, 53–67.

Phil. Trans. (1699). Of the Nature of Silk, as It Is Made in Piedmont. Communicated by William Aglionby, Esq; F.R.S. Vol. 21, 183–186.

18th century

Phil. Trans. (1700). A Letter from Mr John Monro to the Publisher, Concerning the Catacombs of Rome and Naples. Vol. 22, 643–650.

Phil. Trans. (1700). A Letter from Dr P. Silvestre, of the Coll. of Phy. & F. R. S. to the Publisher, Giving an Account of Some New Books and Manuscripts in Italy. Vol. 22, 613–614.

Phil. Trans. (1700). A Letter from Dr Peter Silvestre, F. R. S. to the Publisher, Concerning the State of Learning, and Several Particulars Observed by Him Lately in Italy. Vol. 22, 627–634.

Phil. Trans. (1700). An Abstract of a Letter, Wrote Some Time Since, by Signior John Ciampini of Rome, to Father Bernard Joseph a Jesu Maria, etc. Concerning the Asbestus, and Manner of Spinning and Making an Incombustible Cloath Thereof. Vol. 22, 911–913.

Phil. Trans. (1700). An Account of the Strange Effects of the Indian Varnish. Wrote by Dr Joseph Del Papa, Physician to the Cardinal De Medices, at the Desire of the Great Duke of Tuscany. Communicated by Dr William Sherard. Vol. 22, 947–951.

Phil. Trans. (1700). Extracts of Some Letters to the Publisher, Giving an Account of Some Books Now Printing and Lately Printed in Italy, France, Germany, Holland and Scotland. Vol. 22, 1041–1043.

Phil. Trans. (1700). IV. Account of a Book, viz. Aloysi Ferdinandi comit. Marsigli danubialis operis Prodromus. Ad Regiam Societatem Anglicanam. Fol. Vol. 22, 1038–1041.

Phil. Trans. (1700). Responsio Almi Collegii Romanorum Archiatrorum ad Epistolas Clarissimi D. Raymundi Vieussends Medicinae Doctoris Monspeliensis, in Qua Potissimum Agitur De Existentia Salis Acidi in Sanguine, & De Proportione Principiorum Ejusdem Scripta Per Jo. Mariam Lancisi Olim. S. D. Innocentis XI. Med. a Secretis & Nunc Archiatrum Collegialem, & in Romano Licaeo Theoricae Extraordinariae Professorum. Vol. 22, 599–610.

Phil. Trans. (1700). The Way of Making Several China Varnishes. Sent from the Jesuits in China, to the Great Duke of Tuscany, Communicated by Dr William Sherard. Vol. 22, 525–526.

Phil. Trans. (1701). IV. Account of a book, viz. Aloysi Ferdinandi comit. Marsigli danubialis operis Prodromus. Ad Regiam Societatem Anglicanam. Fol. 1700. Vol. 22, 1038–1041.

Phil. Trans. (1702). An Abstract of Part of a Letter from Dr Bonomo to Sigmor Redi, Containing Some Observations Concerning the Worms of Humane Bodies. By Richard Mead, M. D. Vol. 23, 1296–1299.

Phil. Trans. (1703). Books Imported and Sold by Sam. Smith and Benj. Walford, Printers to the Royal Society, at the Princes Arms in St Paul's Churchyard. Vol. 23, 283.

Phil. Trans. (1706). A Letter from Mr Samuel Dale to Dr Hans Sloane, R. S. Secr. Giving an Account of What Manuscripts Were Left by Mr John Ray, Together with Some Anatomical Observations Made at Padua by the Said Mr Ray. Vol. 25, 2282–2303.

Phil. Trans. (1706). An Extract of a Letter to His Excellency Signior Francisco Cornaro, Ambassador from the Republick of Venice, to the Queen of Great Britain, etc. By Anthony Van Leeuwenhoek, F.R.S. Containing Microscopical Observations of the Salts of Pearls, Oyster-Shells, etc. Vol. 25, 2416–2424.

Phil. Trans. (1706). Observations Made at Rome, by the Late Reverend Mr. John Ray, of the Comet Which Appeared Anno 1664. Communicated to the Publisher by Mr. Samuel Dale. Vol. 25, 2350–2352.

Phil. Trans. (1706). Part of a Letter Written to Signior Antonio Magliabechi, by Mr Anthony Van Leeuwenhoek, F. R. S. Concerning the Particles of Silver Dissolved in Aqua Fortis, etc. Vol. 25. 2425–2432.

Phil. Trans. (1708). Epistola D. Guidonis Grandi, Societatis Regalis Londin. Socii, ad Illust. Comitem D. Laurentium Magalotti, Dictae Societatis Socium, De Natura & Proprietatibus Soni. Vol. 26, 270–288.

Phil. Trans. (1708). Tables of the Barometical Altitudes at Zurich in Switzerland in the Year 1708. Observed by Dr. Joh. Ja. Scheuchzer, F. R. S. and at Upminster in England, Observed at the Same Time by Mr. W. Derham, F. R. S. as Also the Rain at Pisa in Italy in 1707. and 1708. Observed There by Dr. Michael Angelo Tilli, F. R. S. and at Zurich in 1708. And at Upminster in All That Time: With Remarks on the Same Tables, as Also on the Winds, Heat and Cold, and Divers Other Matters Occurring in Those Three Different Parts of Europe. By Mr. W. Derham, Rector of Upminster. Vol. 26, 334-366.

Phil. Trans. (1710). VII. An Account of a Book Intitled, differtatio epistolaris de glandulis conglobatis duræ meningis humanæ, indeque ortis Lymphaticis ad Piam Meningem productis. Authore Antonio Pacchiono. Romæ 1705. 8vo. Vol. 27, 208-211.

Phil. Trans. (1712). Epistola D. Josephi Vallettae Nobilis Neopolitani, ad D. Richardum Waller Armig. Reg. Societ.Sec. de Incendio & Eruptione Montis Vesuvii, Anno MDCCVII. Vol. 28, 22-25.

Phil. Trans. (1714). An Account, or History, of the Procuring the Small Pox by Incision, or Inoculation; As It Has for Some Time Been Practised at Constantinople. Vol. 29, 72-82.

Phil. Trans. (1714). III. An Extract from the Acta Eruditorum for the Month of March, 1713. Pag. 111. Vol. 29(338), 46-49.

Phil. Trans. (1714). Miscellaneous Observations Made about Rome, Naples and Some Other Countries, in the Year 1683 and 1684; and Communicated to the Publisher by Tancred Robinson M. D. R. S. S. Vol. 29, 473-483.

Phil. Trans. (1714). Nova & Tuta Variolas Excitandi per Transplantationem Methodus, Nuper Inventa & in Usum Tracta: Per Jacobum Pylarinum, Venetum, M. D. & Reipublicae Venetae Apud Smyrnenses Nuper Consulem. Vol. 29, 393-399.

Phil. Trans. (1714). VI. An Account of a Book Entituled Methodus Incrementorum, auctore Brook Taylor, LL.D. & R. S. Secr. By the author. II. Ludovici Ferdinandi Marsilii Dissertatio de Generatione Fungorum. Rom. 4to. Vol. 29, 339-352.

Phil. Trans. (1717). A Letter of M. l' Abbe Conti, R. S. S. to the Late M. Leibnitz, Concerning the Dispute about the Invention of the Method of Fluxions, or Differential Method; With M. Leibnitz His Answer. Vol. 30, 923-928.

Phil. Trans. (1717). Extract of a Letter of Mr. Edw Berkeley from Naples, Giving Several Curious Observations and Remarks on the Eruptions of Fire and Smoak from Mount Vesuvio. Communicated by Dr. John Arbuthnot, M D. and R. S. S. Vol. 30, 708-713.

Phil. Trans. (1720). De Peste Constantinopoli Grassante. Auctore Nupero V. Cl. Emanuele Timone, M. D. Hoc Scriptum ab Auctore Clarissimo, Qui Constantinopoli per Multos Annos Medicinam Fecerat, Excellentissimo

Ablegato Britannico. Roberto Sutton, Eq. Aurato, Traditum, Ejusdem Equitis Permissu, cum Societate Regia Communicavit R. Hale, M. D. Vol. 31, 14–21.

Phil. Trans. (1720). Some Remarks on a Late Essay of Mr. Cassini, Wherein He Proposes to Find, by Observation, the Parallax and Magnitude of Sirius, By Edmund Halley, LL. D. R. S. S. Vol. 31, 1–4.

Phil. Trans. (1721). An account of book, intituled, conghietture del Dottor Pietre Anton. Michelotti, Filosofo, e medico d' arco, sopra la natura, cagione e rimedi dell' infermità regnanti ne' animali bovini di molte città, &c. Nell' autunno del' anno cadente, 1711. In Venezia, 1712. Vol. 31(365), 83–86.

Phil. Trans. (1722). An Account of Two Observations upon the Cataract of the Eye; Contain'd in a Letter from Signor Antonio Benevoli, Master-Surgeon in the Hospital of S. Maria Nuova in Florence, to Dr. Valsalva; Printed in Italian at Florence, This Present Year, and Communicated to the Royal Society, at the Desire of the Author, by Sir Thomas Dereham, F. R. S. Vol. 32, 194–196.

Phil. Trans. (1722). Defensio Dissertationis de Motu Aquarum Fluentium, in Actis Philosophicis, No 355. Editae, Contra Animadversiones Viri Cl. Petri Antonii Michelotti. Auctore Jacobo Jurin, M. D. Coll. Med. Lond. Soc. in Theatro Publico Chirurg. Lond. Praelectore Anatomico, & Soc. Reg. Secr. Vol. 32, 179–190.

Phil. Trans. (1722). Florum Geometricorum Manipulus Regiae Societati Exhibitus a D. Guidone Grandi Abbate Camaldulensi, Pisani Lycaei Mathematico, R. S. S. Vol. 32, 355–371.

Phil. Trans. (1724). An Enquiry into a Discovery, Said to Have Been Made by Signor Valsalva of Bologna, of an Excretory Duct from the Glandula Renalis to the Epididymis. By Mr. John Ranby, Surgeon, F. R. S. Vol. 33, 270–271.

Phil. Trans. (1724). An Excretory Duct from the Glandula Renalis. Extracted from the Giornale Di Literati of Venice, for the Year 1719. Vol. 33, 190.

Phil. Trans. (1724). Observatio Ejusdem Cometae ab Illustrissimo Domino Francisco Bianchini Habita Albani Mense Octobri, 1723. & ab Eodem Ulysipponem missa P. Joanni Baptistae Carbone Soc J E S U. Commuuicavit Isaacus Samuda, M. D. Col. Med. Lond. L. S. R. S. Vol. 33, 51–53.

Phil. Trans. (1724). Observations Made in Italy of a Lunar Eclipse, Which Happened the 8th of Sept. 1718. Extracted from the Giornale Di Literati of Venice. Vol. 33, 71–78.

Phil. Trans. (1726). Conspectus Maculae Platonis in Luna Caelo Clarissimo Nocte Sequente Diem 16 Augusti 1725, Hora Prima Post Occasum Solis Romae in Monte Palatino, Per Tubum Opticum Josephi Campani, Palmorum 150 Romanorum, Spectante Eodem Viro Cl. Ex Eadem Epistola. Vol. 34, 181–182.

Phil. Trans. (1726). Observatio Defectus Solis Habita Patavii VII. Cal. Octob. 1726. a Viro Cl. Domino Johanne Poleno, Astr. Prof. R. S. S. Vol. 34, 157.

Phil. Trans. (1727). IV. An Extract of a Letter of Signior Michele Pinelli, Concerning the Causes of the Gout. Translated from the Italian by Joh. Gasp Scheuchzer, M. D. F. R. S. and Coll. Med. Lond. Lic. Vol. 35(403), 491–494.

Phil. Trans. (1727). Lunaris Eclipsis Celebrata die 10. Octob. an. 1726. & in Observatorio Collegii D. Antonii Magni Observata ab Eodem. Vol. 35, 338–342.

Phil. Trans. (1727). Observatio Solaris Deliquii Celebrati Die 25. Septemb. 1726. Habita Ulyssipone in Observatorio Regii Palatii A' P. Jo. Baptista Carbone, Soc. Jes. Vol. 35, 335–338.

Phil. Trans. (1727). Observationes Astronomicae a R. P. Joh. Baptista Carbone Transmissae, Communicante Is. De Seguera Samuda, M. D. R. S. S. & Coll. Med. Lond. Lic. Vol. 35, 471–479.

Phil. Trans. (1727). Observationes Astronomicae Habitae in Observatorio Bononiensi Anno 1727, a Cl. Eustachio Mansredi, R. S. S. Ex Epistola J. Baptistae Carbone ad Isaacum De Sequeyra Samuda M. D. Coll. Med. Lic. & R. S. S. Vol. 35, 534–535.

Phil. Trans. (1727). Observationes Astronomicae Habitae Ulyssipone, Anno 1726. a Rev. P. Joh. Baptista Carbone, Soc. Jes. Communicante Isaaco Sequeyra Samuda, M. D. R. S. S. & Coll. Med. Lond. Lic. Vol. 35, 408–413.

Phil. Trans. (1728). II. Optical Experiments Made in the Beginning of August 1728, before the President and Several Members of the Royal Society, and Other Gentlemen of Several Nations, upon Occasion of Signior Rizzetti's Opticks, with an Account of the Said Book. Vol. 35, 596–629.

Phil. Trans. (1729). An Explanation of the New Chronological Table of the Chinese History, Translated into Latin from the Original Chinese, by Father Johannes Franciscus Foucquet, Soc. Jes. Bishop of Eleutheropolis, and Published at Rome in the Year 1730. Collected from Two Accounts Thereof, Written in French, One Sent from Rome by Sir Tho. Dereham, Bart. to the Royal Society, the Other Sent from Father Foucquet to Father Eustache Guillemeau, a Barnabite at Paris, and by Him Transmitted to Sir Hans Sloane, Bart. Pr. R. S. Vol. 36, 397–424.

Phil. Trans. (1729). De Frigidae in Febribus Usu. Auctore Cl. Nicolao Cyrillo Prim. Med. Prof. Neapol. & R. S. S. Vol. 36, 142–151.

Phil. Trans. (1729). Observatio Ejusdem Defectus Solis Habita Patavii, ab J. Poleno. Vol. 36, 396.

Phil. Trans. (1729). Observatio Lunaris Eclipseos, Ulissipone Habita Die 2 Februarii, An. 1730, N. S. in Collegio Divi Antonii Magni a Rev. P. Joanne Baptista Carbone, Soc. Jes. Ex Ejusdem Cl. Viri Epistola ad Jacobum. De Castro Sarmento, M. D. Coll. Med. Lond. Lic. & R. S. S. Vol. 36, 363–365.

Phil. Trans. (1729). Observationes Caelestes Multifariae Inter Menses Novemb. 1727, & Novemb. 1728, Pekini in SINIS, Habitae & ad Rev. P. Joannem Baptistam Carbone, Soc. Jes. Transmissae. Ex Eadem Epistola Descriptae. Vol. 36, 366–371.

Phil. Trans. (1729). Of the Meteor Called the Ignis Fatuus, from Observations Made in England, by the Reverend Mr. W. Derham, F. R. S. and Others in Italy, Communicated by Sir Tho. Dereham, Bart. F. R. S. Vol. 36, 204–214.

Phil. Trans. (1730). VI. An account of a book entituled, Hesperi & Phosphori Nova Phænomena, &c. Authore Francisco Blanchino. Vol. 36, 158–164.

Phil. Trans. (1731). An Account of an Extraordinary Eruption of Mount Vesuvius in the Month of March, in the Year 1730, Extracted from the Meteorological Diary of That Year at Naples, Communicated by Nichol. Cyrillus, M. D. R. S. S. Vol. 37, 336–338.

Phil. Trans. (1731). VI. An Extract of a Letter from Oliver St. John, Esq; F. R. S. Dated from Florence, November the 30th, 1731, N. S. Communicated by R. Graham, F. R. S. Vol. 37, 256.

Phil. Trans. (1731). VI. De ingenti sanguinis vomitu perquàm gelidissimis brumali tempore potionibus curato, observatio ad Regiam Scientiarum Londinensem Societatem à Petro Antonio Michelotto M. D. R. S. S. transmissa. Vol. 37, 129–145.

Phil. Trans. (1731). Viri Celeberrimi Johannis Marchionis Poleni, R. S. S. ad Virum Doctissimum Jacobum Jurinum, M. D. R. S. S. Epistola, Qua Continetur Summarium Observationum Meteorologicarum per Sexennium Patavij Habitarum. Vol. 37, 201–216.

Phil. Trans. (1733). Aeris Terraeque Physica Historia, Anni Biss. 1732. a Nicolao Cyrillo in Universitate Neapolit. Primar. Med. Profess. & R. S. S. Vol. 38, 184–190.

Phil. Trans. (1733). An Abstract of the Meteorological Diaries, Communicated to the Royal Society, with Re-Marks upon Them, by W. Derham, D. D. Canon of Windsor, F. R. S. [Vide Part III. in Transact. No 433.] Part IV. Vol. 38, 405–412.

Phil. Trans. (1733). Clausula Excerpta, ex Historia Variolarum quae per Incisionem Excitantur, ab E. Timoni, M D. Scripta, R. S. Communicavit Sam. Horseman, M. D. Vol. 38, 296–297.

Phil. Trans. (1733). Historia Terraemotus Apuliam & Totum Fere Neapolitanum Regnum, Anno 1731, Vexantis. A Nicolao Cyrillo, in Regia Universitate Neapolitana, Pr. Med. Prof. & R. S. S. Vol. 38, 79–84.

Phil. Trans. (1733). Nonnullae Jovis Satellitum Eclipses Bononiae, Observatae ab Eustachio Manfredi. Vol. 38, 117–118.

Phil. Trans. (1733). Observatio Ecclipsews Lunaris Romae Habita Die 1 Decembris, 1732, in Aedibus Emiminentiss. de V I A, a D. Didaco Revillas Abbate Hieronymiano, Abbate Joanne Botrario, & Eustachio Manfredio. Vol. 38, 85–88.

Phil. Trans. (1735). A Copy of an Ancient Chirograph, or Conveyance of Part of a Sepulchre, cut in Marble, Lately Brought from Rome, and Now in the Possession of Sir Hans Sloane, Bart. R. S. Pr. with Some Observations upon It by Roger Gale, Esq; V. P. R. & Tr. R. S. Vol. 39, 211–219.

Phil. Trans. (1735). An Account of the Standard Measures Preserved in the Capitol at Rome. By Martin Folkes, Esq; V. P. R. S. Vol. 39, 262–266.

Phil. Trans. (1735). Observatio Eclipseos Telluris Romae Habita in Aedibus Eminentissimi Cardinalis De-Via v. Non. i. e. d. 3 Maij, N. S. Apr. 22. V. S. MDCCXXXIV. per Didacum de Revillas, Abbat. Hieronym. R. S. S. & Andream Celsium, R. S. S. Astrom Profess. Upsal. R. S. Suec. Secr. Vol. 39, 294–296.

Phil. Trans. (1737). Observations of the Transit of Mercury Over the Sun, Oct. 31. 1736. Communicated to the Royal Society. Vol. 40, 102–110.

Phil. Trans. (1737). Viri Celeberrimi Johannis Marchionis Poleni, R. S. Lond. S. ad Virum Doctissimum Jacobum Jurinum, M. D. R. S. S. Epistola, Qua Continetur Summarium Observationum Meteorologicarum per Sexennium Patavii Habitarum. Vol. 40, 239–248.

Phil. Trans. (1739). A Letter from His Excellency Nicolas-Michael d' Aragona, Prince of Cassano, and F. R. S. to the President of the Royal Society, Containing an Account of the Eruption of Vesuvius in May 1737. Translated from the Italian by T. S. M. D. F. R. S. Vol. 41, 237 –252.

Phil. Trans. (1739). Extracts of Two Letters from Sigr. Camillo Paderni at Rome, to Mr. Allan Ramsay, Painter, in Covent-Garden, concerning Some Antient Statues, Pictures, and Other Curiosities, Found in a Subterraneous Town, Lately Discovered Near Naples. Translated from the Italian by Mr. Ramsay, and Sent by Him to Mr. Ward, F. R. S. Prof. Rhet. Gresh. Vol. 41, 484–489.

Phil. Trans. (1739). A Collection of the Observations of the Remarkable Red Lights Seen in the Air on Dec. 5. 1737. Sent from Different Places to the Royal Society. Vol. 41, 583 –606.

Phil. Trans. (1739). The Parabolic Orbit for the Comet of 1739. Observed by Signor Eustachio Zanotti at Bologna. Vol. 41, 809.

Phil. Trans. (1740). A Collection of the Observations of the Remarkable Red Lights Seen in the Air on Dec. 5. 1737. Sent from Different Places to the Royal Society. Vol. 41(459), 583–606.

Phil. Trans. (1740). A Letter from His Excellency Nicolas-Michael d'Aragona, Prince of Cassano, and F. R. S. to the President of the Royal Society, Containing an Account of the Eruption of Vesuvius in May 1737. Translated from the Italian by T. S. M. D. F. R. S. Vol. 41(455), 237–252.

Phil. Trans. (1742). A Letter Written to the Most Reverend Father D. Cla. Fremond Calmad, Publick Professor in the University of Pisa, Giving an Account of the Earthquakes Felt in Leghorn, from the 16th to the 27th of January 1742. With Some Observations Made by the Most Reverend Sig. Pasqual R. Pedini, Principal of the Clergy of the Most Eminent College of the Said City. Communicated to the Royal Society by James Jurin, M. D. F. R. S. &c. Vol. 42, 77–90.

Phil. Trans. (1742). De Calculo Praegrandi a Muliere Cum Urina Excreto Observatio Dni Antonii Leprotti, R. S. S. Pont. Max. Archiat. per Abbatem Didacum de Revillas, R. S. S. ad D. Smart Lethieullier R. S. S. Transmissa. Vol. 42, 363.

Phil. Trans. (1742). Johannis Marchionis Poleni, R. S. S. Dc Novis Quibusdam Cogitationibus ad Explorandum, Num Pendula vi Aliqua Centrifuga Perturbentur, Commentariolum Illustrissimae Societati Regali Londinensi Oblatum. Vol. 42, 298–306.

Phil. Trans. (1744). Extract of a Letter from Dr. Josephus Laurentius Bruni, of Turin, F. R. S. to Mr. Henry Baker, F.R.S. concerning the Bologna Bottles. Vol. 43, 272–273.

Phil. Trans. (1744). An Extract, by Mr. Paul Rolli, F. R. S. of an Italian Treatise, Written by the Reverend Joseph Bianchini, a Prebend in the City of Verona; Upon the Death of the Countess Cornelia Zangari & Bandi, of Cesena. To Which are Subjoined Accounts of the Death of Jo. Hitchell, Who was Burned to Death by Lightning; And of Grace Pett at Ipswich, Whose Body was Consumed to a Coal. Vol. 43, 447–465.

Phil. Trans. (1744). Some Account of a Curious Tripos and Inscription Found Near Turin, Serving to Discover the True Situation of the Ancient City Industria. By David Erskine Baker. Vol. 43, 540–549.

Phil. Trans. (1744). An Account of Some Human Bones Incrusted with Stone, Now in the Villa Ludovifia at Rome: Communicated to the Royal Society by the President, with a Drawing of the Same. Vol. 43, 557–560.

Phil. Trans. (1746). An Extract, by Philip Henry Zollman, Esq; F. R. S. of a Philosophical Account of a New Opinion Concerning the Origin of Petrifactions Found in the Earth, Which Has Been Hitherto Ascribed to the Universal Deluge; As Contained in an Italian Book, Intitled, De Crostacei ed Altri Marini Corpi che se trovano su' Monti, di Anton. Lazzaro Moro, Venice 1740. Communicated together with Several Remarks, by Dr. Balthasar Ehrhart, Physician in Ordinary at Memmingen, and Member of the Acad. Nat.Curios. in High-Dutch at Memmingen, 1745. 4 to. Vol. 44, 163–166.

Phil. Trans. (1746). Extract of a Letter Dated at Rome, Aug. 5. 1747. from Mr. Hoare, a Young Statuary, Now Pursuing His Studies There, to His Brother Mr. Hoare, an Eminent Painter at Bath, Giving a Short Account of Some of the Principal Antique Pictures Found in the Ruins of Herculaneum at Portici, Near Naples: Communicated by the Rev. Mr. Birch. Vol. 44, 567–571.

Phil. Trans. (1746). Extract of a Letter from the Marquis Nicolini F. R. S. to the President, Concerning the Same Mirror Burning at 150 Feet Distance. Vol. 44, 495–496.

Phil. Trans. (1746). The Phaenomena of Venus, Represented in an Orrery Made by Mr. James Ferguson, Agreeable to the Observations of Seignior Bianchini. Vol. 44, 127–146.

Phil. Trans. (1748). A Letter from Mr. Henry Baker F. R. S, to the President, Concerning Several Medical Experiments of Electricity. Vol. 45, 270–275.

Phil. Trans. (1749). Extract of a Letter from the Abbe Nollet, F.R.S. &c. to Charles Duke of Richmond, F. R. S. Accompanying an Examination of Certain Phaenomena in Electricity, Published in Italy, by the Same, and Translated from the French by Mr. Watson, F. R. S. Vol. 46, 368–397.

Phil. Trans. (1749). Some Considerations on the Causes of Earthquakes. By the Rev. Stephen Hales, D.D. and F.R.S. Vol. 46, 669–681.

Phil. Trans. (1749). XI. Part of a Letter from Robert More Esq; to Mr. W. Watson F. R. S. Concerning the Method of Gathering Manna Near Naples. Vol. 46, 470–471.

Phil. Trans. (1749). A Letter from Mr Christopher Maire to the President, Containing Observations Made at Rome of the Eclipse of the Moon, Dec. 23. 1749; And of That of the Sun, Jan. 8. 1750. Vol. 46, 321–323.

Phil. Trans. (1749). A Letter from Robert More Esq; to the President, Containing Several Curious Remarks in His Travels through Italy. Vol. 46, 464–467.

Phil. Trans. (1749). Remarks upon an Antient Roman Inscription, Found in That Part of Italy, Which Formerly Belonged to the Sabines; And Now in the Possession of Richard Rawlinson, LL.D. & F. R. S. By Mr. John Ward, Prof. Rhetor. Gresh. & F. R. S. Vol. 46, 293–304.

Phil. Trans. (1751). An Account of Dr. Bianchini's Recueil d'Experiences Faites a Venise sur le Medicine Electrique; By Mr. William Watson, F. R. S. Vol. 47, 399–406.

Phil. Trans. (1751). An Account of the Eruption of Mount Vesuvius, from Its First Beginning to the 28th of October 1751, in a Letter from Mr. Richard Supple, Communicated by Mr. Benjamin Wilson, F. R. S. Vol. 47, 315–317.

Phil. Trans. (1751). New Discoveries Relating to the History of Coral, by Dr. Vitaliano Donati. Translated from the French, by Tho. Stack, M. D. F. R. S. Vol. 47, 95–108.

Phil. Trans. (1751). Part of a Letter from Mr. John Parker, an English Painter at Rome, to His Father at London, Concerning the Late Eruption of Mount Vesuvius: Communicated by Mr. Henry Baker, F. R. S. Vol. 47, 474–475.

Phil. Trans. (1751). Extract of the Observations Made in Italy, by the Abbe Nollet, F. R. S. on the Grotta de Cani. Translated from the French by Tho. Stack, M. D. F. R. S. Vol. 47, 48–61.

Phil. Trans. (1753). An Account of a Book, Intitled, P. D. Pauli Frisii Mediolanensis, &c. Disquisitio mathematica in causam physicam figuræ et magnitudinis Telluris nostræ; printed at Milan in 1752. inscribed to the Count de Sylva, and consisting of ten sheets and a half in quarto. Vol. 48, 5–17.

Phil. Trans. (1753). An Account of an Extraordinary Disease of the Skin, and Its Cure. Extracted from the Italian of Carlo Crusio; Accompanied with a Letter of the Abbe Nollet, F. R. S. to Mr. William Watson, F. R. S. by Robert Watson, M. D. F. R. S. Vol. 48, 579–587.

Phil. Trans. (1753). Extract of a Letter from Camillo Paderni, Keeper of the Museum Herculaneum, to Tho. Holles, Esq; Dated at Naples, April 27, 1754. Vol. 48, 634–638.

Phil. Trans. (1753). Extract of a Letter from Camillo Paderni, Keeper of the Herculaneum Museum, to Thomas Hollis, Esq; Relating to the Late Discoveries at Herculaneum. Vol. 48, 821–825.

Phil. Trans. (1753). Extract of a Letter from Signor Camillo Paderni, to Dr. Mead, Concerning the Antiquities Dug Up from the Antient Herculaneum, Dated from Naples, Nov. 18, 1752. Translated from the Italian. Vol. 48, 71–73.

Phil. Trans. (1755). An Account of a Work Published in Italian by Vitaliano Donati, M. D. Containing, An Essay towards a Natural History of the Adriatic Sea: By Mr. Abraham Trembley, F. R. S. Translated from the French, by Thomas Birch, D. D. Secret. R. S. Vol. 49, 585–592.

Phil. Trans. (1755). An Account of the Late Discovories of Antiquities at Herculaneum, &c. in Two Letters from Camillo Paderni, Keeper of the

Musaeum Herculanei, to Thomas Hollis, Esq; Translated from the Italian by Robert Watson, M. D. F. R. S. Vol. 49, 490–508.

Phil. Trans. (1755). An Account of What Happened at Bergemoletto, by the Tumbling down of Vast Heaps of Snow from the Mountains There, on March 19, 1755: As Taken by the Intendant of the Town and Province of Cuneo. Received from Dr. Joseph Bruni, Professor of Philosophy at Turin, and F. R. S. Communicated by Mr. Henry Baker, F. R. S. Translated from the Italian. Vol. 49, 796–803.

Phil. Trans. (1755). An Extract of a Letter Written by the Magistrates of the City of Mascali, in Sicily, and Sent from Their Public Office to Naples, Concerning a Late Eruption of Mount Aetna. Translated from the Italian. Vol. 49, 209–210.

Phil. Trans. (1755). Copy of a Letter from a Learned Gentleman of Naples, Dated February 25, 1755, concerning the Books and Antient Writings Dug out of the Ruins of an Edifice Near the Site of the Old City of Herculaneum; To Monsignor Cerati, of Pisa, F. R. S. Sent to Mr. Baker, F. klR. S. and by Him Communicated; with a Translation by John Locke, Esq; F. R. S. Vol. 49, 112–115.

Phil. Trans. (1755). Extract of a Letter from Dr. Vitaliano Donati, Professor of Botany at Turin, to Mr. Abraham Trembley, F. R. S. concerning the Earthquakes Felt at Turin, December 9, 1755, and March 8, 1756. Translated from the Italian. Vol. 49, 612–616.

Phil. Trans. (1755). Extracts of Two Letters to Thomas Hollis, Esq; Concerning the Late Discoveries at Herculaneum. Vol. 49, 109–112.

Phil. Trans. (1757). An Account of the Alterations Making in the Pantheon at Rome: In an Extract of a Letter from Rome to Thomas Hollis, Esq; Communicated by John Ward, LL.D. R.S. Vice-Praes. Vol. 50, 115–117.

Phil. Trans. (1757). An Account of the Late Discoveries of Antiquities at Herculaneum; In an Extract of a Letter from Camillo Paderni, Keeper of the Herculanean Museum, and F. R. S. to Thomas Hollis, Esq; Dated Naples, Dec. 16, 1756. Vol. 50, 49–50.

Phil. Trans. (1757). An Account of the Late Discoveries of Antiquities at Herculaneum, and of an Earthquake There; In a Letter from Camillo Paderni, Keeper of the Museum at Herculaneum, and F. R. S. to Tho. Hollis, Esq; F. R. S. Dated Portici, Feb. 1. 1758. Vol. 50, 619–623.

Phil. Trans. (1757). XCI. An Historical Memoir Concerning a Genus of Plants Called Lichen, by [Pier Antonio] Micheli, Haller, and [Carl] Linnæus; and Comprehended by Dillenius Under the Terms Usnea, Coralloides, and Lichenoids: Tending Principally to Illustrate Their Several Uses. Vol. 50, 652–688.

Phil. Trans. (1759). An Account of the Hot Baths of Vinadio, in the Province of Coni in Piedmont; With the State of the Weather at Turin in the Year 1759: In an Extract of a Letter from Dr. Joseph Bruni, F. R. S. to Mr. Henry Baker, F. R. S. Vol. 51, 839–843.

Phil. Trans. (1759). Conjectures upon an Inedited Parthian Coin. By the Rev. John Swinton, B. D. of Christ-Church, Oxon. F. R. S. Vol. 51, 680–693.

Phil. Trans. (1759). Experiments in Electricity: In a Letter from Father Beccaria, Professor of Experimental Philosophy at Turin, to Benjamin Franklin, L. L. D. F. R. S. Vol. 51, 514–526.

Phil. Trans. (1759). Extracts of Some Letters from Signor Abbate De Venuti, F. R. S. to J. Nixon, A. M. and F. R. S. Relating to Several Antiquities in Italy. Vol. 51, 636–643.

Phil. Trans. (1759). Extracts of Some Letters from Signor Abbate de Venuti, F. R. S. to J. Nixon, A. M. and F. R. S. Relating to Several Antiquities Lately Discovered in Italy. Vol. 51, 201–206.

Phil. Trans. (1759). Some Observations upon a Samnite-Etruscan Coin, Never before Fully Explained. In a Letter from the Rev. John Swinton, B. D. of Christ-Church, Oxon. F. R. S. to the Rev. Thomas Birch, D. D. Secretary to the Royal Society. Vol. 51, 853–865.

Phil. Trans. (1761). An Account of a Printed Memoir, in Latin, Presented to the Royal Society, Intituled, De Veneris ac Solis Congressu Observatio, Habita in Astronomica Specula Bononiensis Scientiarum Instituti, Die 5 Junii 1761. Auctore Eustachio Zanotto, Ejusdem Instituti Astronomo, ac Regiae utriusque Londinensis et Berolinensis Academiae Socio. By Nathanael Bliss, Savilian Professor of Geometry, and F. R. S. Vol. 52, 399–414.

Phil. Trans. (1761). An Account of the Double Refractions in Crystals; By Father John Beccaria, Professor of Experimental Philosophy at Turin. Vol. 52, 486–490.

Phil. Trans. (1761). Extract of a Letter from Mr. Robert Mackinlay, to the Right Hon. the Earl of Morton, F. R. S. Dated at Rome, the 9th January 1761. Concerning the Late Eruption of Mount Vesuvius, and the Discovery of an Antient Statue of Venus at Rome. Vol. 52, 44–45.

Phil. Trans. (1763). An Account of Some Subterraneous Apartments, with Etruscan Inscriptions and Paintings, Discovered at Civita Turchino in Italy [Tab. VII. VIII. IX.]: Communicated from Joseph Wilcox, Esq; F. S. A. by Charles Morton, M. D. S. R. S. Vol. 53, 127–129.

Phil. Trans. (1764). XVI. Observations upon Two Antient Etruscan Coins, Never before Illustrated or Explained. In a Letter to the Rev. Thomas Birch, D.D. Secret. R. S. from the Rev. John Swinton, B.D. F. R. S. Member of the Academy degli Apatisti at Florence, and of the Etruscan Academy of Cortona in Tuscany Vol. 54, 99-109.

Phil. Trans. (1765). An Uncommon Anatomical Observation Addressed to the Royal Society, by John Baptist Paitoni, Physician at Venice: Translated from the Italian. Vol. 55, 79-83.

Phil. Trans. (1765). XXVIII. Extracts of Three Letters of Sir F. H. Eyles Stiles, F. R. S. to Daniel Wray, Esq; F. R. S. Concerning Some New Microscopes Made at Naples, and Their Use in Viewing the Smallest Objects. Vol. 55.

Phil. Trans. (1766). A Report Concerning the Microscope-Glasses, Sent as a Present to the Royal Society, by Father Di Torre of Naples, and Referred to the Examination of Mr. Baker, F. R. S. Vol. 56, 67-71.

Phil. Trans. (1766). Novorum Quorumdam in Re Electrica Experimentorum Specimen, Quod Regiae Londinensi Societati Mittebat Die 14 Januarii, Anni 1766. Joannes Baptista Beccaria ex Scholis Piis, & R. S. Soc. Vol. 56, 105-118.

Phil. Trans. (1767). De Problemate Quodam Algebraico, Deque Evolutione Mechanicae Cujusdam Curvae Inter Infinitas Hypermechanicas, Quae Determinatae Aequationi Satisfaciunt. Auctore Pio Fantoni, Mathematico Bononiensi. Communicated by Sir Horace Mann, His Majesty's Envoy at Florence. Vol. 57, 358-371.

Phil. Trans. (1767). Novorum Quorundam in Re Electrica Experimentorum Specimen, Quod Regiae Londinensi Societati Mittebat Die 26 Aprilis 1766, Joannes Baptista Beccaria, ex Scholis Piis, R. S. Soc. Communicated by M. Maty, Sec. R. S. Vol. 57, 297-311.

Phil. Trans. (1767). Two Letters from the Hon. William Hamilton, His Majesty's Envoy Extraordinary at Naples, to the Earl of Morton, President of the Royal Society, Containing an Account of the Last Eruption of Mount Vesuvius. Vol. 57, 192-200.

Phil. Trans. (1768). A Letter from J. A. Rizzi Zannoni, Member of the Academy of Sciences at Gottingen, and Geographer to His Sicilian Majesty, to the Late Earl of Morton, Pr. R. S. Containing Several Astronomical Observations, Made in Several Parts of the Kingdom of Naples and Sicily; Translated from the French, by Mathew Maty, M. D. Sec. R. S. Vol. 58, 196-202.

Phil. Trans. (1768). An Account of the Eruption of Mount Vesuvius, in 1767: In a Letter to the Earl of Morton, President of the Royal Society, from the Honourable William Hamilton, His Majesty's Envoy Extraordinary at Naples. Vol. 58, 1-14.

Phil. Trans. (1768). An Investigation of the Difference between the Present Temperature of the Air in Italy and Some Other Countries, and What It Was

Seventeen Centuries Ago: In a Letter to William Watson M. D. F. R. S. by the Honourable Daines Barrington F. R. S. Vol. 58, 58–67.

Phil. Trans. (1768). Description of a Punic Coin Appertaining to the Isle of Gozo, Hitherto Attributed to That of Malta, by the Learned. In a Letter to Mathew Maty, M. D. Sec. R. S. from the Rev. John Swinton, B. D. F. R. S. Custos Archivorum of the University of Oxford, Member of the Academy Degli Apatisti at Florence, and of the Etruscan Academy of Cortona in Tuscany. Vol. 58, 261–264.

Phil. Trans. (1768). Extract of a Letter from Rome, to M. Maty, M.D. Sec. R. S. on the Extraordinary Heats Observed There This Last Summer. Vol. 58, 336.

Phil. Trans. (1768). Two Medical Observations by Dr. Joseph Benevuti, Physician at Lucca; Communicated to the Late President of the Royal Society, by Dr. Ch. Allioni of Turin, F. R. S. and Translated from the Latin by Daniel Peter Layard, M. D. Physician to Her Royal Highness the Princess Dowager of Wales, Member of the Royal College of Physicians in London, and of the Royal Societies of London and Gottingen. Vol. 58, 189–191.

Phil. Trans. (1769). A Letter from the Honourable William Hamilton, His Majesty's Envoy Extraordinary at Naples, to Mathew Maty, M. D. Sec.R. S. Containing Some Farther Particulars on Mount Vesuvius, and Other Volcanos in the Neighbourhood. Vol. 59, 18–22.

Phil. Trans. (1769). An Account of an Essay on the Origin of a Natural Paper, Found Near the City of Cortona in Tuscany. In a Letter from John Strange, Esq; F. R. S. to Mathew Maty, M. D. Sec. R. S. Vol. 59, 50–56.

Phil. Trans. (1770). A Letter to Dr. William Watson, F. R. S. Giving Some Account of the Manna Tree, and of the Tarantula: By Dominico Cirillo, M. D. Professor of Natural History at the University of Naples. Vol. 60, 233–238.

Phil. Trans. (1770). An Account of a Journey to Mount Etna, in a Letter from the Honourable William Hamilton, His Majesty's Envoy Extraordinary at Naples, to Mathew Maty, M. D. Sec. R. S. Vol. 60, 1–19.

Phil. Trans. (1770). An Account of Some Very Perfect and Uncommon Specimens of Spongiae from the Coast of Italy: In a Letter to James West, Esquire, President of the Royal Society; From John Strange, Esquire, F. R. S. Vol. 60, 179–183.

Phil. Trans. (1770). De Athmosphaera Electrica Joannis Baptistae Beccariae, R. S. S. ex Scholis Piis ad Regiam Londinensem Societatem, Libellus. Vol. 260, 277–301.

Phil. Trans. (1771). Extract of Another Letter, from Mr. Hamilton, to Dr. Maty, on the Same Subject. Vol. 61, 48–50.

Phil. Trans. (1771). Interpretation of Two Punic Inscriptions, on the Reverses of Two Siculo-Punic Coins, Published by the Prince di Torremuzza, and Never Hitherto Explained. In a Letter to M. Maty, M. D. Sec. R. S. from the Rev. John Swinton, B. D. F. R. S. Custos Archivorum of the University of Oxford, Member of the Academy Degli Apatisti at Florence, and of the Etruscan Academy of Cortona in Tuscany. Vol. 61, 91–103.

Phil. Trans. (1771). Letter from Mr. John Baptist Beccaria, of Turin, F. R. S. to Mr. John Canton, F. R. S. on His New Phosphorus Receiving Several Colours, and Only Emitting the Same. Vol. 61, 212.

Phil. Trans. (1771). Remarks upon the Nature of the Soil of Naples, and Its Neighbourhood; In a Letter from the Honourable William Hamilton, His Majesty's Envoy Extraordinary at Naples, to Mathew Maty, M. D. Sec. R. S. Vol. 61, 1–47.

Phil. Trans. (1771). Remarks upon Two Etruscan Weights, or Coins, Never before Published. In a Letter to Mathew Maty, M. D. Sec. R. S. from the Rev. John Swinton, B. D. F. R. S. Custos Archivorum of the University of Oxford, Member of the Academy Degli Apatisti at Florence, and of the Etruscan Academy of Cortona in Tuscany. Vol. 61, 82–90.

Phil. Trans. (1773). Account of the Effects of a Thunder-Storm, on the 15th of March 1773, upon the House of Lord Tylney at Naples. In a Letter from the Honourable Sir William Hamilton, Knight of the Bath, His Majesty's Envoy Extraordinary at the Court of Naples, and F. R. S. to Mathew Maty, M. D. Sec. R. S. Vol. 63, 324–332.

Phil. Trans. (1773). Farther Remarks upon a Denarius of the Veturian Family, with an Etruscan Inscription on the Reverse, Formerly Considered. In a Letter to Mathew Maty, M. D. Sec. R. S. from the Reverend John Swinton, B. D. F. R. S. Custos Archivorum of the University of Oxford, Member of the Academy Degli Apatisti at Florence, and of the Etruscan Academy of Cortona in Tuscany. Vol. 63, 22–29.

Phil. Trans. (1774). Explication of a Most Remarkable Monogram on the Reverse of a Very Antient Quinarius, Never before Published or Explained. In a Letter to M. Maty, M. D. Sec. R. S. from the Rev. John Swinton, B. D. F. R. S. Custos Archivorum of the University of Oxford, Member of the Academy Degli Apatisti at Florence, and of the Etruscan Academy of Cortona in Tuscany. Vol. 64, 318–327.

Phil. Trans. (1775). An Account of a Curious Giant's Causeway, or Group of Angular Columns, Newly Discovered in the Euganean Hills, Near Padua, in Italy. In a Letter from John Strange, Esq. F. R. S. to Sir John Pringle, Bart. P. R. S. Vol. 65, 418–423.

Phil. Trans. (1775). An Account of Two Giants Causeways, or Groups of Prismatic Basaltine Columns, and Other Curious Vulcanic Concretions, in the Venetian State in Italy; with Some Remarks on the Characters of These and Other Similar Bodies, and on the Physical Geography of the Countries in Which They are Found. Addressed to Sir John Pringle, Bart. P. R. S. by John Strange, Esq. F.R.S. Vol. 65, 5–47.

Phil. Trans. (1775). Extract of a Letter from Dr. John Ingenhousz, F. R. S. to Sir John Pringle, Bart. P. R. S. Containing Some Experiments on the Torpedo, Made at Leghorn, January 1, 1773 (after Having Been Informed of Those by Mr. Walsh). Dated Saltzburg, March 27, 1773. Vol. 65, 1–4.

Phil. Trans. (1775). Extraordinary Electricity of the Atmosphere Observed at Islington on the Month of October, 1775. By Mr. Tiberius Cavallo. Communicated by William Watson, M. D. V. P. R. S. Vol. 66, 407–411.

Phil. Trans. (1777). An Account of Some New Electrical Experiments. By Mr. Tiberius Cavallo: Communicated by Mr. Henley, F. R. S. Vol. 67, 48–55.

Phil. Trans. (1777). New Electrical Experiments and Observations; With an Improvement of Mr. Canton's Electrometer. By Mr. Tiberius Cavallo, in a Letter to Mr. Henly, F. R. S. Vol. 67, 388–400.

Phil. Trans. (1777). Extract of a Letter from John Strange, Esquire, His Majesty's Resident at Venice, to Sir John Pringle, Bart. P. R. S.: With a Letter to Mr. Strange from the Abbe Joseph Toaldo, Professor in the University of Padua, &c. Giving an Account of the Tides in the Adriatic. Vol. 67, 144–161.

Phil. Trans. (1778). New Experiments upon the Leyden Phial, Respecting the Termination of Conductors. By Benjamin Wilson, Esq. F. R. S. Vol. 68, 999–1012.

Phil. Trans. (1779). Account of the Airs Extracted from Different Kinds of Waters; With Thoughts on the Salubrity of Air at Different Places. In a Letter from the Abbe Fontana, Director of the Cabinet of Natural History Belonging to His Royal Highness the Grand Duke of Tuscany, to Joseph Priestley, LL.D. F. R. S. Vol. 69, 432–453.

Phil. Trans. (1779). Experiments and Observations on the Inflammable Air Breathed by Various Animals. By the Abbe Fontana, Director of the Cabinet of Natural History Belonging to His Royal Highness the Grand Duke of Tuscany; Communicated by John Paradise, Esq. F. R. S. Vol. 69, 337–361.

Phil. Trans. (1780). A Conjecture concerning the Method by Which Cardan's Rules for Resolving the Cubic Equation [...]were Probably Discovered by Scipio Ferreus, of Bononia, or Whoever Else Was the First Inventor of Them. By Francis Maseres, Esq. F. R. S. Cursitor Baron of the Exchequer. Vol. 70, 221–238.

Phil. Trans. (1780). An Account of an Eruption of Mount Vesuvius, Which Happened in August, 1779. In a Letter from Sir William Hamilton, K. B. F. R. S. to Joseph Banks, Esq. P. R. S. Vol. 70, 42–84.

Phil. Trans. (1780). An Account of Some New Experiments in Electricity, with the Description and Use of Two New Electrical Instruments. By Mr. Tiberius Cavallo, F. R. S. Communicated by the President. Vol. 70, 15–29.

Phil. Trans. (1780). Memoria sopra il Veleno Americano detto Ticunas. By the Abbe Fontana, Director of the Cabinet of Natural History Belonging to His Royal Highness the Grand Duke of Tuscany; Communicated by John Paradise, Esq. F. R. S. [With a Translation to English Following]. Vol. 70, 163–ix.

Phil. Trans. (1780). Thermometrical Experiments And Observations. By Tiberius Cavallo, F. R. S. Who Was Nominated by the President and Council to Prosecute Discoveries in Natural History, Pursuant to the Will of the Late Henry Baker, Esq. F. R. S. Vol. 70, 585–599.

Phil. Trans. (1781). An Account of Some Thermometrical Experiments; Containing, I. Experiments Relating to the Cold Produced by the Evaporation of Various Fluids, with a Method of Purifying Ether. II. Experiments Relating to the Expansion of Mercury. III. Description of a Thermometrical Barometer. By Tiberius Cavallo, F. R. S. Who Was Nominated by the President and Council to Prosecute Discoveries in Natural History, Pursuant to the Will of the Late Henry Baker, Esq. F. R. S. Vol. 71, 509–525.

Phil. Trans. (1781). Account of a Luminous Appearance in the Heavens. By Mr. Tiberius Cavallo, F. R. S. in a Letter to Sir Joseph Banks, Bart. P. R. S. Vol. 71, 329–330.

Phil. Trans. (1782). Del Modo di Render Sensibilissima la piu Debole Elettricita sia Naturale, sia Artificiale. By Mr. Alexander Volta, Professor of Experimental Philosophy in Como, &c. &c.; Communicated by the Right Hon. George Earl Cowper, F. R. S. Vol. 72, 237–xxxiii.

Phil. Trans. (1782). Relazione di una Nuova Pioggia, Scritta dal Conte De Gioeni Abitante Della 3a Reggione Dell' Etna; Communicated by Sir William Hamilton, K. B. F. R. S. Vol. 72, 1–vi.

Phil. Trans. (1782). An Account of Some Scoria from Iron Works, Which Resemble the Vitrified Filaments Described by Sir Wiliam Hamilton. In a Letter from Samuel More, Esq. to Sir Joseph Banks, Bart. P. R. S. Vol. 72, 50–52.

Phil. Trans. (1783). An Account of the Earthquakes Which Happened in Italy, from February to May 1783. By Sir William Hamilton, Knight of the Bath, F. R. S.; in a Letter to Sir Joseph Banks, Bart. P. R. S. Vol. 73, 169–208.

Phil. Trans. (1783). Account of the Earthquake Which Happened in Calabria, March 28, 1783. In a Letter from Count Francesco Ippolito to Sir William

Hamilton, Knight of the Bath, F. R. S.; Presented by Sir William Hamilton. Vol. 73, 209–vii.

Phil. Trans. (1783). Description of an Improved Air-Pump, and the Account of Some Experiments Made with It. By Mr. Tiberius Cavallo, F. R. S. Vol. 73, 435–452.

Phil. Trans. (1784). Description of a Meteor, Observed Aug. 18, 1783. By Mr. Tiberius Cavallo, F. R. S. Vol. 74, 108–111.

Phil. Trans. (1786). Magnetical Experiments and Observations. By Mr. Tiberius Cavallo, F. R. S. Vol. 76, 62–80.

Phil. Trans. (1786). Some Particulars of the Present State of Mount Vesuvius; With the Account of a Journey into the Province of Abruzzo, and a Voyage to the Island of Ponza. In a Letter from Sir William Hamilton, K. B. F. R. S. and A. S. to Sir Joseph Banks, Bart. P. R. S. Vol. 76, 365–381.

Phil. Trans. (1787). A Letter from the Father Prefect of the Mission in Thibet, F. Joseph da Rovato, Containing Some Observations Relative to Borax. Communicated by Sir Joseph Banks, Bart. P. R. S. Vol. 77, 301–473.

Phil. Trans. (1787). Magnetical Experiments and Observations. By Mr. Tiberius Cavallo, F. R. S. Vol. 77, 6–25.

Phil. Trans. (1788). Description of a New Electrical Instrument Capable of Collecting Together a Diffused or Little Condensed Quantity of Electricity. By Mr. Tiberius Cavallo, F. R. S. Vol. 78, 255–260.

Phil. Trans. (1788). Of the Temperament of Those Musical Instruments, in Which the Tones, Keys, or Frets, are Fixed, as in the Harpsichord, Organ, Guitar, &c. By Mr. Tiberius Cavallo, F. R. S. Vol. 78, 238–254.

Phil. Trans. (1788). Of the Methods of Manifesting the Presence, and Ascertaining the Quality, of Small Quantities of Natural or Artificial Electricity. By Mr. Tiberius Cavallo, F. R. S. Vol. 78, 1–22.

Phil. Trans. (1789). Result of Calculations of the Observations Made at Various Places of the Eclipse of the Sun, Which Happened on June 3, 1788. By the Rev. Joseph Piazzi, C. R Professor of Astronomy in the University of Palermo; Communicated by Nevil Maskelyne, D. D. F. R. S. and Astronomer Royal. Vol. 79, 55–64.

Phil. Trans. (1791). Description of a Simple Micrometer for Measuring Small Angles with the Telescope. By Mr. Tiberius Cavallo, F. R. S. Vol. 81, 283–294.

Phil. Trans. (1793). Account of Some Discoveries Made by Mr. Galvani, of Bologna; With Experiments and Observations on Them. In Two Letters from Mr. Alexander Volta, F. R. S. Professor of Natural Philosophy in the University of Pavia, to Mr. Tiberius Cavallo, F. R. S. Vol. 83, 10–44.

Phil. Trans. (1795). An Account of the Late Eruption of Mount Vesuvius. In a Letter from the Right Honourable Sir William Hamilton, K. B. F. R. S. to Sir Joseph Banks, Bart. P. R. S. Vol. 85, 73–116.

Phil. Trans. (1795). Observations on the Influence, Which Incites the Muscles of Animals to Contract in Mr. Galvani's Experiments. By William Charles Wells, M. D. F. R. S. Vol. 85, 246–262.

19th century

Phil. Trans. (1800). On the Electricity Excited by the Mere Contact of Conducting Substances of Different Kinds. In a Letter from Mr. Alexander Volta, F. R. S. Professor of Natural Philosophy in the University of Pavia, to the Rt. Hon. Sir Joseph Banks, Bart. K. B. P. R. S. Vol. 90, 403–431.

Phil. Trans. (1801). [Davy, Humphry] An Account of Some Galvanic Combinations, Formed by the Arrangement of Single Metallic Plates and Fluids, Analogous to the New Galvanic Apparatus of Mr. Volta. Vol. 91, 397–402.

Phil. Trans. (1806). [Home, Everard] An Account of a Small Lobe of the Human Prostate Gland, Which Has Not before Been Taken Notice of by Anatomists. Vol. 96, 195–204.

Hutton, C., Shaw, G. and Pearson, R. (1809). A New and Safe Method of of Communicating the Small-Pox by Inoculation, lately invented and brought into use. By Jacob Pylarini, M.D. formerly Venetian Consul at Smyrna. No 347, [vol. 29] p. 393. Translated and Abridged from the Latin. In *The Philosophical Transactions of the Royal Society of London, from Their Commencement, in 1665, to the Year 1800; Abridged, with Notes and Biographic Illustrations, by Charles Hutton, Georges Shaw, Richard Pearson ... from 1703 to 1712. London. Printed for C. and R. Baldwin, New Bridge Street, Blackfriars.* Vol. V, 207–210.

Hutton, C., Shaw, G. and Pearson, R. (1809). An Account of the Procuring Small Pox by Incision, or Inoculation; as It Has for Some Time Been Practiced at Constantinople. Being an Extract of a Letter from Emmanuel Timonius, Oxon. and Patav., M.D. S.R.S.. Dated Constantinople, December, 1713. Communicated by John Woodward, M.D. and S.R.S., No 339, p. 72. In *The Philosophical Transactions of the Royal Society of London, from Their Commencement, in 1665, to the Year 1800; Abridged, with Notes and Biographic Illustrations, by Charles Hutton, Georges Shaw, Richard Pearson ... from 1703 to 1712. London. Printed for C. and R. Baldwin, New Bridge Street, Blackfriars.* Vol. V, 88–91.

Hutton, C., Shaw, G. and Pearson, R. (1809). III. An Extract from the Acta Eruditorum for the Month of March, 1713. Pag. 111. [...] In *The Philosophical Transactions of the Royal Society of London, from Their Commencement, in 1665, to the Year 1800; Abridged, with Notes and Biographic Illustrations, by Charles Hutton, Georges Shaw, Richard Pearson ... from 1703 to 1712. London. Printed for C. and R. Baldwin, New Bridge Street, Blackfriars.* Vol. V, 78–80.

Hutton, C., Shaw, G. and Pearson, R. (1809). An Account of a Contagion among the Cattle in the Venetian Territories, in Autumn 1711. By Dr. Peter Antony Michelotti. N. 385 p. 83. In *The Philosophical Transactions of the Royal Society of London, from Their Commencement, in 1665, to the Year 1800; Abridged, with Notes and Biographic Illustrations, by Charles Hutton, Georges Shaw, Richard Pearson ... from 1703 to 1712. London. Printed for C. and R. Baldwin, New Bridge Street, Blackfriars.* Vol. V, 481–483.

Hutton, C., Shaw, G. and Pearson, R. (1809). A collection of Geometrical Flowers, presented to the Royal Society. By Guido Grandi, Abbot of the Camaldules, and Professor of Mathematics at Pisa. No 378, p. 355. In *The Philosophical Transactions of the Royal Society of London, from Their Commencement, in 1665, to the Year 1800; Abridged, with Notes and Biographic Illustrations, by Charles Hutton, Georges Shaw, Richard Pearson ... from 1703 to 1712. London. Printed for C. and R. Baldwin, New Bridge Street, Blackfriars.* Vol. V, 664.

Proceedings. (1815). [Holland, H.] On the Manufacture of the Sulphate of Magnesia at Monte Della Guardia, near Genoa. Vol. 2, 48–49.

Phil. Trans. (1816). [Holland, H.] On the Manufacture of the Sulphate of Magnesia at Monte della Guardia, Near Genoa. Vol. 106, 294–300.

Phil. Trans. (1817). [Wilson Philip A. P.] On the Effects of Galvanism in Restoring the due Action of the Lungs. Vol. 107, 22–31.

Phil. Trans. (1818). [Granville, Augustus Bozzi] XVI. On a Mal-conformation of the Uterine System in Women; and on Some Physiological Conclusions to Be Derived from It. In a Letter to Sir Everard Home, Bart. V. P. R. S. from A. B. Granville, M. D. F. R. S. F. L. S. Physician in Ordinary to H. R. H. the Duke of Clarence; Member of the Royal College of Physicians, and Physician-Accoucheur to the Westminster General Dispensary. Vol. 108, 308–315.

Phil. Trans. (1820). [Granville, Augustus Bozzi] A Case of the Human Foetus Found in the Ovarium, of the Size It Usually Acquires at the End of the Fourth Month. Vol. 110, 101–107.

Phil. Trans. (1825). [Granville, Augustus Bozzi] XIII. An Essay on Egyptian Mummies; with Observations on the Art of Embalming among the Ancient Egyptians. Vol. 115, 269–316.

Phil. Trans. (1826). [South, James] On the Discordances between the Sun's Observed and Computed Right Ascensions, as Determined at the Blackman-Street Observatory, in the Years 1821 and 1822; With Experiments to Show That They did not Originate in Instrumental Derangement. Also a Description of the Seven-Feet Transit with Which the Observations Were Procured, and Upon Which the Experiments Were Made. Vol. 116, 423–483.

Phil. Trans. (1826). [Somerville, Mary] On the Magnetizing Power of the More Refrangible Solar Rays. Vol. 116, 132–139.

Phil. Trans. (1829). [Ritchie, William] An Experimental Examination of the Electric and Chemical Theories of Galvanism. Vol. 119, 361–366.

Proceedings. (1833). [Granville, Augustus Bozzi] An Essay on Egyptian Mummies; with Observations on the Art of Embalming among the Ancient Egyptians. Vol. 2, 240–241.

Proceedings. (1833). On a Mal-conformation of the Uterine System in Women; and on Some Physiological Conclusions to Be Derived from It. In a Letter to Sir Everard Home, Bart. V. P. R. S. from A. B. Granville, M. D. F. R. S. F. L. S. Physician in Ordinary to H. R. H. the Duke of Clarence;Member of the Royal College of Physicians, and Physician-Accoucheur to the Westminster General Dispensary. Vol. 2, 95.

Proceedings. (1833). [Somerville, Mary] On the Magnetizing Power of the More Refrangible Solar Rays. Vol. 2, 263–265.

Proceedings. (1833). A Case of the Human fœtus Found the Ovarium, of the Size It Usually Acquires at the End of the Fourth Month. By A. B. Granville, M. D. F. R. S. In a Letter Addressed to Sir Everard Home, Bart. V. P. R. S. Vol. 2, 123.

Phil. Trans. (1836). [Tiedmann, Frederick] On the Brain of the Negro, Compared with That of the European and the Orang-Outang. Vol. 126, 497–527.

Phil. Trans. (1842). IV. On the Structure and Use of the Malpighian Bodies of the Kidney, with Observations on the Circulation through That Gland. Vol. 132, 57–80.

Proceedings. (1843). Geographical Position of the Principal Points of the Triangulations of the Californias and of the Mexican Coasts of the Pacific, with the Heights of the Principal Points of That Part of the Cordilleras. By the Comte Vincent Piccolomini; in a Letter Addressed to Sir John F. W. Herschel, Bart., V. P. R. S. Communicated by Sir John Herschel. Vol. 4, 196–197.

Proceedings. (1843). On Nobili's Plate of Colours; in a Letter from J. P. Gassiot, Esq., Addressed to J. W. Lubbock, Esq., V. P. and Treasurer R. S. Communicated by J. W. Lubbock, Esq. Vol. 4, 195–196.

Proceedings. (1843). [Carlini, Francesco] Variations de la déclinaison et intensité magnétique horizontale observées à Milan le 28 et 29 Mai, le 23 et 24 Juin, le 21 et 22 Juillet, le 27 et 28 Aout, et le 22 et 23 Septembre 1841. Vol. 4, B.

Phil. Trans. (1845). [Matteucci, Carlo] Electro-Physiological Researches. First Memoir. The Muscular Current. Vol. 135, 283–295.

Phil. Trans. (1845). [Matteucci, Carlo] Electro-Physiological Researches. Second Memoir. On the Proper Current of the Frog. Vol. 135, 297–301.

Phil. Trans. (1845). [Matteucci, Carlo] Electro-Physiological Researches. Third Memoir. On Induced Contractions. Vol. 135, 303–317.

Phil. Trans. (1846). [Matteucci, Carlo] Electro-Physiological Researches. Fourth Memoir. The Physiological Action of the Electric Current. Vol. 136, 483–499.

Phil. Trans. (1847). [Matteucci, Carlo] Electro-Physiological Researches. Fifth Series. Part I. Upon Induced Contractions. Part II. Upon the Phenomena Elicited by the Passage of the Current through the Nerves of a Living Animal, or an Animal Recently Killed, According to the Direction of the Current. Vol. 137, 231–237.

Phil. Trans. (1847). [Matteucci, Carlo] Electro-Physiological Researches. Sixth Series. Laws of the Electric Discharge of the Torpedo and Other Electric Fishes-Theory of the Production of Electricity in These Animals. Vol. 137, 239–241.

Phil. Trans. (1847). [Matteucci, Carlo] Electro-Physiological Researches. Seventh and Last Series. Upon the Relation between the Intensity of the Electric Current, and That of the Corresponding Physiological Effect. Vol. 137, 243–248.

Phil. Trans. (1850). [Matteucci, Carlo] Electro-Physiological Researches. Eighth Series. Vol. 140, 287–296.

Phil. Trans. (1850). [Matteucci, Carlo] Electro-Physiological Researches. On Induced Contraction. Ninth Series. Vol. 140, 645–649.

Phil. Trans. (1850). XXXVI. [Carpenter, B. William] On the Mutual Relations of the Vital and Physical Forces. Vol. 140, 727–757.

Proceedings. (1851). [Carlini, Francesco] Observations de la dèclinaison et intensitè horizontales magnètiques observèes à Milan pendant vingt-quatre heures consécutives le 29 et 30 de Dècembre 1844. Vol. 5, 542.

Proceedings. (1851). [Carlini, Francesco] Variations de la déclinaison et intensité magnetique observées à Milan le 26 et 27 Mai, le 21 et 22 Juin, le 19 et 20 Juillet, le 25 et 26 Aout, le 20 et 21 Septembre, le 18 et 19 Octobre, 1843. Vol. 5.

Proceedings. (1856). Description of the Observatory of the Collegio Romano at Rome. Vol. 8, 83.

Proceedings. (1856). Extract of a Letter to George Rennie, Esq., F.R.S., from P. A. Secchi, Director of the Astronomical Observatory of the Collegio Romano, Containing Explanatory Remarks on a Drawing of the Lunar Spot 'Copernicus,' Presented by Him to the Royal Society. Dated Rome, March 13, 1856. Vol. 8, 72.

Proceedings. (1856). [Philips, John] Notes on the Drawing of 'Copernicus,' Presented to the Royal Society by P. A. Secchi. Vol. 8, 73-75.

Phil. Trans. (1857). [Matteucci, Carlo] Electro-Physiological Researches. Physical and Chemical Phenomena of Muscular Contraction. Tenth Series. Part I. Vol. 147, 129-143.

Phil. Trans. (1858). [Lyell, Sir Charles] XXXII. On the Structure of Lavas Which Have Consolidated on Steep Slopes; with Remarks on the Mode of Origin of Mount Etna, and on the Theory of "craters of elevation". Vol. 148, 703-786.

Proceedings. (1859). [Mallet, Robert] Report to the Royal Society of the Expedition into the Kingdom of Naples to Investigate the Circumstances of the Earthquake of the 16th December 1857. Vol. 10, 486-494.

Proceedings. (1859). [Matteucci, Carlo] XV. Sur la relation entre les courants induits et le pouvoir moteur de l'electricité. Vol. 9, 321-324.

Phil. Trans. (1861). [Matteucci, Carlo] Electro-Physiological Researches. Eleventh Series. On the Secondary Electromotor Power of Nerves, and Its Application to the Explanation of Certain Electro-Physiological Phenomena. Vol. 151, 363-372.

Proceedings. (1864). Comparison of Mr. De la Rue's and padre Secchi's Eclipse Photographs. Vol. 13, 442-444.

Proceedings. (1878). [Brioschi, Francesco] Sur une Equation Differentielle du 3me Ordre. Vol. 27, 126-128.

Proceedings. (1883). [Flight, Walter] Examination of the Meteorite Which Fell on the 16th February, 1883, at Alfianello, in the District of Verolannova, in the Province of Brescia, Italy. Vol. 35, 258-260.

Proceedings. (1887). [Sherington, S. Charles] Note on the Anatomy of Asiatic Cholera as Exemplified in Cases Occurring in Italy in 1886. Vol. 42, 474-477.

Proceedings. (1892). [Mosso, Angelo] Croonian Lecture. — The Temperature of the Brain, Especially in Relation to Psychical Activity. Vol. 51, 83-85.

Phil. Trans. B. (1892). [Mosso, Angelo] VII. Croonian Lecture. — Les phénomènes psychiques et la température du cerveau. Vol. 183, 299-309.

Proceedings. (1897). [Granville, Augustus Bozzi] The Reproduction and Metamorphosis of the Common eel (Anguilla vulgaris). Vol. 60, 359-367.

Complementary papers

Phil. Trans. (1684). An Account of a Sort of Paper made of Linum Asbestinum Found in Wales in a Letter to the Publilher, from Edward LLoyd of Jesus Coll. Oxon. Vol. 14(166), 823–824.

Phil. Trans. (1685). A Discourse Concerning the Incombustible Cloth above Mentioned; Address't in a Letter to Mr. Arthur Bayly Merchant, and Fellow of the Royal Society; and to Mr Nicholas Waite, Merchant of London; by Rob. Plot. LL.D. Vol. 15(172), 1051–1062.

Phil. Trans. (1685). A Letter from Mr. Nich. Waite Merchant of London, to Dr. Rob. Plot; Concerning Some Incombustible Cloth, Lately Exposed to the Fire before the Royal Society. Vol. 15(172), 1059–1051.

Phil. Trans. (1701). A Letter from Mr Willson to the Publisher, Giving an Account of the Lapis Amianthus, Asbestos, or Linum Incombustibile, Lately Found in Scotland. Vol. 22(276), 1004–1007.

Phil. Trans. (1710). VI. Part of a Letter from Mr. Patrick Blair to Dr. Hans Sloane, S. See. Giving an Account of the Asbestos, or Lapis Amiantus, Found in the High-lands of Scotland. Vol. 27(333), 434–437.

Phil. Trans. (1719). [Jurin, Jocobo] III. De motu aquarum fluentium. Vol. 30(355), 748–766.

Phil. Trans. (1759). LXXII. An Account of a late Discovery of Asbestos in France: In a Letter to the Rev. Tho. Birch, D.D. Secretary to the Royal Society, from Mr. Turberville Needham, F. R. S. Vol. 51, 187–189.

Secondary sources

Allen, B., Qin, J. & Lancaster, F. (1994). Persuasive Communities: A Longitudinal Analysis of References in the *Philosophical Transactions of the Royal Society*, 1665–1990. *Social Studies of Science*, 24(2), 279–310.

Andrade, E. N. da C. (1965). The Birth and Early Days of the *Philosophical Transactions. Notes and Records of the Royal Society of London*, 20(1), 9–27.

Atkinson, D. (1992). The Evolution of Medical Research Writing from 1735 to 1985: The Case of the Edinburgh Medical Journal. *Applied Linguistics*, 13(4), 337–374.

Atkinson, D. (1996). *The Philosophical Transactions of the Royal Society of London, 1675–1975: A Sociohistorical Discourse Analysis*. Cambridge: Cambridge University Press.

Atkinson, D. (1999). *Scientific Discourse in Sociohistorical Context. The Philosophical Transactions of the Royal Society of London, 1675–1975*. New York: Routledge.

Bibliography   293

Avramov, I. (1999). An Apprenticeship in Scientific Communication: The Early Correspondence of Henry Oldenburg (1656–63). *Notes and Records of the Royal Society of London*, 53(2), 187–201.

Baldwin, M. (1995). The Snakestone Experiments: An Early Modern Medical Debate. *Isis*, 86(3), 394–418.

Banks, D. (2008a). *The Development of Scientific Writing: Linguistic Features and Historical Context*. Oakville/London: Equinox.

Banks, D. (2008b). The Significance of Thematic Structure in the Scientific Journal Article, 1700–1980. *Systemic Functional Linguistics in Use, Odense Working Papers in Language and Communications*, 29, 1–29.

Banks, D. (2009a). Creating a Specialized Discourse: The Case of the *Philosophical Transactions*. *ASp*, 56, 29–44.

Banks, D. (2009b). Starting Science in the Vernacular. Notes on Some Early Issues of the *Philosophical Transactions* and the *Journal des Sçavans*, 1665–1700. *ASp*, 55, 5–22.

Banks, D. (2010a). The Beginnings of Vernacular Scientific Discourse: Genres and Linguistic Features in Some Early Issues of the *Journal des Sçavans* and the *Philosophical Transactions*. *E-rea*, 8(1), http://journals.openedition.org/erea/1334.

Banks, D. (2010b). Transitivity and Thematic Structure in Some Early Issues of the *Philosophical Transactions*. *ASp*, 58, 57–71.

Banks, D. (2012). How Modality May Function in Some Early Issues of the *Philosophical Transactions*. *Revista de Lingüística y Lenguas Aplicadas*, 7(1), 61–76.

Banks, D. (2017). *The Birth of the Academic Article: Le Journal des Sçavans and the Philosophical Transactions, 1665–1700*. Sheffield/Bristol: Equinox Publishing.

Bazerman, C. (1988). *Shaping Written Knowledge*. Madison: University of Wisconsin Press.

Bednarek, M. (2006). *Evaluation in Media Discourse: Analysis of a Newspaper Corpus*. London: Continuum.

Beer, G. (1990). Translation or Transformation? The Relations of Literature and Science. *Notes and Records of the Royal Society of London*, 44, 81–99.

Beretta, M. (2000). At the Source of Western Science: The Organization of Experimentalism at the Accademia del Cimento (1657–1667). *Notes and Records of the Royal Society of London*, 54(2), 131–151.

Berti, L. (2019). Italy and the Royal Society. Medical Papers in the Early *Philosophical Transactions*. *Token*, 8, 31–60.

Berti, L. (2021) Early Reception of Smallpox Inoculation in Italy: Insights from the Correspondence of the Fellows of the Royal Society. *Diciottesimo Secolo*, 6, 5–18

Bertucci, P. (2013). The In/visible Woman: Mariangela Ardinghelli and the Circulation of Knowledge between Paris and Naples in the Eighteenth Century. *Isis, 104*(2), 226–249.

Berzolari, A. G. (2002). *Alessandro Volta and the Scientific Culture between 1750 and 1850*. Milano: Istituto Lombardo di Scienze e Lettere.

Biagetti, M. T. (2001). *Teoria e Prassi della Catalogazione Nominale: i Contributi di Panizzi, Jewett e Cutter*. Rome: Bulzoni Editore.

Biagioli, M. (1996). Interdependence and Sociability in Seventeenth-Century Science. *Critical Inquiry, 22*(2), 193–238.

Biber, D. (1988). *Variation across Speech and Writing*. Cambridge: Cambridge University Press.

Biber, D. (2013). *A Typology of English Texts*. Amsterdam: Mouton de Gruyter.

Biber, D. & Finegan E. (1989). Drift and the Evolution of the English Style: A History of Three Genres. *Linguistic Society of America, 65*(3), 487–517.

Boccone, P. (1697) *Museo di Fisica e di Esperienze*. Venice.

Boella, A. & Galli, A. (Eds.) (2015). *Philosophia Hermetica*, by Gualdi, F. Rome: Biblioteca Ermetica. Edizioni Mediterranee.

Booth, C. C. (1982). Medical Communication: The Old and the New. The Development of Medical Journals in Britain. *British Medical Journal, 285*(6335), 105–108.

Borgato, M. T. (2016). RICCIOLI, Giovanni Battista. *Dizionario Biografico Treccani*, 87, http://www.treccani.it/enciclopedia/giovanni-battista-riccio li_%28Enciclopedia-Italiana%29/.

Borrelli, A. (1997). Scienza e Accademie negli Stati Italiani del Settecento. *Studi Storici, 38*(2), 571–577.

Boschiero, L. (2002). Natural Philosophizing Inside the Late Seventeenth-Century Tuscan Court. *The British Journal for the History of Science, 35*(4), 383–410.

Boschiero, L. (2010). Translation, Experimentation and the Spring of the Air: Richard Waller's Essays of Natural Experiments. *Notes and Records of the Royal Society of London, 64*(1), 67–83.

Boukala, S. & Wodak, R. (2015). European Identities and Revival of Nationalism in the European Union. A Discourse Historical Approach. *Journal of Language and Politics, 14*(1), 87–109.

Callaway, H. (1992). Does Language Determine Our Scientific Ideas? *Dialectica*, 46(3/4), 225–242.

Camerota, M., Giudice, F. & Ricciardo, S. (2018). The Reappearance of Galileo's Original Letter to Benedetto Castelli. *Notes and Records of the Royal Society of London*, 73(1), 1–18.

Canziani, T., Grego, K. & Iamartino, G. (Eds.) (2014). *Perspectives in Medical English*. Monza: Polimetrica International Scientific Publisher.

Carlino, A. (2010). Tra Antiquaria e Archeologia la Riscoperta dei Templi di Agrigento nell'Opera di G. Pancrazi. *Sicilia Antiqua*, 7, 179–204.

Catenaccio, P. (2017). Negotiating Futures in Socio-Technical Controversies in the Media: Strategies of Opinion Orientation. In Breeze, R., & Olza, I. (Eds.), *Evaluation in Media Discourse* (pp. 121–155). Bern: Peter Lang.

Cavazza, M. (1980). Bologna and the Royal Society in the Seventeenth Century. *Notes and Records of the Royal Society of London*, 35(2), 105–123.

Cavazza, M. (2002). The Institute of Science of Bologna and the Royal Society in the Eighteenth Century. *Notes and Records of the Royal Society of London*, 56(1), 3–25.

Cavazza, M. (2009). Laura Bassi and Giuseppe Veratti: An Electric Couple during the Enlightenment. *Contributions to Science*, 5(1), 115–128.

Cavazza, M. (2010). The Accademia del Cimento and Its European Context, by Beretta, M., Clericuzio, A. and Principe, L.M. [Review] *Isis*, 101(4), 871–872.

Ciancio, L. (1995). *A Calendar of the Correspondence of John Strange, FRS (1732-1799)*. London: The Wellcome Institute for the History of Medicine.

Clericuzio, A. (1992). Promoting Experimental Learning: Experiment and the Royal Society 1660-1727, by M. Boas Hall. [Review]. *Medical History*, 36(3), 343–344.

Clericuzio, A. (2013). Le Accademie Scientifiche del Seicento. Il Contributo Italiano alla Storia del Pensiero. *Scienze, Enciclopedia Treccani*, http://www.treccani.it/enciclopedia/le-accademie-scientifiche-del-seicento_%28Il-Contributo-italiano-alla-storia-del-Pensiero:-Scienze%29/.

Cook, A. (2002). Across the Alps. London and Bologna in the Eighteenth Century. *Notes and Records of the Royal Society of London*, 56(1), 1–2.

Cook, A. (2003). 1703 and Other Anniversaries. *Notes and Records of the Royal Society of London*, 57(1), 1–2.

Cook, A. (2004). Rome and the Royal Society, 1660-1740. *Notes and Records of the Royal Society of London*, 58(1), 3–19.

Costa, G. (1968). Documenti per una Storia dei Rapporti Anglo-Romani nel Settecento. *Saggi e Ricerche sul Settecento*, 371–452.

Crinò, A. M. (1971). *Fatti e Figure del Seicento Anglo-Toscano: Documenti Inediti sui Rapporti Letterari, Diplomatici, e Culturali fra Toscana ed Inghilterra.* Firenze: Olschki.

Crinò, A. M. (1972). *Lorenzo Magalotti: Relazioni d'inghilterra: 1668 e 1688.* Firenze: Olschki.

Crosland, M. (1983). Explicit Qualifications as a Criterion for Membership of the Royal Society: A Historical Review. *Notes and Records of the Royal Society of London, 37*(2), 167–187.

Crosland, M. (2005). Relationships between the Royal Society and the Academie des Sciences in the Late Eighteenth Century. *Notes and Records of the Royal Society of London, 59*(1), 25–34.

D'Amore, M. (2015). Learned Letters from Italy Classical Rome, Vesuvius, and Etna in *Philosophical Transactions* (1665–1700). *Annali di Ca' Foscari. Serie occidentale, 49,* 145–162.

D'Amore, M. (2017). *The Royal Society and the Discovery of the Two Sicilies: Southern Routes in the Grand Tour.* Cham: Palgrave Macmillan.

De Beer, G. R. (1951). John Strange FRS 1732– 1799. *Notes and Records of the Royal Society of London, 9*(1), 96–108.

De Beer, G. R. (1948). Johann Gaspar Scheuchzer, FRS 1702–1729. *Notes and Records of the Royal Society of London, 6*(1), 656–66.

Dereham, T. (1734). *Saggio delle transazioni filosofiche della Società Regia, compendiate da Giovanni Lowthorp; tradotte dall'inglese nell'idioma toscano. A sua Eccellenza la Signora Contessa Donna Clelia Grilla-Borromea,* 5 vols. Napoli: presso il Moscheni.

De Simone, E., Frisullo, F., & Vincenti, P. (2022). *Da Oria a Lisbona. Giovanni Battista Carbone S.J. (1694 –1750) astronomo e diplomatico alla corte dei Braganza.* Società Storia Patria. Lecce: Giorgiani Editore.

Despeaux, S. (2011). Fit To Print? Referee Reports on Mathematics for the Nineteenth-Century Journals of The Royal Society Of London. *Notes and Records of the Royal Society of London, 65*(3), 233–252.

Dooley, B. (1991). *Science, Politics and Society in Eighteenth-century Italy: The Giornale de' letterati d'Italia and Its World.* New York/London: Garland.

Dorris, G. E. (1967). *Paolo Rolli and the Italian Circle in London, 1715–1744.* Paris: Mouton.

Drummond, W. & Monro, W. (1737). An Account of Medical Discoveries, Improvements, and Books Published in the Year 1731 [...]. In *Medical Essays and Observations, The Second Edition Corrected* (pp. 370–381). Philosophical Society of Edinburgh.

Dukes, C. E. (1984). London Medical Societies in the Eighteenth Century. *Proceedings of the Royal Society of Medicine, Section of the History of Medicine*, 53(9), 699–706.

Eccles, J. (1975). Letters from C. S. Sherrington, F. R. S., to Angelo Ruffini between 1896 and 1903. *Notes and Records of the Royal Society of London*, 30(1), 69–88.

Emblen, D. L. (1969). Roget vs Panizzi: A Collision. *Journal of Library History*, 4(1), 9–38.

Emblen, D. L. (1970). *Peter Mark ROGET: The Word and the Man*. London: Longman.

Eriksen, A. (2020). Smallpox Inoculation: Translation, Transference and Transformation. *Palgrave Communications*, 52(6), 1–9.

Fairclough, N. (1992). *Discourse and Social Change*. Cambridge: Polity Press.

Fairclough, N. (2001). Critical Discourse Analysis as a Method in Social Scientific Research. In Wodak, R., & Meyer, M. (Eds.), *Methods of Critical Discourse Analysis* (pp. 121–138). London: Sage Publications, Ltd.

Fairclough, N. (2010). *Critical Discourse Analysis: The Critical Study of Language*. London: Routledge, Taylor & Francis Group.

Fairclough, N. (2015). *Language and Power*. London: Routledge, Taylor & Francis Group.

Ferrari, G. & McConnell, A. (2005). Robert Mallet and the 'Great Neapolitan Earthquake' of 1857. *Notes and Records of the Royal Society of London*, 59(1), 45–64.

Findlen, P. (2009). Founding a Scientific Academy: Gender, Patronage and Knowledge in Early Eighteenth-Century Milan. *Republics of Letters: A Journal for the Study of Knowledge, Politics, and the Arts* 1(1), 1–43.

Fisher, N. (2001). Robert Balle, Merchant of Leghorn and Fellow of the Royal Society (ca. 1640–ca. 1734). *Notes and Records of the Royal Society of London*, 55(3), 351–371.

Fontes Da Costa, P. (2000). The Understanding of Monsters at the Royal Society in the First Half of the Eighteenth Century. *Endeavour*, 24(1), 34–39.

Fontes da Costa, P. (2002a). The Culture of Curiosity at the Royal Society in the First Half of the Eighteenth Century. *Notes and Records of the Royal Society of London*, 56(2), 147–166.

Fontes da Costa, P. (2002b). The Making of Extraordinary Facts: Authentication of Singularities of Nature at the Royal Society of London in the First Half of the Eighteenth Century. *Studies in History of Philosophy of Science*, 33, 265–288.

Fontes Da Costa, P. (2009). *The Singular and the Making of Knowledge at the Royal Society of London in the Eighteenth Century*. Newcastle upon Tyne: Cambridge Scholars Publishing.

Forner, F., Meyer, F., Schwarze, S. (Eds.) (2022). *I Periodici Settecenteschi come Luogo di Comunicazione dei Saperi. Prospettive Storiche, Letterarie e Linguistiche*. Berlin: Peter Lang.

Frati, L. (1913). Un medico Bolognese in Olanda [Rinaldo Duglioli]. *Nuova Antologia*, 5, 144, 310–315.

Fyfe, A. & Moxham, N. (2016). Making Public Ahead of Print: Meetings and Publications at the Royal Society, 1752–1892. *Notes and Records of the Royal Society of London*, 70(4), 1–19.

Fyfe, A., McDougall-Waters, J. & Moxham, N. (2015). 350 Years of Scientific Periodicals. *Notes and Records of the Royal Society of London*, 69, 227–239.

Gascoigne, J. (2003). *Joseph Banks and the English Enlightenment: Useful Knowledge and Polite Culture*. Cambridge: Cambridge University Press.

Gascoigne, J. (2009). The Royal Society, Natural History and the Peoples of the 'New World(s)', 1660–1800. *The British Journal for the History of Science*, 42(4), 539–562.

Gee, J. P. (2014). *How to Do Discourse Analysis: A Toolkit*. London: Routledge.

Generali, D. (2012). Periodici eruditi, carteggi e progetto egemonico della scienza vallisneriana nel «Giornale de' Letterati d'Italia». In Del Tedesco, E. (Ed.), *«Giornale de' letterati d'Italia» trecento anni dopo: scienza, storia, arte, identità (1710–2010)* (pp. 29–48). Pisa/Roma: Fabrizio Serra.

Generali, D. (2016). La Cultura Scientifica a Milano Dal Primo Settecento Sino a Maria Gaetana Agnesi. *Rivista Di Storia Della Filosofia* (1984–) 71(4), 195–214.

Giannini, A. (1936). *I Rapporti Italo-Inglesi. Quaderni dell'Istituto Nazionale Fascista di Cultura*. N.p.

Gibelin, J. (1793). *Compendio delle transazioni filosofiche della Società Reale di Londra Opera compilata, divisa per materie, ed illustrata dal Signor Gibelin dottore di medicina, membro della Società Medica di Londra, ec. ec. e recata in italiano da una società di dotte persone con nuove illustrazioni, e tavole in rame*.

Gómez López, S. (1997). The Royal Society and Post-Galilean Science in Italy. *Notes and Records of the Royal Society of London*, 51(1), 35–44.

Gotti, M. (2006). Disseminating Early Modern Science: Specialized News Discourse in the *Philosophical Transactions*. In Brownlees, N. (Ed.), *News Discourse in Early Modern Britain* (pp. 41–70). Bern: Peter Lang.

Gotti, M. (2011). The Development of Specialised Discourse in the *Philosophical Transactions*. In Taavitasainen, I. & Pahta, P. (Eds.), *Medical Writing in Early Modern English* (pp. 205-219). Cambridge: Cambridge University Press.

Gotti, M. (2014). Scientific Interaction within Henry Oldenburg's Letter Network. *Journal of Early Modern Studies, 3*, 151-171.

Gray, B., Biber, D. & Hiltunen, T. (2011). The Expression of Stance in Early (1665-1712) Publications of the *Philosophical Transactions* and other Contemporary Medical Prose: Innovations in a Pioneering Discourse. In Taavitasainen, I. & Pahta, P. (Eds.), *Medical Writing in early Modern English* (pp. 220-247). Cambridge: Cambridge University Press.

Greco, G. & Rosa, M. (Eds.). (2013). *Storia degli Antichi Stati Italiani*. Roma: Editori Laterza.

Greenstone, G. (2009). The Roots of Evidence-Based Medicine. *BC Medical Journal, 51*(8), 342-344.

Gross, A. G., Harmon, E. J., & Reidy, M. S. (2000). Argument and 17th Century Science: A Rhetorical Analysis with Sociological Implications. *Social Studies of Science, 30*(3), 371-396. Sage Publications Ltd.

Gross, A., Harmon, E. J. & Reidy, M.S. (2002). *Communicating Science. The Scientific Article from the 17th Century to the Present*. West La Fayette, IN: Parlor Press.

Gross, A. G. (2001). *Scientific Discourse in Sociohistorical Context*. The Philosophical Transactions of the Royal Society of London, 1675-1975, by Atkinson, D. [Review]. *Isis, 92*(3), 576-577.

Gunnarsson, B.-L. (Ed.). (2011). *Languages of Science in the Eighteenth-Century*. Berlin: Mouton De Gruyter.

Hall, A. R. & Hall, M. B. (Eds.). (1966). *The Correspondence of Henry Oldenburg. Vol. III 1666-1667*. Madison, WI: University of Wisconsin Press.

Hall, A. R. & Hall, M. B. (Eds.). (1967). *The Correspondence of Henry Oldenburg. Vol. IV 1667-1668*. Madison, WI: University of Wisconsin Press.

Hall, A. R. & Hall, M. B. (Eds.). (1971). *The Correspondence of Henry Oldenburg. Vol. VIII 1671-1672*. Madison, WI: University of Wisconsin Press.

Hall, A. R. & Hall, M. B. (Eds.). (1986). *The Correspondence of Henry Oldenburg. Vol. XIII 1676-1681*. London: Taylor and Francis.

Hall, J. (1992). Where History and Sociology Meet: Forms of Discourse and Sociohistorical Inquiry. *Sociological Theory, 10*(2), 164-193.

Hall, M. B. (1975). The Royal Society's Role in the Diffusion of Information in the Seventeenth Century. *Notes and Records of the Royal Society of London, 29*(2), 173-192.

Hall, M. B. (1982). The Royal Society and Italy 1667–1795. *Notes and Records of the Royal Society of London, 37*(1), 63–81.

Hall, M. B. (1984). *All Scientists Now: The Royal Society in the Nineteenth-Century*. Cambridge: Cambridge University Press.

Hall, M. B. (1991). *Promoting Experimental Learning: Experiment and the Royal Society, 1660–1727*. Cambridge: Cambridge University Press.

Hall, M. B. (1992). *The Library and Archives of the Royal Society 1660–1990*. London: The Royal Society.

Halliday, M. A. & Matthiessen, C. M. (2014). *Halliday's Introduction to Functional Grammar* (4th ed.). Abingdon: Routledge.

Hamilton, W. (2015). Fiery fields – Volcanoes as Never Seen Before. In Furlong G. (Author), *Treasures from UCL* (pp. 118–121). London: UCL Press.

Hannesdóttir, A. H. (2011). From Vernacular to National Language: Language Planning and the Discourse of Science in Eighteenth-Century Sweden. In Gunnarsson, B.-L. (Ed.), *Languages of Science in the Eighteenth Century* (pp. 107–122). Berlin: De Gruyter Mouton.

Henderson, F. (2013). Faithful Interpreters? Translation Theory and Practice at the Early Royal Society. *Notes and Records of the Royal Society, 67*, 101–122.

Henderson, F. (2017). Translation in the Circle of Robert Hooke. In Fransen, S., Hodson, N., & Enenkel, K. A. E. (Eds.) *Translating Early Modern Science* (pp. 17–40). Leiden/Boston: Brill.

Hunter, M. (2014). John Ray in Italy: Lost Manuscripts Rediscovered. *Notes and Records of the Royal Society of London, 68*, 93–109.

Hunting, P. (2003). The Medical Society of London. *Postgraduate Medical Journal, 80*, 350–354.

Iamartino, G. (2001). La Contrastività Italiano-Inglese in Prospettiva Storica. *Rassegna Italiana di Linguistica Applicata, 33*, 7–126.

Iamartino, G. (2002). Non Solo Maccheroni, Mafia e Mamma Mia!: Tracce Lessicali dell'Influsso Culturale Italiano in Inghilterra. In San Vicente F. (Ed.), *L'inglese e le Altre Lingue Europee. Studi sull'Interferenza linguistica* (pp. 23–49). Bologna: CLUEB.

Ilardi, V. (1976). Eyeglasses and Concave Lenses in Fifteenth-Century Florence and Milan: New Documents. *Renaissance Quarterly, 29*(3), 341–360.

Johnson, J. (1794). *Medical Facts and Observations*, Vol. 5. London: Printed for J. Johnson, No 72, Saint Paul's Church Yard.

Kader, N. A. (2016). A Critical Analysis of Anti-Islamisation and Anti-immigration Discourse: The Case of the English Defence League and Britain First. *International Journal for Innovation Education and Research, 4*, 26–53.

Keeler, C. R. (2011). Three Hundred Fifty Years of the Royal Society: Fellows of Vision. *Arch Ophthalmol, 129*(10), 1361-1365.

Keller, R. (2011). The Sociology of Knowledge Approach to Discourse (SKAD). *Human Studies, 34*(1), 43-65.

Kipnis, N. (1987). Luigi Galvani and the Debate on Animal Electricity, 1791-1800. *Annals of Science, 44*, 107-142.

Knowles Middleton, W. E. (1969). The Title of the "Saggi". *The British Journal for the History of Science, 4*(3), 283-286.

Knowles Middleton, W. E. (1975). Science in Rome, 1675-1700, and the Accademia Fisicomatematica of Giovanni Giustino Ciampini. *The British Journal for the History of Science, 8*(2), 138-154.

Knowles Middleton, W. E. (1977). What Did Charles II Call the Fellows of the Royal Society? *Notes and Records of the Royal Society of London, 32*(1), 13-16.

Knowles Middleton, W. E. (1979). Some Italian Visitors to the Early Royal Society. *Notes and Records of the Royal Society of London, 33*(2), 157-173.

Knowles Middleton, W. E. (1980). *Lorenzo Magalotti at the Court of Charles II: His Relazione D'Inghilterra of 1668*. Waterloo, Ont.: Wilfrid Laurier University Press.

Koyré, A. (1955). A Documentary History of the Problem of Fall from Kepler to Newton: *De Motu Gravium Naturaliter Cadentium in Hypothesi Terrae Motae. Transactions of the American Philosophical Society, 45*(4), 329-395.

Kronick, D. A. (1994). Medical Publishing Societies in Eighteenth-Century Britain. *Bulletin of the Medical Library Association, 82*(3), 277-82.

List of Fellows of the Royal Society 1660 - 2007. (2007). London: Royal Society.

Locke, T. (2004). *Critical Discourse Analysis*. London: Continuum.

Lonati, E. (2013). Health and Medicine in 18th Century England: A Sociolinguistic Approach. In Kermas, S. & Christiansen, T. (Eds.), *The Popularisation of Specialised Discourse and Knowledge Across Communities and Cultures*. Bari: Edipuglia.

Lonati, E. & Grego, K. (2012). Reasoning, Rhetoric and Dialogue in Galileo's Mathematical Discourses. In Mazzon, G., & Fodde, L. (Eds.), *Historical Perspectives on Forms of English Dialogues* (pp. 181-207). Milan: FrancoAngeli.

Lyons, H. (1944). *The Royal Society, 1660-1940. A History of Its Administration under Its Charters*. Cambridge: Cambridge University Press.

MacLeod, R. (2010). The Royal Society and the Commonwealth: Old Friendships, New Frontiers. *Notes and Records of the Royal Society of London, 64*, S 137-S 149.

Marples, A. (2019). Scientific Administration in the Early Eighteenth Century: Reinterpreting the Royal Society's Repository. *Historical Research*, 92, 183–204.

Martin, J. R. & White, P. R. R. (2005). *The Language of Evaluation: Appraisal in English*. Basingstoke: Palgrave.

Mazzotti, M. (2013). Il Newtonianesimo e la Scienza del Settecento. Il Contributo Italiano alla Storia del Pensiero – *Scienze, Enciclopedia Treccani*, http://www.treccani.it/enciclopedia/il-newtonianesimo-e-la-scienza-del-settecento_%28Il-Contributo-italiano-alla-storia-del-Pensiero:-Scienze%29/.

McConnell, A. (1986). L. F. Marsigli's Voyage to London and Holland, 1721–1722. *Notes and Records of the Royal Society of London*, 41(1), 39–76.

McConnell, A. (1993). L. F. Marsigli's Visit to London in 1721, and His Report on the Royal Society. *Notes and Records of the Royal Society of London*, 47(2), 179–204.

McCutcheon, R. (1924). The *Journal Des Scavans* and the *Philosophical Transactions of the Royal Society*. *Studies in Philology*, 21(4), 626–628.

Meli Bertoloni, D. (2008). The Collaboration Between Anatomists and Mathematicians in the Mid-Seventeenth Century with a Study of Images as Experiments and Galileo's Role in Steno's Myology. *Early Science and Medicine* 13, 665–709.

OED. *Oxford English Dictionary*. https://www-oed-com.pros.lib.unimi.it:2050/.

Olivari, E. & Tornaghi, P. (2012). Dialogues and Scientific Writing in the 16th Century: Cyprian Lucar's Translation of Niccolò Tartaglia. In Mazzon, G. & Fodde L. (Eds.), *Dialogic/Dialogue Forms in One Thousand Years of English Texts from Old English to Late Modern English* (pp. 229–247). Milano/Roma: FrancoAngeli.

Ortore, M. (2022). L'astronomia nel primo Settecento tra carteggi e riviste: albori dell'articolo scientifico. In Forner, F., Meyer, F., Schwarze, S. (Eds.), *I Periodici Settecenteschi come Luogo di Comunicazione dei Saperi. Prospettive Storiche, Letterarie e Linguistiche*. Berlin: Peter Lang.

Pahta, P. & Taavitasainen, I. (2011). An Interdisciplinary Approach to Medical Writing in Early Modern English. In Taavitasainen, I., & Pahta, I. (Eds.), *Medical Writing in Early Modern English* (pp. 1–8). Cambridge: Cambridge University Press.

Panizzi, A. (1836). *Catalogue of the Scientific Books in the Library of the Royal Society*. London: Richard & John E. Taylor.

Parent, A. (2004). Giovanni Aldini: From Animal Electricity to Human Brain Stimulation. *The Canadian Journal of Neurological Sciences*, 31(4), 576–584.

Patterson, E. C. (1983). *Mary Somerville and the Cultivation of Science, 1815-1840*. Boston: Martinus Nijhoff.

Peiffer, J., Conforti, M., Delpiano, P. (Eds.). (2013). L'Europe des journaux savants (xviie-xviiie siècles). Communication et construction des savoirs / Scholarly journals in early modern Europe. Communication and the construction of knowledge, n° spécial des Archives internationales d'histoire des sciences, fasc. 170-171, vol. 63.

Pesaresi, F., & Ascari, M. (2015). *La scoperta dell'Inghilterra: Epistolari e diari dei viaggiatori italiani del settecento*. Verona: QuiEdit.

Pinnavaia, L. (2001). *The Italian Borrowings in the Oxford English Dictionary: A Lexicographical, Linguistic and Cultural Analysis*. Roma: Bulzoni.

Placanica, A. (1985). *Il filosofo e la catastrofe: un terremoto del Settecento*. Torino: Einaudi.

Plescia, I. (2011). "Strangers to our Nation" Anglo-Italian Relations and Linguistic Encounters in Two Early Modern Scientific Translations. *Textus*, 24(3), 559-578.

Poliakoff, M. (2015). The Royal Society, the Foreign Secretary, and International Relations. *Science & Diplomacy*, 4(1), http://www.sciencediplomacy.org/letter-field/2015/royal-society-foreign-secretary-and-international-relations.

Pomata, G. (2007). Malpighi and the Holy Body: Medical Experts and Miraculous Evidence in Seventeenth-Century Italy. *Renaissance Studies*, 21(4), 568-586.

Praz, M. (1939). Fortuna della Lingua e della Cultura Italiana in Inghilterra. *Romana*, 8, 465-82.

Praz, M. (1944). *Ricerche Anglo-Italiane*. Roma: Edizioni di Storia e Letteratura.

Preiano', A. (2016). Manfredo Settala, Accumulatore Seriale di meraviglie. *Milanoplatinum.com: Luxury Web Magazine*, https://www.milanoplatinum.com/manfredo-settala-accumulatore-seriale-di-meraviglie.html.

Quinn, T. (2005). Editorial: An Italian Correspondence, an Italian Earthquake and the Homes of the Royal Society. *Notes and Records of the Royal Society of London*, 59(1), 1-4.

Ray, J. (1673). *Observations Topographical, Moral, & Physiological; Made in a Journey Through Part of the Low-Countries, Germany, Italy, and France: with a Catalogue of Plants Not Native of England, Found Spontaneously Growing in Those Parts, and Their Virtues*. London: Printed for John Martyn, printer to the Royal Society.

Rebora, P. (1936). *Civiltà Italiana e Civiltà Inglese: Studi e Ricerche*. Firenze: Le Monnier.

Rebora, P. (1938). *Civiltà Italiana nel Mondo. In Inghilterra*. Roma: Società Nazionale Dante Alighieri.

Reisigl, M. & Wodak, R. (2016). The Discourse-Historical Approach (DHA). In Wodak, R., & Meyer, M. (Eds.), *Methods of Critical Discourse Analysis* (pp. 23–62). London: Sage.

Renaldo, J. (1976). Bacon's Empiricism, Boyle's Science, and the Jesuit Response in Italy. *Journal of the History of Ideas, 37*(4), 689–695.

Rivington, F. (1793). Account of Some Discoveries Made by Mr. Galvani, of Bologna; With Experiments and Observations on Them. In Two Letters from Mr. Alexander Volta, F. R. S. Professor of Natural Philosophy in the University of Pavia, to Mr. Tiberius Cavallo, F. R. S. In *The British Critic and Quarterly Theological Review, 2*, 89–91.

Rivington, C. A. (1984). Early Printers to the Royal Society 1663–1708. *Notes and Records of the Royal Society of London, 39*(1), 1–27.

Robertson, J. (2000). Enlightenment and Revolution: Naples 1799. *Transactions of the Royal Historical Society, 10*, 17–44.

Rotta, S. (1968). Bianchini,Giuseppe. In *Dizionario Biografico degli Italiani*, vol. 10. Treccani. https://www.treccani.it/enciclopedia/giuseppe-bianchini_%28Dizionario-Biografico%29/.

Rotta, S. (1990). L'Accademia Fisico-Matematica Ciampiniana: Un'Iniziativa di Cristina?. In Di Palma, W., & Bovi, T. (Eds.), *Cristina di Svezia. Scienza ed Alchimia nella Roma Barocca* (pp. 99–186). Bari: Dedalo.

Rusnock, A. (1996). *The Correspondence of James Jurin (1648–1750): Physician and Secretary to the Royal Society*. Amsterdam: Rodopi.

Rusnock, A. (1999). Correspondence Networks and the Royal Society, 1700–1750. *The British Journal for the History of Science, 32*(2), 155–169.

Sabba, F. (2018). I periodici bibliografici italiani dalle origini: prospettive di un censimento. *Bibliothecae.it, 7*(2), 8–55.

Scala, G. E. & Schullian, D. M. (Eds.). (1974). An Index of Proper Names in Thomas Birch, the History of the Royal Society (London, 1756–1757). *Notes and Records of the Royal Society of London, 28*, 263–330.

Schickore, J. (2010). Trying Again and Again: Multiple Repetitions in Early Modern Reports of Experiments on Snake Bites. *Early Science and Medicine, 15*(6), 567–617.

Schofield, R. S. (1953). John Wesley and Science in 18th Century England. *Isis, 44*(4), 331–340.

Shapin, S. (1988). The House of Experiment in Seventeenth-Century England. *Isis, 79*(3), 373–404.

Shapin, S. (1996). *The Scientific Revolution*. Chicago: The University of Chicago Press.

Shapiro, B. J. (2002). Testimony in Seventeenth-Century English Natural Philosophy: Legal Origins and Early Development. *Studies in History and Philosophy of Science, 33*(2), 243–263.

Simpson, P. (2003). *Language, Ideology and Point of View*. London: Routledge.

Skouen, T. (2011). Science Versus Rhetoric? Sprat's History of the Royal Society Reconsidered. *Rhetorica: A Journal of the History of Rhetoric, 29*(1), 23–52.

Sorrenson, R. & King Hele, D. G. (1996). Towards a History of the Royal Society in the Eighteenth Century. *Notes and Records of the Royal Society of London, 50*, 29–46.

Taavitsainen, I. & Pahta, P. (2011). *Medical Writing in early Modern English*. Cambridge: Cambridge University Press.

Taddia, E. (2009). Corpi, cadaveri, chirurghi stranieri e ceroplastiche: l'ospedale di Pammatone a Genova tra Sei e Settecento. *Mediterranea. Ricerche Storiche, VI*, 155–192.

Teleman, U. (2011). The Swedish Academy of Sciences: Language Policy and Language Practice. In Gunnarsson, B.-L. (Ed.), *Languages of Science in the Eighteenth Century* (pp. 63–88). Berlin: De Gruyter Mouton.

Testa, S. (2015). *Italian Academies and Their Networks, 1525–1700: From Local to Global*. London: Palgrave Macmillan.

Turner, A. (2008). An Interrupted Story: French Translations from "Philosophical Transactions" in the Seventeenth and Eighteenth Centuries. *Notes and Records of the Royal Society of London, 62*(4), 341–354.

Udías, A. (2014). *Jesuit Contribution to Science: A History*. Cham/ New York: Springer.

Vicentini, A. (2019). Popularizing and Translating Science in 18th century Europe: Francesco Algarotti's *Newtonianismo per le dame* and its English-language Editions (1737–1772). *Status Quaestionis, 17*, 235–262.

Waller. (2012). Lorenzo Magalotti in England, 1668-9. *Italian Studies, 1*(2), 49–66.

White, P. R. R. (2004). Subjectivity, Evaluation and Point of View in Media Discourse. In Coffin C., Hewings, A. & O'Halloran, K. (Eds.), *Applying English Grammar* (pp. 229–246). London: Hodder Arnold.

White, P. R. R. (2015). Appraisal Theory. In Tracy, K., Sandel, T., & Ilie, C. (Eds.), *The International Encyclopedia of Language and Social Interaction* (pp. 1–7). Hoboken, New Jersey: John Wiley & Sons, Inc.

Wills, H. (2019). Charles Blagden's Diary: Information Management and British Science in the 18th Century. *Notes and Records of the Royal Society of London*, *73*, 61–81.

Wilson, F. (1935). English Letters and the Royal Society in the Seventeenth Century. *The Mathematical Gazette*, *19*(236), 343–354.

Wis, M. (1996). Lorenzo Magalotti e la sua relazione di Svezia: I. Origini del testo. *Neuphilologische Mitteilungen*, *97*(4), 343–363.

Wis, M. (1997). Lorenzo Magalotti e la Sua Relazione di Svezia: II manoscritti e edizione. *Neuphilologische Mitteilungen*, *98*(4), 351–370.

Wis, R. (1970). Lorenzo Magalotti e la sua relazione del grande viaggio di Cosimo De' Medici. *Neuphilologische Mitteilungen*, *71*(3), 451–454.

Wodak, R. & Meyer, M. (Eds.) (2016). *Methods of Critical Discourse Analysis*. London: Sage.

Index of names

Acta Eruditorum 35, 143, 145, 271, 288
AGLIONBY, WILLIAM 86, 269
AGNESI, MARIA GAETANA 36, 108
ALBERTINI, GIAMBATTISTA 104, 240
ALDINI, GIOVANNI 130, 132, 201, 249, 302
ALESSI, DARIO 31
ALGAROTTI, FRANCESCO 101, 105, 237, 305
ALLIONI, CARLO 38, 102, 240, 282
AMALDI, EDOARDO 31
AMICI, GIOVANNI BATTISTA 136–137
ANDREANI, PAOLO 106–107
ARDINGHELLI, MARIA LUISA 141, 294
ARGENTATI, RAFFAELE 130, 132, 250
AUZOUT, ADRIAN 78, 82, 152, 168, 170–172, 177, 261–262
AVERANI, GIUSEPPE 105, 111–112, 232

BABBAGE, CHARLES 30
BACCHINI, BENEDETTO 35
BACON, FRANCIS 17, 23–24, 304
BADILY, WILLIAM 85
BAGLIVI, GIORGIO 74, 94, 229
BAINES, THOMAS 34, 84
BAKER, HENRY 110, 116–117, 182, 190, 246, 276–281, 285
BALDINI, GIOVANNI ANTONIO 104, 232
BALLE, CHARLES 85
BALLE, ROBERT 17, 83, 85, 152, 297
BALLE, WILLIAM 84

BANKS, JOSEPH 19, 24, 27–28, 107–108, 126–127, 130, 201, 285–287, 293
BAROZZI, JACOPO 78
BARTOLI, CARLO 78
BASSI, LAURA 116, 295
BAYARDI, ANTONIO OTTAVIO 105, 238
BECCARI, JACOPO BARTOLOMEO 102–103, 112, 120, 235
BECCARIA, GIOVANNI BATTISTA 103, 117, 238, 280–281, 283
BELGIOIOSO, COUNT LUDOVICO DI 102, 242
BELLINI, LORENZO 17, 33, 87, 93–94, 109, 163, 173, 263
BENEVOLI, ANTONIO 106, 244, 246, 272
BENVENUTI, GIUSEPPE 107
BIANCHI, VENDRAMINO 104, 232
BIANCHINI, FORTUNATO 106, 116, 205
BIANCHINI, FRANCESCO 33, 103, 114, 203, 232, 244, 272, 276–278
BIANCHINI, GIUSEPPE 112–113
BIDDELL AIRY, GEORGE 29
BLAGDEN, CHARLES 104, 106–108, 306
BOCCONE, PAOLO 76, 87, 94–95, 162–163, 167–168, 180, 230, 265, 268, 294
BONANNI, FILIPPO 78, 94
BONAPARTE, NAPOLEON 39–40, 243
BONFIGLIOLI, SILVESTRO 74, 229
BONOMO, GIOVANNI COSIMO 76, 106, 117, 176, 243, 270

Index of names

BORELLI, GIOVANNI ALFONSO 17, 33, 74, 76, 85, 87, 92–93, 109, 154, 169, 229, 266
BORGHESE, MARCANTONIO 75, 228
BORROMEO, CLELIA 38, 296
BOSCOVICH, GIUSEPPE RUGGERO 38
BOTTONI, DOMENICO 74, 85, 118, 229
BOVI, ROCCO 106, 244, 304
BOYLE, ROBERT 34, 82, 90, 155, 174, 266, 304
BOZZI GRANVILLE, AUGUSTUS. *See* Granville, Augustus Bozzi
BRADLEY, JAMES 110
BRIGOLI, CANONICO 107
BRIOSCHI, FRANCESCO 130, 132–133, 138, 249, 291
BROMLEY, WILLIAM 109
BRUGNATELLI, LUIGI VALENTINO 107
BRUNI, GIUSEPPE LORENZO 102–103, 116–117, 182, 238, 276, 279–280
BURNET, GILBERT 44

CALANDRUCCIO, SALVATORE 133
CALDANI, LEOPOLDO MARCO ANTONIO 102, 240
CAMPANI, GIUSEPPE 33, 76–78, 80, 82–83, 85, 87, 154, 157, 168, 170–172, 176, 230, 261–262, 272
CANNIZZARO, STANISLAO 40, 129, 248
CAPELLO, PIETRO ANDREA 104, 238
CAPUA, LEONARDO DI 34, 77, 94
CAPUANI, P. 78
CARACCIOLI, DOMENICO 104, 240
CARAFA, GIOVANNI 104, 240
CARBONE, GIOVANNI BATTISTA 102–103, 112, 114, 235, 272–274

CARBURI, GIOVANNI BATTISTA 102, 240
CARDELLI, LUCA 31
CARLINI, FRANCESCO 128–129, 138, 147, 217, 247, 290
CARNAVALLIA, MARQUIS 107
CASSINI, GIOVANNI DOMENICO 42, 74–75, 78, 82–83, 85, 87, 114, 140, 152, 157, 170–172, 176–177, 228, 262–263, 265–268, 272
CASTAGNA, MARCO ANTONIO 89–90, 139, 177–179, 230, 264–265
CASTELLI, BENEDETTO 18, 295
CAVALLI SFORZA, LUIGI LUCA 31
CAVALLO, TIBERIO 42, 103, 107, 117–118, 185–186, 191–193, 199, 214, 242, 284–286, 304
CELESIA, PIETRO PAOLO 104, 240
CERATI, GASPARE 104, 237, 279
CERVI, JOSEPH 102, 236
CESI, FEDERICO 33, 91
CESTONI, GIACINTO 77
CHARAS, MOYSE 89, 168, 172, 265
CHRISTINA, QUEEN OF SWEDEN 33, 44, 93, 176
CIAMPINI, GIOVANNI GIUSTINO 33, 76, 93, 141, 176, 180, 231, 267, 269, 301
CIGNA, GIOVANNI FRANCESCO 102–103, 240
CIRILLO, DOMENICO 106, 126
CIRILLO, NICOLA 36, 118, 198, 244, 282
COCCHI, ANTONIO 102, 237, 246
Collegio Romano 248, 290–291
COLLINS BRODIE, BENJAMIN 29
COMPTON, SPENCER 29
CONCLUBET, ANDREA 34
CONFIGLIACHI, LUIGI 217
CONTI, ANTONIO SCHINELLA 103, 198, 232, 271
CORNARO, FRANCESCO 104, 232, 270

Cornelio, Tommaso 34, 80–81, 94, 151, 162, 204, 264
Corsini, F. 107, 143, 192
Count of Barbiano di Belgioioso, Luigi Carlo Maria 104
Crawford, James 85
Cremona, Luigi 128–129, 248
Crivelli, Giovanni Francesco 103, 106, 236
Crusio, Carlo 106, 244, 278
Cuzzoni, Abbot 107

D'Albini, Maffeo 107
D'Aragona, Nicolo Alerbo 104, 118, 186–187, 209, 235
D'Este, Francesco Maria 104, 236
D'Orsi, Giovanni Giuseppe 105
Da Rovato, Giuseppe 106, 286
Dalton Hooker, Joseph 29
Daniell, John Frederic 131
Davia, Giovanni Antonio 35, 77
Davy, Humphry 29, 129, 287
De Angeli *See* Degli Angeli
De Baillou, Cavalier 104
De Gioeni, Count 107
De Guasco, Ottavio 105
De la Rue, Warren 137, 291
De Medici, Cosimo 45, 231, 243–244, 306
De Paoli, Pasquale 42
De Revillas, Didacus 111, 186, 275–276
De' Medici, Cardinal Leopold 33, 74
Degli Angeli, Stefano 36, 168–169
Del Bene, Tommaso 75, 229
Del Papa, Giuseppe 94, 106, 244, 269
Del Riccio, Luigi 173

Della Torre, Giovanni Maria 106, 189–190, 203, 244
Dereham, Thomas 17, 38, 42, 44, 106, 109, 111–113, 115–116, 119–120, 141–142, 193, 198, 272–274
Dereham, William 119–120
Desaguliers, John Theophilus 197–198, 201
Di Napoli, Anello 94, 240
Di Paoli, Filippo 105
Di Petrizzi, Antonio 108
Di Tommaso, Carron Count of Briancon 102, 104, 232
Divini, Eustachio 78, 80, 87, 154, 168, 171–172
Dodington, John 78, 80, 85, 94, 151, 176, 264–265
Dohrn, Anton 136
Donati, Vitaliano 102, 114, 118, 191, 240, 278–279
Doni, Giovanni Battista 96, 99, 230, 263
Donzelli, Tommaso 94
Doria, Paolo Mattia 77, 195, 201–202
Douglas, James 27
Dulbecco, Renato 31
Duliolo, Rinaldo 102, 232

Erskine Baker, David 121
Eyles-Stiles, Francis 190, 203

Fabbroni, Giovanni 37
Fabretti, Raffaele 94
Fabri, Honorè 74, 85
Fabris, Pietro 108
Fagnano dei Toschi, Giulio Carlo 103, 234
Falconer, Mr. 135
Falconieri, Paolo 45, 74, 76, 173
Fano, Ugo 31
Fantoni, Pio 107, 281

Index of names

FARADAY, MICHAEL 129
FERDINAND II, Grand Duke of
 Tuscany 33, 88, 119, 175
FERMI, ENRICO 31
FERRARI, DOMENICO 20, 42, 105,
 234, 297
FINCH, JOHN 34, 74, 81, 83-84
FLAMSTEED, JOHN 177, 265
FOLKES, MARTIN 27, 41, 100, 120,
 124, 275
FONTANA, FELICE 37, 106, 118, 183-
 184, 186, 243-244, 246, 284-285
FONTANA, GREGORIO 103
FORNASARI, IPPOLITO 75, 229
FORTIS, ALBERTO 102, 105, 108,
 243, 270
FOSCARINI, MARCO 105, 240
FRACASSATI, CARLO 33, 57-58, 76,
 88, 166-168, 229, 262
FRISI, PAOLO 103-105, 192, 240
FULLER, JOHN 186-188

GALE, ROGER 121, 275
GALIANI, CELESTINO 36-37, 105-
 106, 141, 198, 236
GALILEI, GALILEO 18, 32-33, 35, 41,
 44, 295, 301-302
GALVANI, LUIGI 40, 106, 116, 131-
 132, 137-138, 192, 198-201, 222,
 244, 258, 286-287, 301, 304
GARNIERI, GIOVANNI BATTISTA 94
GASSENDI, PIERRE 34
GATTA, XAVERIO 108
GATUCCI 105, 232
GEMELLI CARERI, GIOVANNI
 FRANCESCO 93
GIACOMELLI,
 MICHELANGELO 105, 237
GIANNETI, PASCASIO 94
GIANOTTI, FABIOLA 31
GIBELIN 142, 298
GILBERT, DAVIES 29

Giornale de' Letterati 10, 13, 23, 33,
 35, 42, 73, 111, 141, 145, 153, 176-
 177, 196
GIUNTINI, JERONIMO 102, 236
GIZLANZONI, SIGNOR 198
GORI, ANTONIO FRANCESCO 105,
 108, 237
GORNIA, GIOVANNI BATTISTA 76,
 173, 231
GOTTIGNIES, GILLES FRANCOIS 85,
 92, 172
GRANDI, GUIDO 103, 111-112, 138,
 198, 216
GRANDI, JACOBUS 74, 87, 154, 176,
 228, 232, 263, 270, 272
GRANVILLE, AUGUSTUS BOZZI 30,
 129-130, 146-147, 214, 247, 288-
 289, 291
GRASSI, GIOVANNI BATTISTA 130,
 133, 135, 217, 250
GREGORY, JACOB 168-169, 262
GREW, NEHEMIAH 25, 180
GRIMALDI, FRANCESCO MARIA 160
GRIMANI, PIETRO 104, 232
GUASCONI, BERNARDO 74-75, 86,
 228, 231
GUGLIELMINI, DOMENICO 35, 75,
 94, 164, 229

HALES, STEPHEN 110, 141, 209, 277
HALLEY, EDMOND 26, 44, 108,
 202, 272
HAMILTON, WILLIAM 42, 110-111,
 118-119, 122, 125-126, 142, 185,
 208, 281-283, 285-287, 300
HARTOP, MARTIN 86, 90, 268
HARVEY, WILLIAM 34
HERMAN, JACOB 36
HILL, JOHN 27
HOARE, MR. 121, 277
HOLLAND, HENRY 136, 269, 288, 302
HOLLIS, THOMAS 110, 278-279

Index of names

Hooke, Robert 25, 120, 162, 169, 300
Howard, Henry 85
Huxley, Thomas Henry 29

Imperato, Ferrante 108
Ippolito, Francesco 75, 106, 118, 229, 244, 285

James III 111
Jattica, Jacobus 102, 106, 236
John, Oliver 121
Johnson, Samuel 153, 199, 300
Journal de Sçavans 24, 35, 293
Jurin, James 17, 26, 44, 106, 108–109, 111–112, 115, 119–120, 195–196, 201, 208, 272, 276, 292, 304
Justel, Henri 82

King, William 26

Lagrangia, Giuseppe Ludovico 102–103, 243
Lana, Francesco 76, 89, 153, 156, 230, 264–265
Lancisi, Giovanni Maria 87–88, 93, 102, 163, 176, 232, 268, 270
Langrangia, Giuseppe Luigi 38
Leibniz, Gottfried Wilhelm 44, 103, 198, 232
Leopold de Medici *See* De Medici, Cardinal Leopold
Leopold II, Grand Duke of Tuscany 129, 137, 247
Leprotti, Antonio 102, 236, 276
Lercari, Abbot 107
Leti, Gregorio 42, 75, 228, 231
Levi Montalcini, Rita 31
Levi-Civita, Tullio 31
Lind, James 199
Lister, Joseph 29, 163, 267, 269
Lorenzini, Lorenzo 112, 173

Lorenzini, Stefano 107, 116
Lorgna, Antonio Maria 38, 104, 106, 242
Lyell, Charles 135, 138, 291

Mackinlay, Robert 118, 280
Maffei, Francesco Scipione 105, 112–113, 192, 204, 236
Magalotti, Lorenzo 33–34, 45, 74–76, 80–81, 86, 89, 105, 150, 173–174, 230, 232, 270, 296, 301, 305–306
Magliabechi, Antonio 94, 106, 243, 270
Maitland, Charles 115
Malaspina, Luigi 104, 107, 242
Mallet, Robert 135, 138, 291, 297
Malpighi, Marcello 17, 21, 33, 41–42, 74, 77–78, 80, 87–88, 91, 93–94, 109, 115, 139, 163, 191, 228, 231, 258, 262, 264, 268, 303
Mancini, Pasquale Stanislao 130, 132, 250
Manetti, Saverio 102, 240
Manfredi, Eustachio 35, 43, 74, 103, 112, 114, 169, 198, 235, 275
Manfredi, Michele 169
Maraldi, Giovanni Domenico 107
Marchetti, Antonio 163
Maria Theresa of Austria 39
Marinoni, Giovanni Giacomo 102–103, 237
Marsigli, Luigi Ferdinando *See* Marsili, Luigi Ferdinando
Marsili, Luigi Ferdinando 35, 37, 75, 229, 271
Marsili, Antonio Felice 35, 37, 77–78, 163, 229, 240
Marsili, Giovanni 102
Matani, Antonio 102, 240

312 Index of names

MATTEUCCI, CARLO 130–131, 137–138, 213–217, 249, 290–291
MATY, MATTHEW 26, 28, 108, 142, 281–283
MAY, JOSEPH 83, 135, 276, 285, 293
MAZZOCCHI, CANONICO 107
MELLONI, MACEDONIO 40, 128–129, 248
MERCATOR, NICHOLAS 177, 263
MICHELI, PIER ANTONIO 107, 279
MICHELOTTI, PIETRO ANTONIO 102, 142, 195–196, 201, 234, 272, 288
MOLINELLI, PIER PAOLO 102, 238
MONFORTE, ANTONIO DI 93
MONRO, JOHN 121, 269, 296
MONTAGU, MARY WORTLEY 115
MONTANARI, GEMINIANO 42, 76, 230
MORE, ROBERT 27, 78–79, 109–110, 123–124, 158, 162, 277, 285, 289
MORGAGNI, GIOVANNI BATTISTA 102, 109, 197, 234
MORICHINI, DOMENICO PINO 129, 138, 217, 247, 250
MORO, LAZZARO 106, 202, 244, 277
MOROSINI, FRANCESCO LORENZO 104, 240
MOROZZO, CARLO LUIGI 106
MORTIMER, CROMWELL 26–27, 77, 108
MORTON, CHARLES 27, 105, 121, 280–281
MOSCATI, PIETRO 199
MOSSO, ANGELO 130–132, 249–250, 291
MURATORI, LUDOVICO ANTONIO 105, 234

NARDUCCI, TOMMASO 141
NAZZARI, FRANCESCO 23, 33, 42–43, 141, 176
NERI, ANTONIO 78

NEWTON, HENRY 42
NEWTON, ISAAC 26–27, 36, 43, 101, 103, 141, 169, 192, 196–198, 202, 232, 237, 301
NICOLINI, ANTONIO 104, 238, 277
NOBILI, LEOPOLDO 130–131, 138, 215–216, 249, 289
NOLLET, JEAN ANTOINE 111, 117, 124, 203–206, 277–278

OCCHIALINI, GIUSEPPE 31
OLDENBURG, HENRY 17, 24–25, 43, 74–83, 91, 108–109, 139–141, 150–151, 165–167, 172–173, 177, 219, 293, 299
OLIVIERI, ANNIBALE 107
ORIANI, BARNABA 40, 103, 243
OSORIO, CAVALIERE 104, 238

PACCHIONI, ANTONIO 191
PACICHELLI, GIOVANNI BATTISTA 74–75, 228
PADERNI, CAMILLO 105, 110, 118, 120, 240, 275, 278–279
PAITONI, GIOVANNI BATTISTA 106, 244, 281
PANCIATICHI, LORENZO 45
PANCRAZI, GIUSEPPE MARIA 105, 240, 295
PANIZZI, ANTONIO 130, 133–134, 249, 294, 297, 302
PARKER, GEORGE 27, 118, 278
PAROLINI, ALBERTO 136
PARRINELLO, MICHELE 31
PARSONS, WILLIAM 29, 104
PARTENIO GIANNETASIO, NICOLA 94
PASSERI, GIOVANNI BATTISTA 105, 238
PEDINI, PASQUALE 106, 118, 208–209, 244, 276
PETER LEOPOLD, GRAND DUKE OF TUSCANY 37

Pettini, Max 31
Piazzi, Joseph 106, 128, 244, 247, 250, 286
Piccolomini, Count Vincent 130, 132, 249, 289
Pighi, Giacomo 74, 228
Pilarini, Giacomo 102, 106, 115, 246, 271, 287
Pinelli, Flaminio 106, 115, 145, 193–194, 244, 273
Piranesi, Giovanni Battista 107
Pivati, Giovanni Francesco 107, 116–117, 192, 203–205
Plana, Giovanni Antonio Amedeo 128, 247
Planta, Joseph 28
Platt, Thomas 85, 89, 172, 265
Plot, Robert 25, 178, 292
Poleni, Giovanni 36, 103, 112, 114, 119–120, 186–187, 198, 232, 274–276
Poli, Giuseppe Saverio 102, 242
Pope, Walter 85, 96, 261
Posi, Paolo 107
Pozzan, Tullio 31
Priestley, Joseph 191, 246, 284
Pringle, John 27, 41, 283–284
Pucci, V. 106, 244
Pulleyn, Octavian 86, 162, 268

Ramazzini, Bernardino 107, 112
Ramsay, Mr. 121, 275
Ranby, John 197, 272
Rappuoli, Rino 31
Ray, John 34, 44, 95, 99, 109, 163, 178, 270, 300, 303
Recanati, Giambattista 105, 234
Recupero, Giuseppe 108
Redi, Francesco 33, 74, 76, 85, 88–89, 91, 107, 152, 159–160, 164, 168, 172, 174, 229, 265, 269–270

Rezzonico, Abondio 104, 240
Riccardi, Francesco 45
Riccati, Jacopo 36, 198
Ricci, Michelangelo 33, 74, 85
Riccioli, Giovanni Battista 76, 168–169, 230, 262
Ricolvi, Paolo 107
Righi, Augusto 31
Rinuccini, Folco 104, 238
Ripa, Ludovico 102, 234
Ritchie, William 216, 289
Rivantella, Antonio 107
Rizzardo, Ezio 31
Rizzetti, Giovanni 36, 192, 197–198, 201, 273
Rizzi Zannoni, Giovanni 106, 244, 281
Robinson, Tancred 96, 98, 110, 122–123, 154, 267, 271
Roget, Mark 134, 297
Rolli, Paolo 42, 77, 105, 112–113, 195, 202, 235, 276, 296
Rooke, Lawrence 23
Rubbia, Carlo 31
Ruffini, Angelo 130–131, 249, 297
Rutty, William 26, 108, 111

Sabine, Edward 29
Sacchetti, Giulio 105, 237
Sacchi, Pompeio 94
Saggi di Naturali Esperienze 34, 45–46, 76, 85, 262, 295, 301
Saluzzo, Giuseppe Angelo 38, 103–104, 240
Salvemini, Giovanni Francesco Melchiorre 103, 238
Salvetti, Pietro 173
Salvini, Antonio Maria 84, 105, 234
Sandoro, Ottavio 94
Santorelli, Antonio 108

SAROTTI, GIOVANNI AMBROSIO 75, 86, 96, 98, 159, 162, 228, 231, 266–267
SAROTTI, PAOLO 75
SBARAGLI, GIANGIROLAMO 93–94
SCARPA, ANTONIO 40, 102, 243
SCHEUCHZER, JOHAN JAKOB 191, 270
SCHEUCHZER, JOSEPH CASPAR 142, 193–194, 270, 273
SCHIAPPARELLI, GIOVANNI 128
SCOTTI, EMMANUEL 108
SECCHI, PADRE ANGELO 128–129, 137–138, 215–216, 248, 291
SEGNI, ALESSANDRO 45
SEPTALLA, MANFRED. *See* Settala, Manfredo
SETTALA, MANFREDO 76, 80–81, 85, 96, 99, 178, 179, 180, 230, 303
SHERARD, WILLIAM 196, 269–270
SHERRINGTON, CHARLES 131, 297
SHORT, JAMES 139, 186, 188, 266, 277
SILVESTRE, PIERRE 92–93, 109, 269
SKIPPON, PHILIPP 34
SLOANE, HANS 25–27, 82, 85, 106, 108, 111, 113, 270, 273, 275, 292
SOMERVILLE, MARY 129, 136–137, 212–213, 289, 303
SORANZO, CAVALIER 94
SOUTHWELL, ROBERT 85
SPALLANZANI, LAZZARO 38, 40, 102, 240, 246
SPILLANTINI, MARIA GRAZIA 31
SPOLETI, FRANCESCO 74, 94, 229
SPOTTISWOODE, WILLIAM 29
SPRAT, THOMAS 81, 84, 305
STACK, THOMAS 142, 278
STELLUTI, FRANCESCO 78
STENO, NICOLAUS 33, 78, 81, 173, 302
STOKES, GEORGE GABRIEL 29

STRANGE, JOHN 82, 110–111, 118, 125, 192, 266, 269, 282–284, 295–296
STRATICO, SIMONE 103, 240
STROZZI, ABBOT 173
SUSSEX, AUGUSTUS FREDERICK, Duke of 29
SWINTON, JOHN 110, 121, 142–143, 191–192, 202, 280, 282–283

TACCHINI, PIETRO 128–129, 248
TAGLINI, CARLO 105, 112, 236
TARGIONI TOZZETTI, GIOVANNI 37, 108
TATA, ABBE 108
TESTA, GIUSEPPE ANTONIO 106
THOMSON, WILLIAM 29
TILLI, MICHELANGELO 102, 111–112, 232, 244, 271
TIMONE, EMANUELE 102, 115, 232, 271
TOALDO, GIUSEPPE 104, 242, 284
TORTI, FRANCESCO 102, 234
TOZZI, MONSIGNOR LUCA 94, 102, 232
TRAVAGINO, FRANCESCO 74, 80–81, 164–166, 168, 176, 228
TRIONFETTI, GIOVANNI BATTISTA 93
TRONI, NICOLO 104, 232

UBALDINI, CARLO 75, 228

VALETA, JOSEPH. *See* Valletta, Giuseppe
VALLETTA, GIUSEPPE 94, 106, 244
VALLISNERI, ANTONIO 36, 38, 102, 112, 232, 234
VALSALVA, ANTONIO MARIA 78, 93, 107, 115–116, 196–197, 246, 272
VENANZIO D'AQUINO, FRANCESCO MARIA 104, 242

VENTURA ROSETO, GIOVANNI 78
VENTURI, MARSILIO 102, 238
VENUTI, FILIPPO 105
VENUTI, RIDOLFINO 105, 120, 143, 192, 240, 280
VERATTI, GIUSEPPE 107, 116, 204, 295
VERNON, FRANCIS 85, 96–97
VIVIANI, VINCENZO 33, 74–75, 94, 229
VOLTA, ALESSANDRO 38, 40, 103, 117, 138, 185, 192, 198–201, 215–217, 243, 285–287, 294, 304

WALLER, RICHARD 20, 25, 45–46, 77, 85, 180, 271, 294, 305
WALMESLEY, CHARLES 110
WARD, JOHN 121, 275, 277, 279
WARING, EDWARD 110
WATSON, ROBERT 142
WATSON, WILLIAM 110, 142, 205, 277–279, 282, 284
WELLS, WILLIAM CHARLES 200, 287
WEST, JAMES 27, 157, 187–188, 282, 299
WILCOX, JOSEPH 121, 280
WILLIS, THOMAS 34
WILLOUGHBY, FRANCIS 34, 231
WOLLASTON, WILLIAM HYDE 29
WROTTESLEY, JOHN 29

ZACCHIROLI, FRANCESCO 37
ZAMBECCARI, GIUSEPPE 94
ZANI, ERCOLE 45
ZANOTTI, EUSTACHIO 103, 105, 114
ZANOTTI, FRANCESCO MARIA 43, 105, 186–187, 237–238, 276
ZOPPIO, MELCHIORRE 35

EUROPA PERIODICA
STUDIES ON PERIODICALS AND NEWSPAPERS

Edited by
Patrizia Delpiano, Fabio Forner, Giovanni Iamartino, Sabine Schwarze and Corrado Viola

Volume 1 Fabio Forner / Franz Meier / Sabine Schwarze (Ed.): I periodici settecenteschi come luogo di comunicazione dei saperi. Prospettive storiche, letterarie e linguistiche. 2022.

Volume 2 Lucia Berti: Scientific Crosscurrents between Italy and England. Italian Contributions to the *Philosophical Transactions of the Royal Society*, Seventeenth to Nineteenth Centuries. 2023.

www.peterlang.com

www.ingramcontent.com/pod-product-compliance
Ingram Content Group UK Ltd.
Pitfield, Milton Keynes, MK11 3LW, UK
UKHW041912140426
5217IPUK00002B/11